# RESEARCH ON ACCOUNTING ETHICS

*Volume 4* • 1998

# RESEARCH ON ACCOUNTING ETHICS

*Managing Editor:* LAWRENCE A. PONEMON
*Price Waterhouse LLP*

*Associate Editors:* MARC J. EPSTEIN
*Rice University*

JAMES GAA
*University of Alberta*

VOLUME 4 • 1998

JAI PRESS INC.

*Stamford, Connecticut* *London, England*

*100 Prospect Street*
*Stamford, Connecticut 06901-1640*

*JAI PRESS LTD.*
*38 Tavistock Street*
*Covent Garden*
*London WC2E 7PB*
*England*



*ISBN: 0-7623-0339-5*

*Manufactured in the United States of America*

# CONTENTS

# LIST OF CONTRIBUTORS


| | |
|---|---|
| *Dennis M. Bline* | Accounting Department<br>Bryant College |
| *Peter Clarke* | Department of Accountancy<br>University College<br>Dublin, Ireland |
| *Charles P. Cullinan* | Accounting Department<br>Bryant College |
| *Patricia Casey Douglas* | Department of Accounting<br>Loyola Marymount University |
| *Cynthia Firey Eakin* | Eberhardt School of Business<br>University of the Pacific |
| *Terry J. Engle* | School of Accountancy<br>University of South Florida |
| *Lois Deane Etherington* | Faculty of Business Administration<br>Simon Fraser University, Canada |
| *Sharon H. Fettus* | School of Business and Management<br>University of Maryland, College Park |
| *Dale L. Flesher* | School of Accountancy<br>University of Mississippi |
| *Martin Freedman* | School of Management<br>State University of New York at Binghamton |
| *G. William Glezen* | Department of Accounting<br>University of Arkansas |

| | |
|---|---|
| *Lawrence P. Grasso* | School of Accountancy<br>Arizona State University |
| *Nancy Thorley Hill* | School of Accountancy<br>DePaul University |
| *Thomas G. Hodge* | Department of Accounting<br>Northeast Louisiana University |
| *Steven E. Kaplan* | School of Accountancy<br>Arizona State University |
| *Joshua Krausz* | Sy Syms School of Business<br>Yeshiva University |
| *Alfred R. Michenzi* | Department of Accounting<br>Loyola College in Maryland |
| *Krishnamurty Muralidhar* | School of Management<br>University of Kentucky |
| *Robert J. Nagoda* | Department of Accounting<br>Webber College |
| *Karen Paul* | Department of Marketing and Business Environment<br>Florida International University |
| *Moses L. Pava* | Sy Syms School of Business<br>Yeshiva University |
| *Lawrence A. Ponemon* | Price Waterhouse LLP |
| *Bernadette M. Ruf* | Department of Accounting<br>University of Delaware |
| *Allen G. Schick* | School of Business and Management<br>Morgan State University |
| *Bill N. Schwartz* | Department of Accounting<br>Virginia Commonwealth University |

| | |
|---|---|
| *Terry L. Sincich* | Department of Information Systems and Decision Sciences<br>University of South Florida |
| *A.J. Stagliano* | Department of Accounting<br>Saint Joseph's University |
| *Kevin T. Stevens* | School of Accountancy<br>DePaul University |
| *James H. Thompson* | Department of Accounting and Finance<br>Oklahoma City University |
| *Linda Thorne* | School of Business<br>York University<br>Canada |
| *Dan R. Ward* | Department of Accounting<br>University of Southwestern Louisiana |
| *Suzanne Pinac Ward* | Department of Accounting<br>University of Southwestern Louisiana |
| *John P. Wendell* | School of Accountancy<br>University of Hawaii at Manoa |
| *P. Richard Williams* | Department of Accounting and Finance<br>Austin Peay State University |
| *Thomas E. Wilson, Jr.* | Department of Accounting<br>University of Southwestern Louisiana |
| *Gail B. Wright* | Accounting Department<br>Bryant College |
| *Kristi Yuthas* | Department of Accounting<br>Anderson School of Management<br>University of New Mexico |

# ANNUAL REPORTS AS A MEDIUM FOR VOLUNTARILY SIGNALING AND JUSTIFYING CORPORATE SOCIAL RESPONSIBILITY ACTIVITIES

Moses L. Pava and Joshua Krausz

## ABSTRACT

The primary motivation for this study is to examine empirically why some companies disclose more information about corporate social responsibility activities than other companies. To this end, we measure annual report disclosures of social responsibility activities for a sample of firms which had been previously identified as meeting corporate social responsibility criteria and compare these results to an analysis of disclosures for a control sample. In addition, we also examine the "rhetoric of justification." More specifically, through the use of content analysis of the president's letter to shareholders, we examine whether or not management provides a formal justification, using strategic or nonstrategic language for its selection of socially responsible actions.

**Research on Accounting Ethics, Volume 4, pages 1-27.**

**ISBN: 0-7623-0339-5**

## INTRODUCTION

The accounting literature is deeply divided about the question of what influences some companies to choose to disclose more information about corporate social responsibility (CSR) activities than others. One perspective assumes that social responsibility disclosures are attempts to satisfy the demands of outside "pressure groups." We label this perspective "bluffing." An alternative view suggests that corporate social responsibility disclosures reflect managers' attempt to communicate accurately their choice for engaging in CSR activities through the use of the annual report. This perspective is labelled here as "signaling."

Accordingly, the primary motivation for this study is to examine empirically this question. To this end, we measure annual report disclosures of social responsibility activities for a sample of firms which had previously been identified as meeting corporate social responsibility criteria. The companies were identified by the Council on Economic Priorities (CEP). We compare these results to an analysis of social responsibility disclosures for a control sample.

In addition, given the existence of at least some meaningful disclosures about social responsibility, we also examine the "rhetoric of justification." In this study, we examine whether or not management provides a formal justification, using strategic or nonstrategic language, for its selection of socially responsible activities. Specifically, we analyze the text of the president's letter to shareholders, through the use of content analysis, for any differences between articulated goals and disclosures of specific activities related to CSR.

Documenting the existence of nonstrategic justifications is important as a purely descriptive exercise. As far as the authors are aware, no study to date has addressed itself to this issue. We believe that the existence (or nonexistence) of nonstrategic justifications in the annual report is an important missing piece in our understanding of the nature of corporate social responsibility disclosures. Before development of a full-fledged "theory" of social responsibility disclosure, it is necessary to describe accurately specific characteristics of the reporting environment. It is a fairly straightforward question which has received little or no attention in the previous literature.

There are numerous reasons for examining the president's letter to shareholders. First, the letter is systematically made available to all shareholders and other interested parties. The president's letter is a highly visible document which is available on an annual basis. Second, it is not difficult to understand. In a recent survey of individual investors' readership and use of the annual report, Epstein and Pava (1993) concluded that the president's letter to shareholders was the least difficult item to understand. Third, the letter is specifically targeted to shareholders (although this is not universally the case). This is important, given that our research focus is primarily concerned with how managers disclose and justify corporate social responsibility activities to corporate shareholders. Finally, and perhaps most importantly, the president's letter provides an open-ended and relatively unrestricted medium for managers to voluntarily convey information to interested parties. While this is also true for the Management, Discussion & Analysis section (MD&A) (Pava and Epstein 1993), our initial comparison between CSR disclosures in the MD&A and the president's letter strongly indicated management's propensity to convey CSR information through the president's letter, rather than the MD&A. Wolfe (1991) similarly concluded that the management discussion focused solely on technical explanation of financial results; and reference to CSR issues was absent.

## DISCLOSURE RATIONALES: BLUFFING, SIGNALING, AND THE RHETORIC OF JUSTIFICATION

As stated above, at least two diametrically opposed reasons have been offered as to why some companies disclose more information about CSR activities than others. We label the first reason "bluffing" and the second "signaling." In this section, we explore both of these perspectives. Finally, in the third part of this section, we discuss issues related to the rhetoric of justification.

### A: Bluffing

Underlying the bluffing perspective are the following two assumptions: First, investors, acting in their role as investors, are assumed to attach value exclusively to the expectation of future cash flow, and no value to the creation of positive or negative externalities. This assumption has been described as the presumptive shareholder desire to maximize profit (Engel 1979). Second, the perspective assumes that there is a negative association between CSR activities and financial performance. Thus, it is assumed that higher levels of CSR activities lead to lower levels of financial performance. These two assumptions, taken together, lead to the strong conclusion that companies rated high in terms of CSR, in an environment which allows for strictly voluntary disclosures, will have significant incentives in place not to disclose this information, especially through the president's letter to shareholders. In addition, companies rated low in terms of CSR would be indifferent to this type of disclosure and would not have much to report, in any case.

Abbott and Monsen (1979) have summarized and defended this view as follows:

> There are theoretical reasons to expect the corporation to under report its social involvement activities. Since social involvement activities are also costs, reading of such social activities by shareholders can be taken to mean that the firm's management is failing to put highest priority on the interests of the shareholders by not maximizing income available to be distributed as dividends (p. 506).

More recently, a similar position was articulated by Cowen, Ferreri, and Parker (1987), who wrote:

> A company may be highly involved in social responsibility actions but may not choose to disclose such actions in the annual reports. Conversely, some companies may have little concern with societal welfare but may make numerous disclosures of relatively trivial activities to enhance their corporate image (p. 121).

By far the clearest and most complete explication of this view has been offered by George J. Benston in a controversial and thought-provoking article titled "Accounting and Corporate Accountability" (1982). Because of the importance of this paper, we review it extensively. Benston, applying arguments borrowed from the efficient market school, suggested that managers possess little or no latitude in terms of choosing CSR activities.

To build his case, Benston explicitly asserted that consumers, shareholders, and management demonstrate little or no demand for CSR activities. Given these restrictive assumptions, Benston builds a strong analytical case suggesting the impossibility for the existence of substantial CSR activities. Benston lists four general reasons why managers are limited in terms of their choices: the market for goods and services; the markets for finance and for corporate control; the market for managerial services; and finally, internal and external monitoring systems.

Given that Benston's consumers are "rarely interested" in CSR activities, the market for goods and services serves to constrain managerial decisions. Managers are unable to expend additional resources on environmental projects and pass along the costs of the CSR activities to consumers. Consumers simply will not pay more to help defray the costs of environmental clean-up.

Similarly, since shareholders put no value on CSR activities, the markets for finance and corporate control also limit the ability of managers to engage in CSR activities. Benston asserts that the "poor stewardship" associated with decisions to engage in socially responsible activities will inevitably lead to lower share prices. And, in turn, "the decline in the market price of corporations' shares increases the likelihood of a corporate takeover to displace the managers" (p. 91). Therefore, managers concerned about their own jobs have a strong built-in incentive to avoid CSR activities.

The market for managerial services, like the markets discussed above, also constrains managers. At this point in the argument, Benston equates CSR activities to "self-serving decisions" (p. 91). Thus, the more a manager is perceived as one who engages in CSR activities the more difficult he or she will find obtaining a new managerial position. In Benston's words, "As is the case in markets generally, other producers as well as consumers have incentive to learn about and provide information on the value of the product" (p. 91). To the extent that a manager is known as a CSR manager, his or her value will decline.

Finally, and according to Benston most importantly, the existence of internal and external monitoring systems will prevent managers from "misusing the shareholders' resources" (p. 92). According to Benston, most managers (like consumers and shareholders) have no desire to engage in CSR activities. Managers thus find it in their own interest to set up and design internal and external monitoring systems to attest to the fact that they are not engaging in unprofitable activities like CSR. In the absence of external auditors, for example, managements' "compensation would be reduced by the amount of resources that they would be expected to divert from the shareholders" (p. 92). Simply put, it is in the interest of managers to hire auditors to attest to the fact that no CSR activity has been undertaken.

Benston's conclusion is clearly stated. Managers have little or no discretion to act other than in the interests of shareholders. Since it has been assumed that shareholders have no taste for CSR activities, managers cannot and will not engage in these activities. As part of the empirical evidence to support his position, Benston concludes this part of the discussion by prematurely noting the virtual failure of three CSR-screened mutual funds.

Benston, however, is well aware of the existence of some CSR disclosures and, therefore, must explain the phenomenon in light of his efficient market view.

To explain CSR disclosures, Benston introduces the notion of "pressure groups" into the analysis. Benston suggests that CSR disclosure is an attempt to mitigate "the costs that pressure groups might be able to impose" (p. 99). Managers strategically disclose selected bits of information which serve the interests of the corporation by pacifying the various pressure groups concerned with employee, consumer, and government issues. In other words, managers attempt to bluff pressure groups. Benston continues that since measuring CSR activities is a subjective process, it is not surprising that the reports generally favor the companies. Benston believes that even in a regulatory environment which required CSR disclosures, the best we could hope for would be reports about social responsibility which would "have little chance of being other than public relations or other self-serving exercises" (p. 100).

Benston's position is that real CSR activity is a virtual impossibility. Nevertheless, companies have an incentive to portray themselves as if they are socially responsible. It is the "homage which vice pays to virtue" (Lindsay 1962, p. 97). Benston's pressure groups represent the single class of stakeholders with a demand for socially responsible actions.

Ullmann (1979) understands CSR disclosures in Germany as a result of managers trying to preempt union demands for increased wages. It thus provides a clear example of Benston's self-serving disclosures. Ullmann describes the German reporting environment as follows:

> Today CSR is capable of serving a very different strategic purpose within the context of a drastically changed economic environment, namely to prevent demands for high wage increases. In periods of slow economic growth and unemployment, distribution conflicts intensify, corporate social reports which indicate that the largest share of the wealth created by corporations is already going to employees, can facilitate management arguments in discussion with labor unions. As the Federation of German Employers' Associations pointed out, CSR "can help reduce utopian concepts and to harmonize the social demands with the true capacity of the economy" (p. 127).

Benston's suggestion that disclosure decisions are the result of self-serving calculations on the part of managers can certainly explain many of the reporting failures documented in the national media. For example, the *New York Times* (December 17, 1993) recently gave page one coverage to the indictment of Con Edison and two retired officials for allegedly concealing information that an explosion released large amounts of asbestos into a highly populated area of Manhattan. According to the *New York Times* report, following the explosion, the company gave assurances that the blast did not result in the release of asbestos. On the basis of this allegedly false information, residents returned back to their apartments. It is certainly plausible that Con Ed management made the disclosure decision as a result of an informal or implicit cost benefit analysis. Management, at the time of the explosion, may have believed it was in their interest, at least in the short run, to convey assurances to residents that all was normal.

Similarly, the following episode, as described by the *New York Times*, also suggests a strong element of strategic behavior on the part of management:

> For New York and the 8.2 million people insured by Empire, questions remain. Federal and local law-enforcement agencies are investigating two sets of books kept by Empire over the last several years. Empire officials acknowledged that they supplied false financial data to the New York State Insurance Department, and used the data in their successful lobbying campaign for changes in insurance law to shore up Empire's finances (*New York Times*, July 11, 1993).

The common thread in both of these cases is an attitude on the part of management that information can be manipulated in such a way as to further either individual or corporate goals, or both simultaneously. In each case, management allegedly failed to disclose information in an open and neutral manner.

Neither of these cases is unique. Therefore, an approach like Benston's which explicitly recognizes managers' disclosure decisions as "self-serving" possesses great intuitive appeal.[1,2] Nevertheless, we believe it is open to a number of severe criticisms. First, the argument hinges on some seemingly restrictive assumptions. Consumers, shareholders, and even managers, are assumed to be indifferent toward social responsibility. While these extreme assertions may be true, Benston's empirical evidence is less than completely convincing. Second, the analysis leaves unanswered the question of why pressure groups (who apparently do value CSR activities) are satisfied with social responsibility reports that are merely self-serving documents. If the author of the model is aware that CSR reports are nothing more than public relations, why are pressure groups not aware of this? Analytically, it is not satisfying to conclude that the author of the model has information that some of the assumed economic players are missing. Third, the theory itself is somewhat over-advertised. It is by no means a dramatic conclusion that social responsibility reports will be self-serving, given the starting position which assumes all managerial activity as self-serving.

Yet, in spite of our reservations, Benston's paper undoubtedly provides the clearest model and the most precise statement of the assumptions underpinning the bluffing perspective. It therefore represents an important contribution to understanding CSR disclosures.

## B: Signaling

An alternative view to Benston's begins with the assumption that in addition to pressure groups, in some situations and for some issues, shareholders, consumers, and managers would be willing to incur costs to promote CSR goals.

For example, during the 1950s, Richardson-Merrell began developing and testing a new drug, MER/29, which the company hoped would help reduce cholesterol. According to Christopher D. Stone (1975) in his book *Where the Law Ends*:

> While its top officers were enthusing about the drug, and preparing for a major marketing campaign, other parts of the organization were receiving bad news. In one laboratory test all the female rats on a high dosage died within six weeks. In a subsequent test, all rats on a somewhat lower dosage had to be destroyed halfway through the experiment. On autopsy, it was revealed that they had suffered abnormal blood changes. Corneal opacities were observed in other animals. Monkeys suffered blood changes and weight loss. . . . When the company finally filed a new drug application with the FDA, seeking permission to place MER/29 on the market, its application contained many false statements (p. 54).

Stone further suggests that regardless of the perceived benefits to the corporation of falsifying the data, "It is hard to believe that a representative sampling of board members, fully informed of what was happening to animals in Richardson-Merrell's laboratories, would have voted to push ahead with MER/29." In explaining his view, Stone suggests that not to have done so would have run counter to "moral views that are held by the general public and general business community" (p. 135).

This explanation leads us to an exploration of the signaling perspective. According to a recent study by Wolfe (1991), one of the most important issues concerning annual reports is whether variations in annual report disclosures are simply reflections of variations of communication of espoused values and behaviors or variations of values and behaviors in use. The signaling perspective maintains the latter view.

This view suggests that firms might signal their identity as socially responsible companies through the annual report. It does not imply a one to one correspondence between CSR performance and CSR disclosures. It merely states that, on average, companies which are perceived as having met CSR criteria will be more likely to report this information in the annual report.

Observers agree that CSR disclosure in the United States is predominately voluntary. Wolfe describes the disclosure environment:

> Though considerable resources have been spent debating and studying it, social reporting in the United States remains predominately voluntary... and is expected to remain so. The SEC considers that social information falls outside of its area of responsibility, unless it has material economic consequences (p. 290).

Given this voluntary environment, why would one accept a signaling perspective? Gibbins, Richardson, and Waterhouse (1990), in their study on the management of corporate disclosures, concluded that among executives involved in making disclosure decisions, the credibility of the disclosures is of primary concern. Basing their conclusions on extensive interviews with corporate managers in Canadian corporations, the authors wrote as follows:

> Credibility seems central to effective disclosure. While credibility may be enhanced, for example, by employing external agents to attest to information, it also rests on the firm's disclosure reputation (p. 138).

To the extent that Gibbins, Richardson, and Waterhouse have drawn the correct conclusion from their data, they provide a strong a priori reason to believe that even in the area of CSR, managers cannot afford to be perceived as manipulating disclosures merely to serve their own ends. If shareholders believe that corporate disclosures are merely self-serving attempts to manage a hostile environment, the credibility of the disclosures will soon come into question.

The two most important assumptions upon which the signaling perspective is based can be stated as follows:

1. Unlike Benston's view discussed above, shareholders, consumers, and managers believe that some CSR activity is desirable; therefore, real CSR activity exists.
2. Managers attempt to communicate information to users in a neutral and unbiased way.

The second assumption concerning the communication of neutral information to users is a bedrock concept in accounting. According to the Financial Accounting Standards Board's (FASB's) Concept Project (1980), corporate executives must satisfy their obligation to report *all* information that is of sufficient importance to influence the judgement and decisions of an informed user. More specifically, the FASB has warned that "the primary concern should be the relevance and reliability of the information that results." In addition, the Board explains:

> Accounting information must report economic activity as faithfully as possible, without coloring the image it communicates for purpose of influencing behavior in *some particular direction* (emphasis in the original).

This well-known criterion is called "neutrality."

While the first assumption above remains somewhat controversial, an increasing number of observers accept it. For example, Anderson and Frankle (1980) concluded their empirical study as follows:

> Thus the "ethical investor" may exist and, in fact, dominate the market. Whether this finding can be attributed to altruistic or economic motivations on the part of investors has not yet been resolved (p. 477).

Similarly, survey studies of investors such as Epstein and Pava (1993) have suggested that though the stereotype is that shareholders are worried only about profits, this is clearly not the case. For example, when investors were asked to rank their preferences as to how corporate funds were to be allocated, pollution and product safety were ranked higher than dividends.

Numerous researchers have speculated about the degree to which CSR disclosures correspond to actual performance and have provided reasoning similar to that offered here.

Bowman (1984), in defending his use of CSR disclosure as a surrogate for actual CSR performance, wrote that he would not expect "unusual puffery" on issues like corporate social responsibility. Because annual reports are written "essentially to shareholders" (p. 63), it is unlikely that managers will invent or exaggerate CSR activities.

Similarly, Wolfe believes that in spite of its voluntary nature, there are pressures on management to report honestly and accurately. According to Wolfe:

> Increased public scrutiny, competitive forces, the independent financial press, audits and anti-fraud laws, potential impacts on pubic confidence, and moral obligation motivate corporate managers to provide honest annual report information (p. 290).

Abbott and Monsen (1979) agree that there are good reasons to believe that managers will disclose CSR activities in an unbiased and honest way. Their emphasis lies less on the perceived altruism of shareholders or Wolfe's "moral obligations" but rather in the need for managers to legitimate the business enterprise. According to Abbott and Monsen:

> stockholders have a vested interest in the stability and legitimacy of the entrepreneurial institution and the autonomy of that institution from state control. Aware, then, of the criticisms that have been made of the corporation, reading of its progressive views on social responsibilities in the annual report can enhance confidence of the politically savvy shareholder in managements' policies (p. 506).

Fair and honest disclosure of CSR activities might, thus, preempt government intrusion and safeguard the autonomy of corporate managers.

Based on our review of both the bluffing perspective and the signaling perspective we can formulate the following testable hypothesis:

**Hypothesis 1.** Although some disclosures about CSR activities exist, there is no statistical association between perceived CSR performance and CSR disclosures.

Acceptance of Hypothesis 1 would lend strong credence to the bluffing perspective. By contrast, rejection of Hypothesis 1 would lend support to a signaling perspective. Under the bluffing perspective, CSR activity is assumed to be a virtual impossibility. Therefore, it is highly unlikely that any third party assessments of

CSR behavior—which did not use the president's letter as an input into formulating the ratings (CEP ratings)—would correlate with CSR disclosures. On the other hand, if CSR activities is measurable, and managers attempt to communicate in a neutral way—as the signaling perspective holds—one would anticipate rejection of Hypothesis 1.

Prior empirical evidence is mixed on this question. At least three studies have concluded that in the area of environmental pollution there is little or no association between disclosures and performance (Ingram and Frazier 1980; Freedman and Jaggi 1982; Wiseman 1982). Some evidence in support of a positive association between disclosures and performance, where social responsibility is viewed in more general terms, has also been documented (Abbott and Monsen 1979; Bowman 1984; Wolfe 1991).

## C: The Rhetoric of Justification

Not only is there a positive association between CSR performance and disclosures, but some managers will defend CSR activities by using nonstrategic language in the annual report. In other words, firms engaging in CSR activities claim to do so not only to serve the financial interests of shareholders by maximizing profits but to meet the legitimate moral and ethical claims of other stakeholder groups. This possibility, which of course is more extreme than the signaling perspective reviewed above, has received virtually no attention in the literature on CSR disclosures. This is surprising. The existence (or nonexistence) of nonstrategic justifications in the annual report is an important missing piece in our understanding of the nature of corporate social responsibility disclosures. Before development of a full-fledged "theory" of social responsibility disclosure, it is necessary to describe accurately specific characteristics of the reporting environment. This is a fairly straightforward question which has received little or no attention in the previous literature. We formally articulate this perspective as follows:

**Hypothesis 2.** Some managers justify CSR activities in nonstrategic terms.

For the purposes of this study, we utilize Kenneth E. Goodpaster's (1991) distinction between strategic and nonstrategic decision making. According to Goodpaster:

> A management team, for example, might be careful to take positive and (especially) negative stakeholder effects into account for no other reason than that offended stakeholders might resist or retaliate (e.g., through political action or opposition to necessary regulatory clearances). It might not be *ethical* concern for the stakeholders that motivates and guides such analysis, so much as concern about potential impediments to the achievement of strategic objectives. Thus positive and negative effects on relatively powerless stakeholders may be ignored or discounted (p. 57).

Management, adopting a strategic perspective, views all groups other than stockholder groups "instrumentally, as factors potentially affecting the overarching goal of optimizing stockholder interests" (p. 58). Goodpaster believes that weighing stakeholder considerations as potential sources of goodwill or retaliation as a practical matter "is morally neutral" (p. 57). Expanding on this point, Goodpaster continues:

> The point is simply that while there is nothing necessarily wrong with strategic reasoning about the consequences of one's actions for others, the kind of concern exhibited should not be confused with what most people regard as moral concern (p. 60).

An excellent example of strategic decision making in the area of corporate social responsibility is the decision of Wal-Mart Stores, Inc., to stop selling handguns in its stores, a move hailed by gun control advocates (Wal-Mart will continue to sell handguns by catalogue). As the *Wall Street Journal* noted (December 23, 1993), this decision "coincides with the recent heightening of the nation's long-running debate over guns and violence. The controversy has been fueled by a spate of multiple shootings and the passage of the Brady Bill, which mandates a five-day waiting period for handgun purchases." A spokesman for Wal-Mart carefully, and no doubt purposely, explained the decision in purely strategic terms. According to the spokesman, Don Shinkle, "A majority of our customers tell us they would prefer not to shop in a retail store that sells handguns." The firm makes no moral or ethical claims. Wal-Mart, at least in this instance, would prefer that this decision be understood in terms of cost benefit analysis only. Wal-Mart apparently would like to avoid making any kind of moral or ethical stand on a visceral and hotly debated public issue.

By contrast, nonstrategic decision making explicitly incorporates the legitimate rights of even the least powerful stakeholders. Goodpaster suggests that "moral concern would avoid injury or unfairness to those affected by one's actions because it is wrong, regardless of the retaliatory potential of the aggrieved parties" (p. 60).

Is there any evidence that public companies defend corporate activities through the use of nonstrategic language, as Hypothesis 2 predicts? While there is little systematic empirical evidence, some anecdotal evidence exists. For example, Goodpaster quotes the following memo from the CEO of a major public company to middle management. The purpose of the memo was to explain and justify the firm's decision to put significant resources behind their affirmative action program. The CEO's letter reads in part:

> I am often asked why this is such a high priority at our company. There is, of course, the obvious answer that it is in our best interest to seek out and employ good people in all sectors of our society. And there is the answer that enlightened self-interest tells us that more and more of the younger people, whom we must attract as future employees, choose companies by their social records as much as by their business prospects. *But the one overriding reason for this emphasis is because it is right.* Because this company has always set for itself the objective of

assuming social as well as business obligations. Because that's the kind of company we have been. And with your participation, that's the kind of company we'll continue to be (p. 65, emphasis in the original).

This memo provides an unusually stark example of what is meant by nonstrategic decision making; it contrasts sharply with the Wal-Mart example above. The CEO articulates an unambiguous justification of the firm's affirmative action program. The CEO implies that even in the absence of financial gains to the corporation, the corporation has social obligations to society at large. The CEO explicitly recognizes business obligations as well as "overriding" social obligations.

Based on the above discussion, a major goal of this study is to explore whether or not this anecdotal evidence suggests an important component of CSR disclosures? While it is obvious that no research methodology will be able to assess the credibility of nonstrategic disclosures, documenting the mere existence of these types of disclosures is an important task. The rhetoric and language of justification is itself an important variable in fully understanding CSR disclosures.

## METHODOLOGY: MEASURING CSR PERFORMANCE AND DISCLOSURES

Our single most important methodological issue is the selection of empirical proxies to measure the two key variables: CSR performance and CSR disclosures. Our estimates of CSR performance are based on an analysis completed by the CEP et al. (1991). To develop a proxy for CSR disclosures, we utilized a methodology known as content analysis. This methodology has been advocated both in the CSR literature and beyond.[3]

### CSR Performance

In particular, we examine CSR disclosures for a group of 33 firms which have been identified by the CEP as meeting social responsibility criteria (Group 1), and compare their 1989 disclosures to a control sample matched by both industry and size (Group 2). The Group 2 firms were selected from the same industries and were selected on the basis of sales revenues. Specifically, the Group 2 firms are those firms in the same industry as the Group 1 firms ranking closest in terms of sales revenue (1989). Appendix A provides the names of the 66 companies. CEP's original list consisted of 53 companies. Due to data requirements, 20 companies were eliminated from our analysis. Ingram (1978), Trotman and Bradley (1981), Cowen, Ferreri, and Parker (1987), and Roberts (1992) suggest the importance of both industry and size as factors affecting CSR disclosures.

Drawing on the holdings listed in the prospectuses of the socially responsible mutual funds and based on their own analyses, the CEP described the companies

***Table 1.*** Characteristics of Groups 1 and 2 Firms

| *Characteristics* | *Group 1* *33 Firms* | *Group 2* *33 Firms* |
|---|---|---|
| Domini 400 Social Index | 26 \| 79% | 7 \| 21% |
| 100 Best Companies to Work For | 15 \| 45% | 0 \| 0% |
| 75 Best Companies For Working Mothers | 5 \| 23% | 0 \| 0% |
| 50 Best Places for Blacks to Work | 8 \| 24% | 0 \| 0% |
| Best Companies For Women(50) | 6 \| 18% | 0 \| 0% |
| More Than 20% Employee Ownership | 3 \| 9% | 0 \| 0% |
| Top 100 Defense Department Contractors | 2 \| 6% | 0 \| 0% |
| Direct Investment in South Africa | 1 \| 3% | 3 \| 9% |
| Top 50 Manufacturers Releasing Toxic Chemicals | 1 \| 3% | 1 \| 3% |
| Top 100 Nuclear Weapons Contractors | 0 \| 0% | 1 \| 3% |
| Tobacco Companies | 0 \| 0% | 1 \| 3% |

in Group 1 as "ethical" portfolio companies. The advantages of choosing the CEP firms for our study are as follows:

1. The CEP is highly regarded as a credible source of information on CSR. Numerous published studies have used previous CEP studies as the basis for forming measures of CSR. We concur with Shane and Spicer (1983), who concluded, "The most detailed, consistent, and comparable data bearing on corporate social performance has been published by the CEP. It appears to be the most active external producer of information in this area" (p. 522).
2. The CEP ratings are not unique. The firms included in Group 1 tend to be rated high in terms of CSR by numerous external groups.[4] Table 1 summarizes some characteristics of the Group 1 and Group 2 firms and provides additional support for the CEP ratings. There is significant overlap between the Group 1 firms, as identified by the CEP, and firms included in the Domini 400 Social Index. Of the 33 firms identified by the CEP as ethical, 26 firms are included in the Domini Index. Only seven of the Group 2 firms were included in the Domini Index. About half the Group 1 firms (15 firms) were rated among the "100 Best Companies to Work For," while none of the Group 2 firms were included on this list. Further, five of the Group 1 firms were among the "75 Best Companies for Working Mothers," while none of the Group 2 firms were identified among the "75 Best Companies for Working Mothers." Table 1 also indicates that few of the Group 1 firms are listed among the "Top 100 Defense Department Contractors" or the "Top 50 Manufacturers Releasing Toxic Chemicals."
3. To achieve the goals of this study, we needed an aggregate measure of CSR, as opposed to a measure of one or more of the components of CSR. The

CEP ratings, based on an assessment of 12 specific CSR components, provided a convenient and well-respected third party assessment. Further, we believe that the CEP ratings provide a more precise measure of CSR, per se, than those obtained from the next best competitor, *Fortune* magazine's annual survey of "corporate reputations."

4. The Group 1 firms were selected from diverse industries, thus enhancing the generalizability of the results.
5. Neither the mutual funds from which the firms were originally identified nor the CEP uses social responsibility disclosure in the annual report as a criterion for evaluating CSR behavior. Mutual funds which advertise themselves as socially responsible have begun to define the practice with more and more exactness. Table 2 summarizes both the positive and negative screens used by nine of the most important and influential socially responsible mutual funds. Issues like environmental performance, South Africa, weapons production, and employee relations were cited by almost all of the mutual funds examined.

To conclude this discussion, many of the 33 firms in Group 1 have been described as socially responsible by a wide variety of outside evaluators. The CEP is one of the most highly regarded external producers of social responsibility information. The 33 firms represent a diverse sample of companies. The sample thus provides an important, and inherently interesting, point of departure.

## CSR Disclosures

Content analysis is a relatively new methodology which has proved useful in analyzing written texts. Bowman (1984), in emphasizing the use and importance of the methodology, suggested that content analysis provided a "gestalt, not readily available through other methods" (p. 62). The method consists of codifying qualitative information in literary form (the raw data) into meaningful categories, where categories are defined as precisely as possible with regard to the hypotheses being tested. The results of the analysis can then be transformed into quantitative scales, allowing for comparison among different literary documents (Abbott and Monsen 1979; Ingram and Frazier 1980).

The most important methodological steps involved in content analysis are the following: (1) determine the sampling unit, (2) determine the recording units, and (3) determine the themes and categories to be used in the coding (Wolfe 1991).

The sampling unit chosen for this study was the president's letter to shareholders (step 1).[5] Each paragraph of the letter was viewed as a recording unit (step 2). Our choice of paragraphs rather than sentences or words was influenced both by our research objectives and by a test code of sample texts which indicated that using paragraphs would enhance reliability.

**Table 2.** Social Responsibility Screens Used by Nine Mutual Funds

| *Negative Screens* | *Number of Funds Using Screen* |
|---|---|
| South Africa | 8 |
| Weapons | 7 |
| Nuclear Power | 6 |
| Tobacco, Alcohol, Gambling | 3 |
| EPA Violations, Polluters | 1 |
| *Positive Screens* | |
| Environmental Issues | 8 |
| Employee Relations | 6 |
| Corporate Citizenship | 4 |
| Product Quality and Safety | 4 |
| Altenative Energy | 3 |

***Notes:*** This table reports the number of mutual funds that explicitly cited the above social responsiblity screens in the fund prospectuses. It is based on the following nine mutual funds: 1—Calvert-Ariel Appreciation Fund, 2—Calvert Social Investment Fund, 3—Domini Social Index Trust, 4—Dreyfus Third Century, 5—New Alternatives, 6—Parnassus Fund, 7—Pax World Fund, 8—Rightime Social Awareness Fund, and 9—Schield Progressive Environmental Fund

***Source:*** Social Investment Forum (updated August).

Within the president's letter, paragraphs were identified as belonging to one of two major classificatory themes: (1) general corporate goals, or (2) specific CSR activities (step 3).

## *General Corporate Goals*

All paragraphs in the presidents' letters were identified in which a general articulation of corporate goals was provided. Specifically, we categorized each of these paragraphs as follows.

First, the paragraph suggests that among the most important goals of this corporation are:

A. Maximizing shareholder value;
B. Increasing shareholder value;
C. Aachieving well-defined financial targets (for example, return on equity, return on sales, market share, earnings per share, or market return);
D. Producing highest quality products or services for customers;
E. Satisfying the demands of important stakeholders. Stakeholders can include shareholders and customers but must also include at least one of the following: creditors, employees, suppliers, government, community, or society at large.

An example of each should help clarify:

A. *Maximizing Shareholder Value:*

> These objectives and strategies remain part of the corporation's culture, committing us to maximizing value for our shareholders—as a successful, independent marketer of leading consumer brands—by balancing competitive current returns with investment in profitable growth for the future.

B. *Increasing Shareholder Value:*

> We pledge to continue increasing both the value of your company and the dependable flow of dividend payments that has now risen each year for 15 consecutive years.

C. *Achieving Well-Defined Financial Targets:*

> The focus of the entire organization is on achieving target profitability. And this year should be much improved. . . . We continue to target 5% return on sales as the chief measure of our success.

D. *Producing Highest Quality Products:*

> We believe much of this growth is a reflection of our commitment to 100% customer satisfaction. We can now say with confidence that the progress we have made during the past year has firmly established our company as one of the few viable participants in the air express industry.

E. *Satisfying the Demands of Important Stakeholders:*

> Throughout this year and beyond, we hope to "refine the bottom line" and look for more and more ways of running an innovative business for the benefit of our stockholders, our customers, our employees, and our community.

### *2. CSR Activities*

In addition to classifying corporate goals, all paragraphs were identified in which the author explicitly indicated that the company was engaged in one or more of the following socially responsible activities:

A. Donates corporate funds to charitable causes;
B. Is actively engaged in meeting environmental concerns;
C. Has an employee stock ownership program;
D. Links employee performance with rewards (salary, benefits, promotion, etc.);
E. Promotes increased employee autonomy and responsibility;
F. Seeks out women and/or minorities for hiring and/or promotion;

G. Is helping to solve national social problems in addition to environmental concerns (crime, unemployment, education, health care, etc.);
H. Is meeting local community needs (unemployment, education, day care, museums, etc.);
I. Avoids nuclear energy;
J. Avoids military contracts;
K. Other socially responsible activities.

These activities were chosen based on a review of the CSR literature. We chose a broad definition of CSR activities rather than a more restrictive definition of CSR in order to ensure a complete list of potential CSR activities.

Once a paragraph was identified as articulating a specific CSR activity, two independent raters were asked to determine whether or not a formal justification for the activity was provided. (Raters did not know the identity of the companies, as references to company names were eliminated.)

Raters were asked to carefully consider the context of the statement and indicate whether or not justification for the activity (or avoidance of activity ) is explicitly provided. Second, if the author provided justification, raters were asked if the justification is of a strategic or nonstrategic nature, or both? Following Goodpaster (as discussed above), a strategic justification explicitly links the activity to enhanced corporate financial performance. A strategic justification views the activity instrumentally, as a means toward achieving a financial objective. Examples include: increasing profits, boosting sales, raising margins, enhancing operating efficiency, or expanding operating flexibility. A nonstrategic justification would be a formal statement explaining a given activity which did not link it to financial performance. Key words or phrases include: social responsibility, community responsibility, obligations, ethics, morality, or fairness. This methodology resulted in a 91% interrater reliability.

Examples of both strategic and nonstrategic justifications should once again help clarify. One of the companies in our sample justified its employee stock ownership program as follows:

> We anticipate the ESOP will enhance stockholder values by more closely aligning the interests of our personnel and company stockholders, and by providing us with a cost-efficient mechanism to make matching contributions under the savings plan.

This company explicitly linked the employee stock ownership program to financial performance. The statement is clear and direct. By implementing this new compensation practice, the firm expects to enhance stockholder values. Therefore, this statement is characterized as strategic.

Additional examples of strategic justifications follow:

*Increasing Employee Autonomy:*

The development of an entrepreneurial spirit within the Company will provide the flexibility to assure profitability growth in the years ahead.

*Seeks out Women and Minorities:*

The Hudson Institute has predicted that 85% of all entry-level jobs between now and the year 2000 will be filled by women, minorities, and immigrants. Those companies that become attractive places for these people to work will be the companies that succeed in the 1990s.

*Engaged in Meeting Environmental Concerns:*

While we saw significant potential for parts cleaner services in the industrial market, we also saw opportunities to help industrial plants with other hazardous waste disposal problems. Tens of thousands of industrial plants use solvents, paint thinners, lubricating oils, coolants and a variety of other fluids. Many of these fluids are considered to be hazardous and must be disposed of in accordance with Federal and State Environmental Regulation. Companies that generate small quantities of these fluids find coping with the environmental regulations and proper disposal to be particularly cumbersome, as they might not have in-house expertise in all of the various environmental regulations.

The language used to justify these activities contrasts with the nonstrategic language employed in the following examples:

Donates Corporate Funds to Charitable Causes:

Last year, we celebrated the ten year anniversary of the opening of our doors in the old gas station in Burlington. In looking back over what we have accomplished in our first ten years, we probably take most pride in our growing reputation as one of the most socially responsible businesses in America. The Council on Economic Priorities recently honored our Company with the "Corporate Conscience Award for Corporate Giving." The award was given in recognition of our Company's policy of denoting 7 1/2% of our pretax income to social service agencies and community causes through the corporate foundation.

*Is Helping to Solve National Social Problems:*

Our efforts to relocate manufacturing and increasingly, R & D internationally are designed to make an ongoing contribution to every society in which we operate...It also demonstrates the corporation's desire to help move international trade toward greater equilibrium.

*Seeks Out Women and/or Minorities:*

As we move ahead, we renew our commitment to equal opportunity and diversity throughout the corporation as expressed on the cover of this report. That commitment runs top to bottom. We believe in it.

*Links Employee Performance With Rewards:*

> The Company recognizes that many employees invest significant portions of their lives in their jobs and that they make many significant contributions to its growth and success. Believing that it is in the best interests of all employees, the Company has adopted a "Security of Employment Plan" for the purpose of providing some financial stability to employees whose employment is terminated following a change of control in the Company.

These examples illustrate the distinction between strategic and nonstrategic justifications. In the first example, the company explicitly uses the term "social responsibility." The second example justifies its increased international activities by stating that it fulfills the corporate "desire to move international trade toward greater equilibrium." The third example defends the company's stance of equal employment opportunities and diversity in the workplace by simply noting, "We believe in it." Finally, notice that in the last example, the letter to shareholders simply takes it as self-evident that because employees invest significant portions of their lives working for the corporation, the corporation has an obligation to do what is in the "best interest of all employees." In each of the cases, the corporations justifies a particular activity not because it will enhance corporate profits but for nonstrategic reasons. The authors of these letters assume that the reader will understand and accept corporate responsibilities beyond financial and legal concerns.

## RESULTS: DO SOCIALLY RESPONSIBLE FIRMS SIGNAL THEIR IDENTITY?

The annual report to shareholders, and specifically the president's letter, provides a potential medium for management to voluntarily signal and justify CSR activities. In order to test Hypotheses 1 and 2 above, we ask and answer each of the following specific questions:

1. Are companies which have been perceived as meeting CSR criteria more likely than non-socially responsible companies to communicate CSR activities? Specifically, which types of activities do the two groups disclose?
2. Do companies which have been perceived as meeting CSR criteria communicate corporate goals in ways which can be distinguished from non-socially responsible firms?
3. Are companies justifying CSR activities using language which be can be described as nonstrategic?

Our results show that companies identified as meeting CSR criteria (Group 1) are more than twice as likely to disclose activities related to CSR than the control sample (Group 2). Group 1 firms reported 56 activities versus 25 for the Group 2 firms. This difference is statistically significant using the appropriate $t$-test. In

## *Table 3.* CSR Activities

| | Group 1: CSR Firms | | | | | Group 2: Control Firms | | | | |
|---|---|---|---|---|---|---|---|---|---|---|
| *CRS Activity* | *N.J.* | *S.J.* | *N-S. J.* | *Both* | *Totals* | *N.J.* | *S.J.* | *N-S. J.* | *Both* | *Totals* |
| Donates corporate funds to charitable causes | 1 | 0 | 2 | 0 | 3 | 0 | 0 | 1 | 0 | 1 |
| Is actively engaged in meeting environmental concerns | 0 | 3 | 0 | 1 | 4** | 0 | 0 | 0 | 0 | 0 |
| Has an employee stock ownership program | 0 | 3 | 0 | 3 | 6 | 0 | 2 | 1 | 0 | 3 |
| Links employee performance with rewards (salary, benefits, promotion, etc.) | 1 | 1 | 1 | 1 | 4 | 0 | 1 | 1 | 4 | 6 |
| Promotes increased employee autonomy and responsibility | 4 | 9 | 0 | 2 | 14 | 1 | 7 | 0 | 0 | 8 |
| Seeks out women and/or minorities for hiring and/or promotion | 0 | 3 | 1 | 0 | 4 | 0 | 0 | 1 | 0 | 1 |
| Is helping to solve national social problems in addition to environmental concerns (crime, unemployment, education, health care, etc.) | 2 | 3 | 5 | 4 | 14*** | 1 | 0 | 2 | 1 | 4 |
| Is meeting local community needs (unemployment, education, museums, etc.) | 1 | 0 | 2 | 2 | 5* | 0 | 1 | 0 | 0 | 1 |
| Avoids nuclear energy | 0 | 0 | 0 | 0 | 0 | 0 | 0 | 0 | 0 | 0 |
| Other socially responsible activities | 1 | 2 | 0 | 0 | 3 | 0 | 0 | 0 | 1 | 1 |
| TOTALS | 10 | 24 | 11 | 11 | 57*** | 2 | 11 | 6 | 6 | 25 |

***Notes:*** N.J.: No justification for activity provided.
S.J.: Strategic justification for activity provided.
N.S.J.: Nonstrategic justification for activity provided.
Both: Both strategic and nonstrategic justification for activity provided.
***Group 1 result is significantly higher than Group 2 result at the .01 level.
**Group 1 result is significantly higher than Group 2 result at the .05 level.
*Group 1 result is significantly higher than Group 2 result at the .10 level.

addition, we examine the data using a nonparametric sign test. Again, the differences between the two samples are significant at the .01 level. These results are shown in Table 3.

A positive association characterizes the relationship between CSR disclosures (through the president's letter to shareholders) and CSR performance. This conclusion is neither driven by a small number of Group 1 firms reporting many CSR activities, nor is it driven by Group 1 firms focusing only on a small number of CSR activities. In fact, a careful examination of the results indicates the opposite is true. Comparing the 33 matched pairs shows that in 20 cases, the Group 1 firm have more disclosures than the Group 2 firms. In only six cases do the Group 2 firms have more disclosures than the Group 1 firms. In seven cases, the Group 1 and 2 firms reported the same number of CSR activities. Further, the single activity—out of 11 potential activities—in which the Group 2 firms have more disclosures than the Group 1 counterparts is "Links employee performance with reward," arguably the least relevant category in terms of "pure" corporate social responsibility.

Examining specific types of disclosures reveals that the most important differences between the two groups are in the following three areas: "Is helping to solve national social problems in addition to environmental concerns" (Group 1: 14 disclosures, Group 2: 4 disclosures, significant at .01 level), "Is actively engaged in meeting environmental concerns" (Group 1: 4 disclosures, Group 2: 0 disclosures, significant at .05 level), and "Is meeting local community needs" (Group 1: 5 disclosures, Group 2: 1 disclosure, significant at .10 level).

In addition to these statistically significant results, the Group 1 firms were more likely to report information about charitable donations (Group 1: 3 disclosures, Group 2: 1 disclosure), the existence of employee stock ownership programs (Group 1: 6 disclosures, Group 2: 3 disclosures), promotion of increased employee autonomy (Group 1: 14 disclosures, Group 2: 8 disclosures), and seeking out women and minorities for hiring and promotion (Group 1: 4 disclosures, Group 2: 1 disclosure). Thus, the evidence unambiguously confirms that managers of CSR firms are much more likely to report about CSR activities than the control sample. Hypothesis 1 articulated above can thus be rejected. The results in favor of a positive association confirm the "signaling" perspective.

Table 4 allows us to draw an even stronger conclusion. Examination of our tabulation of articulated corporate goals indicates an important and consistent distinction between the two groups of firms. One of the most important findings of this study is the following: More than half of the Group 1 firms (17 of 33) articulated that an important goal of the corporation was "Satisfying the demands of important stakeholders." Stakeholders could include shareholders and customers but must also include at least one of the following: creditors, employees, suppliers, government, or community or society at large. By contrast, only five of the Group 2 firms explicitly articulated this objective. This difference is significant at the .01

***Table 4.*** Corporate Goals

| *Articulated Corporate Goal* | *Group 1 Totals* | *Group 2 Totals* |
|---|---|---|
| Maximizing shareholder value | 3 | 1 |
| Increasing shareholder value | 26 | 23 |
| Achieving well-defined financial targets (for example, return on equity, return on sales, market share, earnings per share, or market return) | 8* | 3 |
| Producing highest quality products or services for customers | 17 | 14 |
| Satisfying the demands of important stakeholders; stakeholders could include shareholders and customers, but must also include at least one of the following: creditors, employees, suppliers, government, community, or society at large | 17*** | 5 |

***Notes:*** ***Group 1 result is significantly higher than Group 2 result at the .01 level.
*Group 1 result is significantly higher than Group 2 result at the .10 level.

level. The result corroborates the results reported in Table 3 above and, again, lends strong credence to the rejection of Hypothesis 1.

In addition to the statistically significant result above, the Group 1 firms are also more likely to describe "well-defined financial targets (for example, return on equity, return on sales, market share, earnings per share, or market return)" as an important objective (Group 1: 8, Group 2: 3, significant at .10 level). This result, in line with the results above, suggests more precise objectives on the part of the CSR firms than the control sample.

Interestingly, only four firms out of the entire sample of 66 explicitly indicated "maximizing shareholder value" as a corporate objective, whereas 49 firms explicitly indicated "increasing shareholder value" as an objective. Regardless of whether a firm has been identified as having met CSR criteria or not, managers are apparently extremely reluctant to invoke the language of maximization. If one takes these results at face value, they conflict sharply with the traditional economic assumption of managerial behavior. Even if one continues to suggest, however, that there need be no link between statements located in the president's letter to shareholders and actual managerial actions, the findings suggest an interesting puzzle; if managers are really attempting to maximize profits, why don't they admit it to shareholders?

The results discussed thus far lead us to conclude against Hypothesis 1 stated above. We now turn to a discussion of how firms justify CSR activities. This additional evidence lends support to Hypothesis 2. We note the following four observations:

1. Of the 81 CSR activities disclosed (Groups 1 and 2 combined), 21% were justified using nonstrategic language (17 out of 81). Further, removing the

effects of the 12 activities in which no justification was offered, the data show that 25%, or one out of four disclosures, were justified using purely nonstrategic rationales (17 out of 69). These data do not necessarily suggest that 25% of the activities were originally conceived and undertaken for nonstrategic reasons. Many of the activities may well have been selected to increase financial returns, that is, for strategic reasons. Rather, what the data do tell us is that after the fact, the president's letter to shareholders justifies the socially responsible activities using a significant amount of nonstrategic language.

2. Managers will use nonstrategic language to justify corporate activities. This finding is strengthened if we include those activities in which the letter used both strategic and nonstrategic language to justify the same activity. Twenty-one percent of the total CSR activities were justified using a combination of reasons (17 out of 81). Therefore, combining nonstrategic justifications with justifications using both strategic and nonstrategic language we find that 42% of all CSR activities were justified using some nonstrategic language. We can push further and also state that of the 69 activities in which some justification was provided, just under half used at least some nonstrategic justification (as opposed to pure strategic justification).
3. Of the 81 CSR activities disclosed, the most likely set of activities which are justified using either pure nonstrategic language or both nonstrategic and strategic language combined are those activities designed to help solve national social problems. Forty-one percent of the pure nonstrategic justifications are associated this category (7 out of 17).
4. Comparing the results of Groups 1 and 2 shows that Groups 1 firms are almost twice as likely to justify activities using nonstrategic language (Group 1: 11 nonstrategic justifications, Group 2: 6 nonstrategic justifications). Similarly, the Group 1 firms were almost twice as likely to justify activities using both strategic and nonstrategic justifications (Group 1: 11 combination justifications, Group 2: 6 combination justifications). These results should not be overstated. While it is certain that Group 1 firms are more likely to use some nonstrategic justification, this may simply be due to the finding that they are much more likely to disclose CSR activities. (For example, if we look at nonstrategic justifications as a percentage of all CSR activities, the Group 2 firms are actually more likely to use nonstrategic language.)

The above evidence supporting the existence of nonstrategic justifications for CSR activities can be interpreted in at least two distinct ways. First, it may simply indicate a sensitivity and a response to outside pressure groups. If we maintain, as Benston would have it, that shareholders, managers, and customers are opposed to CSR and view it, at best, as a drain on corporate resources, then these nonstrategic justifications are merely "self-serving exercises" (p. 100), to quote Benston once again. The nonstrategic language is used to bluff pressure groups. According to this interpretation, the nonstrategic justifications are meaningless ornaments, and are understood as such by managers and shareholders in the know. Hence, the nonstrategic justifications really constitute a different level of strategic justification. Strategically speaking, it is in the interests of the corporation to use "nonstrategic" language. Undoubtedly, this view probably can explain some of the nonstrategic justifications. The following example of employee related activity would seem to support such an interpretation:

> In the fourth quarter, we established a provision of $78 million after tax to cover closing of the Chicago plant and certain operating departments at other U.S. plants in connection with the consolidation of their production at other manufacturing locations. The provision will cover the write-off of assets as well as retraining, relocation, separation allowances, and other personnel-related costs where applicable. Our paramount concern will be to ease and dislocation burden to our employees and to assist them during the transition period.

Nevertheless, we believe that this is not a full explanation; the strategic use of "nonstrategic language" would seem to be a poor strategy. Why would pressure groups be expected to accept the bluff? Noting that the position assumes that managers and shareholders are not fooled, would pressure groups not possess equal sophistication? Over time, at least, the strategy would become self-defeating.

By contrast, the examples of nonstrategic justification in the president's letter to shareholders is also consistent with the view articulated above; real CSR activity exists and managers "signal" this information to shareholders and other interested parties. This constitutes a second interpretation of the results. Nonstrategic justifications exist because managers believe that some nonstrategic justifications represent legitimate corporate concerns. While clearly an alternative to the classical view of the corporation, this explanation avoids the major limitation noted above.

Based on the four observations outlined above and regardless of how one interprets the data, the empirical results provide strong evidence in favor of accepting Hypothesis 2. Managers justify at least some CSR activities using nonstrategic language. Further, Group 1 firms are more likely than Group 2 firms to invoke nonstrategic language.

## CONCLUSION

We began this study with a question: Do socially responsible firms "signal" their identity through the annual report? Our evidence suggests that the answer to this question is yes. There exists a positive association between CSR performance (based on CEP analysis) and CSR disclosures (based on content analysis of president's letter to shareholders). We do not find these results surprising. At least some executives of major corporations perceive CSR activities as legitimate endeavors. Therefore, they conclude both that it is in their own interests and that they have obligations to carefully craft and prepare the president's letter to shareholders in order to effectively communicate corporate responsibilities and activities undertaken to satisfy them.

# APPENDIX A

***Table A.1.*** Socially Screened

| | *Group 1: Socially Screened* | *Group 2: Control* |
|---|---|---|
| 1 | Baxter International Inc | Smithkline Beecham Plc—Ads |
| 2 | Ben & Jerry's Homemde—Cl A | Dreyer's Grand Ice Cream Inc |
| 3 | Clorox Co-Del | NCH Corp |
| 4 | Cummins Engine | Brunswick Corp |
| 5 | Delta Air Lines Inc. | AMR Corp—Del |
| 6 | Federal Express Corp | Airborne Freight Corp |
| 7 | Fuller (H.B.) Co | Loctite Corp |
| 8 | Gannett Co | Times Mirror Co—Del, -Ser A |
| 9 | Hawaiian Electric Industries | Puget Sound Power & Light |
| 10 | Heinz (H.J.) Co | Cpc International Inc |
| 11 | Hershey Foods Corp | Savannah Foods & Industries |
| 12 | Houghton Mifflin Co | Western Publishing Group Inc |
| 13 | Huffy Corp | Harley-davidson Inc |
| 14 | Kellogg Co | American Maize—products, Cl A |
| 15 | Knight-Ridder Inc | New York Times Co—Cl A |
| 16 | Eastman Kodak Co | Canon Inc—ADR |
| 17 | Lifeline Systems Inc | Pico Products Inc |
| 18 | Maytag Corp | Whirlpool Corp |
| 19 | Merck & Co | American Home Products Corp |
| 20 | Miller (Herman) Inc | Kimball International—Cl A |
| 21 | Penney (J.C.) Co | Ito Yokado Co Ltd—ADR |
| 22 | Pitney Bowes Inc | General Binding Corp |
| 23 | Polaroid Corp | Ricoh Co Ltd—ADR |
| 24 | Procter & Gamble Co | Colgate-Palmolive Co |
| 25 | Rouse Co | Vornado Inc |
| 26 | Rubbermaid Inc | Illinois Tool Works |
| 27 | Ryder System Inc | Rollins Truck Leasing |
| 28 | Quaker Oats Co | Borden Inc |
| 29 | Safety-Kleen Corp | Sotheby's Holdings—CL A |
| 30 | Stride Rite Corp | Wolverine World Wide |
| 31 | Tennant Co | Tokheim Corp |
| 32 | Tootsie Roll Industries | Mei Diversified Inc |
| 33 | Weyerhaeuser Co | Georgia-Pacific Corp |

## NOTES

1. A similar, and perhaps more radical, view than Benston's has recently been put forth by Baruch Lev. Accounting researchers are often reticent about offering normative guidelines to corporate managers. Most accounting research is self-described as a branch of "positive" economics. Therefore, it is with great anticipation and interest that one approaches the recent article by Berkeley Professor Baruch Lev where he adopts an unabashedly normative tone (1992).

In this article, Lev suggested that corporate executives involved in financial reporting activities need to evaluate their information disclosure decisions using cost-benefit analysis. He labeled his program "Information Disclosure Strategy." Throughout his analysis, Lev emphasized the need to view disclosure decisions as fundamentally similar to other corporate activities. Accordingly, he began his paper by noting:

> Most importantly, disclosure activity does not differ in principle from other corporate activities, such as investment, production, and marketing. Disclosure shares with these activities the fundamental characteristics of providing benefits and incurring costs, and it therefore warrants the careful attention and long-term planning accorded to any major corporate activity. Hence the need for an information disclosure strategy (p. 10).

2. Ingram and Frazier (1980) point out that not only are CSR disclosures generally voluntary, but they are usually unaudited. "Few efforts have been made to monitor firms' social activities or to validate their disclosures so that motivation may exist for management to distort voluntary disclosures, to the extent that these disclosures reflect aspects of managements' relative performances" (p. 614).

3. For interesting examples of the use of content analysis outside the area of CSR, see Ingram and Frazier (1983), Staw, McKenzie, and Pufffer (1983), and Bowman (1984).

4. An analysis of ownership structure of the Group 1 and 2 firms revealed no statistically significant differences. Specifically, we tested the percentage of insider ownership for the two groups.

5. The rationale for our choice of the president's letter as the sampling unit was provided in the introduction.

## REFERENCES


Abbott, W.F., and R.J. Monsen. 1979. On the measurement of corporate social responsibility: Self-reported disclosures as a method of measuring corporate social involvement. *Academy of Management Journal.* 22(3): 501-515.

Anderson, J.C., and A.W. Frankle. 1980. Voluntarity social reporting: An iso-beta portfolio analysis. *The Accounting Review* (LV (3): 467-479.

Benston, G.J. 1982. Accounting and corporate accountability. *Accounting, Organizations and Society* 7(2): 87-105.

Bowman, E.H. 1984. Content analysis of annual report for corporate strategy and risk. *Interfaces* 14(1): 61-71.

Council on Economic Priorities, M. Alperson, A.T. Marlin, A. Tepper, J. Schorch, and R. Will. 1991. *The Better World Investment Guide.* New York: Prentice Hall.

Cowen, S.S., L.B. Ferreri, and L.D. Parker. 1987. The impact of corporate characteristics on social responsibility disclosure: A typology and frequency-based analysis. *Accounting, Organizations and Society* 12(2): 111-122.

Engel, D.L. 1979. An approach to corporate social responsibility. *Stanford Law Review* 32(1): 1-97.

Epstein, M.J., and M.L. Pava. 1993. *Studies in Managerial and Financial Accounting,* Vol. 2: *The Shareholder's Use of Corporate Annual Reports.* Greenwich, CT: JAI Press.

Financial Accounting Standards Board (FASB) 1980. *Statedment of Financial Accounting Concepts No. 2: Qualitative Characteristics of Accounting Information.* Stamford, CT: FASB.

Freedman, M., and B. Jagi, 1982. Pollution disclosures, pollution performance and economic performance. *The International Journal of Management Sciences* 10(2): 167-176.

Gibbins, M., A. Richardson, and J. Waterhouse. 1990a. The management of corporate financial disclosure: Opportunism, ritualism, policies, and processes. *Journal of Accounting Research* 28(1): 121-143.

Goodpaster, K.E. 1991. Business ethics and stakeholder analysis. *Quarterly Journal of Business Ethics* 1: 53-73.

Ingram, R.W. 1978. An investigation of the information content of (certain) social responsibility disclosures. *Journal of Accounting Research* 16(2): 270-285.

Ingram, R.W., and K.B. Frazier. 1980. Environmental performance and corporate disclosure. *Journal of Accounting Research* 18(2): 614-622.

Ingram, R.W., and K.B. Frazier. 1983. Narrative disclosures in annual reports. *Journal of Business Research* 11: 49-60.

Lev, B. 1992. Information disclosure strategy. *California Management Review* (Summer): 9-30.

Lindsay, A.D. 1962. *The Modern Democratic State.* New York: Oxford University Press.

Pava, M.L., and M.J. Epstein. 1993. How good is MD&A as an investment tool? *Journal of Accountancy* 175(3): 51-53.

Roberts, R.W. 1992. Determinants of corporate social responsibility disclosure: An application of stakeholder theory. *Accounting, Organizations and Society* 17(6): 595-612.

Shane, P., and B. Spicer. 1983. Market response to environmental information produced outside the firm. *The Accounting Review* 58(3): 521-538.

Staw, B.M., P.I. McKenzie, and S.M. Puffer. 1983. The justification of organizational performance. *Administrative Science Quarterly* 28: 583-598.

Stone, C.D. 1975. *Where the Law Ends.* New York: Harper & Row.

Trotman, K.T., and G.W. Bradley. 1986. Associations between social responsibility disclosure and characteristics of companies. *Accounting, Organizations and Society* 6(4): 355-362.

Ullmann, A.A. 1979. Corporate social reporting: Political interests and conflicts in Germany. *Accounting, Organizations and Society* 4(1/2): 123-133.

Wiseman, J. 1982. An evaluation of environmental disclosures made in corporate annual reports. *Accounting, organizations and Society* 7(1): 53-63.

Wolfe, R. 1991. The use of content analysis to assess corporate social responsibility. In *Research in Corporate Social Performance and Policy,* Vol. 12, ed. J.E. Post, 281-307. Greenwich, CT: JAI Press.

# RECOGNIZING ETHICAL ISSUES:
## THE JOINT INFLUENCE OF ETHICAL SENSITIVITY AND MORAL INTENSITY

Gail B. Wright, Charles P. Cullinan, and Dennis M. Bline

**ABSTRACT**

Models of ethical reasoning include four dimensions: recognizing an ethical issue, making a moral judgment, and creating an intention leading to a behavior. Much research to date has focused on the second stage—making a moral judgment—to the exclusion of other stages. Moral reasoning, as described by Kohlberg (1969, 1976) underlies moral judgment and is often measured in research studies by Rest's Defining Issues Test (DIT) (1986). Recognizing the ethical dimensions of an issue is critical to employing morally based reasoning. The present study extends the research of Shaub et al. (1993) and Jones (1991) by combining ethical sensitivity and moral intensity into a model of ethical issue recognition, the first component of the ethical reasoning model, and by examining the relationship between ethical issue recognition and moral judgment. Based on an educational intervention, the study concludes that ethical issue recognition is a function of both the subject's ethical sensitivity and the moral intensity of the issue. Interventions based on stakeholder theory increased subjects' ability to recognize ethical issues but did not change their levels of moral

**Research on Accounting Ethics, Volume 4, pages 29-52.**

**ISBN: 0-7623-0339-5**

reasoning. No relationship was found between changes in ethical issue recognition and changes in levels of moral reasoning measured by the DIT.

# RECOGNIZING ETHICAL ISSUES: THE JOINT INFLUENCE OF ETHICAL SENSITIVITY AND MORAL INTENSITY

Models of ethical reasoning generally have several common elements, including recognizing an ethical issue, decision making, resulting intentions and behavior. Much of the published research centers on the second stage of the model (decision making), measuring the level of moral development employed in the decision making process. The emphasis on moral reasoning may result from the availability of a widely accepted instrument, the Defining Issues Test (DIT) (Rest 1986), to measure moral reasoning. Recent research has focused on understanding what happens in the first stage of the model when a moral agent faces an ethical dilemma.

Shaub, Finn, and Munter (1993) examined the first stage of recognizing an ethical issue, which they labeled ethical sensitivity. The present study extends the work of Shaub et al. (1993) by examining ethical issue recognition as a two part process. Ethical sensitivity is examined in light of a second aspect of recognition, the moral intensity of the issue (Jones 1991). This study used an educational intervention to investigate: (1) whether moral intensity of issues and a process for increasing subjects' awareness of the ethical dimensions of an issue improves their ability to recognize ethical issues and (2) whether changes in ethical issue recognition are associated with changes in moral reasoning.

The next section of the paper presents the literature relevant to the study. The subsequent section presents the methods. The final two sections present the results and conclusions.

## LITERATURE REVIEW

### Theoretical Framework of Ethical Behavior Models

Two fundamental approaches to the study of ethics are deontology and teleology. Deontology focuses on the goodness of actions or behaviors themselves, examining decision making at the *individual* level. Teleology emphasizes the consequences of actions or behaviors, looking for the best balance of good over evil, examining consequences of decisions at the *societal* level. Most models of ethical decision making in business (e.g., Trevino 1986; Hunt and Vitell 1986; Ferrell and Gresham 1985) are positive and descriptive and are developed along deontological lines to examine individual decisions about behaviors. Hunt and Vitell (1986) also incorporated a teleological focus on consequences to society.

Rest (1986) presents a four-component model of *individual* behavior to describe the psychological process that occurs when a moral agent faces an ethical dilemma. He states that a "person must have performed at least four basic psychological processes" (p. 3) and he stresses that these are processes, not traits. These processes are: (1) recognize that a moral issue exists, (2) make a moral judgment, (3) establish moral intent, and (4) act on the intent. Rest states that the four processes "are presented in a logical sequence, as an analytical framework for depicting what must go on for moral behavior to occur" although some research "suggests complicated interactions among the components" (p. 5). Both earlier and more recent models have been compared to and fit into the four-component model (Jones 1991).

Although decision making is an individual process, decision makers, especially in business, should consider the consequences of their actions. Stakeholder theory builds a bridge between the individual's decisions and the consequences of their behaviors to society. Stakeholder theory takes a teleological approach and relates the business entity to groups (*macro* level) that the entity affects, including employees, shareholders, customers, suppliers, and public stakeholders. A focus on stakeholder groups rather than on society as a whole, is supported by Clarkson (1995, 92) who observes that "corporations manage relationships with stakeholder groups rather than with society as a whole."

The core of stakeholder theory lies in a subset of ethical principles which guide the organization and its leaders in decisions which affect various groups. Donaldson and Preston (1995), among others, credit Freeman (1984) with providing the foundation for the descriptive accuracy, instrumental power and normative validity of stakeholder theory and returning it to prominence within management theories. Jones (1995) posits that when management actively considers the needs and desires of its stakeholders, significant competitive advantages accrue to the entity they manage.

## Measuring Moral Development

The second stage of Rest's model (make a moral judgment) is anchored in the decision maker's level of moral development and the work of Kohlberg (1969, 1976). Kohlberg's (1969) theory of cognitive moral development is the cornerstone of much empirical research in ethical behavior. His framework identifies justice as the universal principle that reaches to the highest level of moral development. Kohlberg identified three levels of moral reasoning, each divided into two stages. The preconventional level (Stages 1 and 2) is concerned with personal consequences, incorporating first a punishment and obedience orientation, then an instrumental relativist orientation where decisions are made in relative terms rather than in absolute terms. In the conventional level, the individual's reasoning is governed by "good boy—nice girl" thinking in Stage 3 and a concern for law and order in Stage 4. It is in Stage 4 that the orientation turns notably away from self to consider the greater good. At the postconventional level, reasoning is abstracted, so that moral principles are defined outside societal or peer pressure. Stage 5 exhibits a social-contract legalistic orientation, while Stage 6 is the stage of universal ethical principles.

Progression through the stages occurs as an invariant sequence. As a result of education, life experiences, age, and so forth, individuals may progress to higher stages or levels of moral reasoning but will do so without skipping stages; termination of development may occur at any stage (Weber 1990). Whereas individuals do not regress to lower stages than they have achieved, they may employ reasoning at a stage above or below their dominant stage in different circumstances and depending upon the issues involved (Elm and Weber 1994). Stages of development represent ways of reasoning about moral dilemmas rather than attitudes about ethical issues.

Kohlberg (1969) based his theory on the works of Piaget (1932) and a 15-year longitudinal study of 75 boys. Through later studies, Kohlberg and his associates developed and refined their research instrument, the Moral Judgment Interview (MJI), and its scoring (Colby et al. 1983; Kohlberg 1976). This instrument contains a structured interview for subjects who respond to questions about several moral dilemmas. The MJI has enjoyed limited use in the study of ethics in business because of the complex interview process it employs (see, e.g., Weber 1990).

Rest (1986) developed the Defining Issues Test (DIT) to study moral judgment and changes in moral reasoning, the second component of the model. Rest (1986) applied Kohlberg's cognitive moral development theory to evaluate the agent's stage of development in the moral judgment (second) component of his four component model. The DIT is an instrument that uses six moral dilemmas, including Kohlberg's Heinz dilemma, to measure moral development.

Questions have arisen about the appropriate application of instruments such as the DIT and the MJI to measure moral reasoning in business situations. In a study of the moral reasoning levels of managers conducted in the manner of Kohlberg's (1976) MJI, Weber (1990) combined two moral dilemmas anchored in business situations with the Heinz dilemma (Kohlberg 1976). The levels of moral reasoning employed by his subjects were higher for the Heinz dilemma than for the dilemmas set in a business context. Weber noted the importance of the role that the issue respondents confront plays in evoking moral reasoning and he attributed the difference in levels of moral reasoning to the issues involved and their consequences. These findings support contentions that, while the DIT may reflect the dominant level of moral reasoning, other stages above or below the dominant stage may be employed when circumstances are of limited scope, such as business-related situations.

## Measuring Ethical Issue Recognition: Moral Intensity and Ethical Sensitivity

A common element implicit in all the models discussed above is the recognition of the issue's ethical dimensions (Ferrell et al. 1989; Hunt and Vitell 1986; Rest

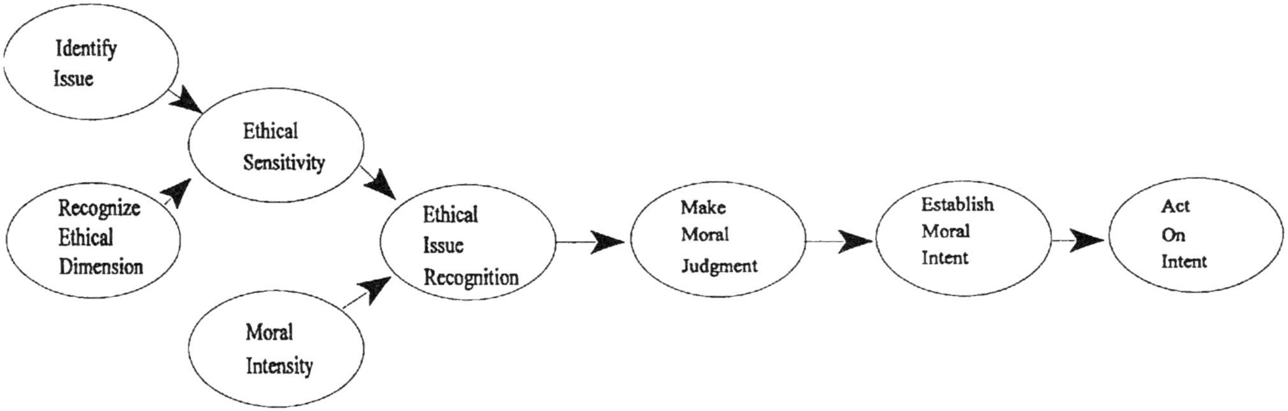


*Figure 1.* Model of Ethical Issue Recognition

1986; Trevino 1986; Ferrell and Gresham 1985). Ferrell et al. (1989, 59) state: "The individual first must perceive the situation as a problem having *ethical* content before the model can be applied" (emphasis added). Jones (1991, 380) explicitly recognized the implications and consequences of this component of the decision model for the study of ethics: "a person who fails to recognize a moral issue will fail to employ moral decision-making schemata and will make the decision according to other schemata, economic rationality, for example." A model of ethical issue recognition, as an elaboration of the first stage of the four component model, is presented in Figure 1.

Recognition of the ethical dimensions of an issue is a significant, if often overlooked, first element common to all models of ethical decision making which derives from two separate constructs: the ability of the moral agent to recognize the ethical dimension (Shaub et al. 1993) and the moral intensity of the issue itself (Jones 1991). Ethical sensitivity is the moral agent's ability to recognize that an issue exists and that it has ethical dimensions. Taken together with the moral intensity of the issue, these two constructs comprise ethical issue recognition. Figure 1 portrays the two components of ethical issue recognition: an individual's ethical sensitivity and the issue's moral intensity.

Definitions of the term "ethical sensitivity," as used in previous research, have been inconsistent. Shaub et al. (1993) conducted a study of public accountants on the relationship among ethical sensitivity, ethical orientation, and the auditor's commitment to organization and profession. They emphasized the nature of the issue involved and defined ethical sensitivity as the ability to "recognize the ethical nature of a situation in a professional context" (p. 146). Hebert et al. (1990, 141) noted that "one aspect of ethical sensitivity [is] the ability to recognize ethical issues." Thus, both Shaub et al. (1993) and Hebert et al. (1980) define ethical sensitivity in relation to the first component of Rest's four-component model. Cohen et al. (1992, 1) focused on the actions taken (step four of the four component model) and described ethical sensitivity as "the evaluation of, or belief concerning whether a particular action is considered to be ethical."

The measurement of ethical sensitivity has also varied. Hebert et al. (1990, 1992) and Shaub et al. (1993) measured the ability of various groups to identify ethical issues. These studies used vignettes embedded with ethical issues to evaluate respondents' ethical sensitivity. Hebert et al. (1990, 1992), in studies using medical students, asked students "to list all the ethical issues related to the case they received" (1990, 141), thus ensuring that respondents focused only on ethical issues. The instructions given by Shaub et al. (1993) asked participants to identify issues that the respondents believed were important in the context of the vignette developed by the authors, without sensitizing them to the study's focus on ethics. As portrayed in Figure 1, we view ethical sensitivity as comprised of two distinct components: identifying an issue which requires resolution (labeled "Identify Issue" in Figure 1), and being aware that the issue has an ethical dimension (described as "Recognize Ethical Dimension in Figure 1).

The second construct of ethical issue recognition involves the severity of characteristics of the issue itself. Jones (1991) constructed an issue-contingent model of ethical decision making to study the effects of moral issues by superimposing moral intensity on Rest's four-component model. Jones (1991, 371) notes that "characteristics of the moral issue itself, collectively called moral intensity, are important determinants of ethical decision making and behavior." The six elements of moral intensity are magnitude of consequences, social consensus, probability of effect, temporal immediacy, proximity, and concentration of effect. The concept of moral intensity indicates that consequences of behavior bear on the human response or recognition of the moral elements of the issue.

Jones (1991, 373) considers moral intensity to be a pervasive construct, "derived from the normative arguments of moral philosophers who differentiate levels of moral responsibility based on proportionality." This suggests a direct relationship between moral intensity and the moral agent's responsibility to consider consequences to others in society when an ethical issue has been recognized. Moral intensity is the sum total of issue characteristics that make its moral dimensions transparent to the observer because of salience and vividness. Jones defines salience as the extent to which an issue stands out from its background. Vividness is the ability to provoke an emotional interest, concrete imagery, and feelings of proximity in sensory, temporal or spatial ways (Nisbett and Ross 1980). Thus, Jones' model emphasizes the significance of the *issue* and its ability to be recognized.

The moral intensity of issues is incorporated in Figure 1 as a factor influencing ethical issue recognition. For a given level of ethical sensitivity, a person is more likely to recognize issues having greater moral intensity. Thus, the likelihood of a person realizing the need to apply a moral framework to an issue (ethical issue recognition) is a function both of the *individual's* ethical sensitivity, and the moral intensity of the *issue* itself.

Newstrom and Ruch (1975, 1976) developed a scale to measure managers' perceptions of the severity or moral intensity of 17 discrete and nonrepetitive behaviors found in the workplace. Akaah and Lund (1994) subsequently validated the original scale and reduced it to six ethical behavior subscales. Newstrom and Ruch (1975) found that their subjects listed all behaviors as basically unethical. They provide some evidence about managers' perceptions of the seriousness (moral intensity) of the 17 issues in their rank ordering of mean values.

## Ethics and Education

Beginning in the mid-1980s, practitioners called for an increase in instruction in ethics, particularly for students in accounting. Cohen and Pant (1989) reported that there was a significant increase in the instruction in ethics for accountants, although not as much as practitioners desired. The American Accounting

Association's Committee on the Future Structure, Content and Scope of Accounting Education (Bedford Committee 1986), the National Commission on Fraudulent Financial Reporting (Treadway Commission 1987), and the American Assembly of Collegiate Schools of Business (1988) are among the many groups who have recommended integrating ethics into the accounting curriculum. Loeb (1988, 321) states that the goals of ethics education in accounting are to stimulate a student to "b) *recognize issues* in accounting that have *ethical* implications, [and] c) develop a sense of moral obligation or responsibility" (emphasis added).

Langenderfer and Rockness (1989) argue that integrating ethics into the accounting curriculum to educate accounting majors about the ethical dilemmas faced by professionals in the business world might increase students' awareness of significant, ethical dimensions of decisions they will be called on to make. Their suggested approach to teaching ethics cases involves seven discussion steps. The steps are: (1) identify the facts; (2) identify the ethical issues; (3) identify the norms, principles, and values related to the case; (4) name alternative courses of action; (5) choose the best course of action consistent with the norms, values, and principles; (6) relate the consequences to each possible course of action; and (7) make a decision. Since stakeholders' interests are implicit when identifying ethical issues and considering the consequences of each course of action, the seven step approach recommended by Langenderfer and Rockness is an application of the stakeholder framework to accounting and auditing.

In terms of Rest's (1986) four-component model, much of the research using ethical interventions in education to date has emphasized the second component, making a moral judgment (e.g., Armstrong 1993; Ponemon 1992; Ponemon and Gabhart 1990; Ponemon and Glazer 1990; St. Pierre, Nelson, and Gabbin 1990; Armstrong 1987), using Rest's DIT as the measurement instrument to identify levels and/or changes in moral reasoning. Only recently have studies focused on the ability of respondents to identify the ethical dimension of an issue to ensure application of the proper schemata for evaluation (e.g., Shaub et al. 1993; Hebert et al. 1992; Hebert et al. 1990).

Ethical interventions have generally used cases or vignettes as a vehicle to sensitize students to ethical issues (Claypool, Fetyko, and Pearson 1990). Reasoning models generally use either a stakeholder theory framework (Langenderfer and Rockness 1989) or a moral reasoning approach (Armstrong 1993). To date, findings are inconclusive as to whether the levels of moral development exhibited by accountants and accounting students are affected by such interventions. Some studies report positive results, that is, increases in levels of moral reasoning or moral reasoning above that of comparable groups (Armstrong 1993; Hiltebeitel and Jones 1991). Other studies report no increase or levels of moral reasoning below that of liberal arts students (Ponemon 1993; St. Pierre, Nelson, and Gabbin 1990). Implicitly, these authors assume that instruction in the decision making process provides adequate sensitization to issues as having an ethical dimension so that the appropriate schemata can be employed.

## HYPOTHESES

This study examines the first step in the Rest (1986) four-component model, ethical issue recognition and its components, ethical sensitivity and moral intensity. The study also considers the relationship between ethical issue recognition and moral judgment.

The model depicts the need to identify an ethical issue. Since issue identification is a necessary step to applying a moral framework, an ethical intervention which increases issue identification would lead to a greater likelihood of an agent applying a moral framework when an ethical issue arises. Thus, the first hypothesis is:

**Hypothesis 1.** Intervention based on stakeholder theory will enhance subjects' ability to identify issues with ethical dimensions.

Ethical sensitivity is the ability to recognize an issue and its ethical dimensions, when applicable. The stakeholder theory framework recommended by Langenderfer and Rockness (1989) begins by identifying the facts followed by identifying ethical issues. The next step in our process incorporates the subjects' beliefs about whether the issues identified are ethical. Even if intervention does not result in ability to recognize more ethical issues, it may result in the ability to better recognize that the ethical issues have an ethical dimension. This leads to the second hypothesis:

**Hypothesis 2.** Intervention based on stakeholder theory will enhance subjects' ethical sensitivity.

Ethical issues will have differing levels of consequences to stakeholders and/or differing numbers of stakeholders. Jones (1991) concluded that the moral intensity of an issue, including its salience and vividness, determines whether the issue is identified along an ethical dimension. As the intervention based on stakeholder theory emphasizes the consequences of ethical decisions in terms of both intensity and number of stakeholders affected, one may expect participants would be more likely to recognize issues and their ethical dimensions when action resulting from the issue would have the most negative consequences to the greatest number of stakeholders. Thus, the third hypothesis is:

**Hypothesis 3.** Intervention based on stakeholder theory will enhance subjects' ethical issue recognition, giving consideration to the moral intensity of the issues and the ethical sensitivity of the subjects.

Rest (1986, 5) observed that the four components of his model "have distinctive functions" and noted that interactions have been found among the components. It has been noted that changes in moral development result from age, experience, and education and are not always observable because of the ability to employ levels of moral reasoning above or below the dominant stage in certain circumstances (Elm and Weber 1994). The intervention conducted here was limited to ethical issues relevant to business and specifically to the practice of auditing and there is no expectation that it will affect the level of broader moral development of subjects. Therefore, Hypothesis 4 is:

**Hypothesis 4.** Intervention based on stakeholder theory will not change the level of moral reasoning measured by the DIT.

Rest's four-component model begins with recognition of an ethical issue to which a moral judgement is applied. However, this model is sequential, not causal (Rest 1986), and therefore no relationship is implied between changes in the ability to recognize ethical dimensions of an issue and changes in the level of moral reasoning other than Rest's observation that all four basic processes of the model must be performed. When the decision-making environment is limited in scope, as in the present study, changes in ethical issue recognition, while increasing the likelihood of applying ethical schemata, would not necessarily relate to changes in moral reasoning. Thus, changes in ethical issue recognition are expected to be independent of changes in the subject's level of moral reasoning as measured by the DIT. This can be represented in the following hypothesis:

**Hypothesis 5.** There is no correlation between changes in ethical issue recognition for business dilemmas and changes in moral development measured by the DIT.

## RESEARCH METHODS

### Research Design

Data were gathered in the current study using a pretest, posttest design to examine changes in ethical issue recognition and levels of moral development. Senior accounting majors in two sections of auditing at a private college in the northeastern United States were subjects for an educational intervention. One of the researchers distributed and collected the questionnaires; participation was voluntary.

Partners at two Big Six accounting firm offices reviewed the vignettes developed for this project to ensure that the issues embedded in the vignettes were ones that might be encountered by an auditor in the context of an audit engagement (see Appendix A.) Subsequently, to minimize the possibility of misinterpreting questions, the scales were pilot tested in an auditing class that did not participate in the study. Students were asked to provide narrative feedback on the instrument and the

**Table 1.** Moral Intensity of Ethical Issues

| *Behavior Categories*[*] *(Newstrom and Ruch, 1975; Akaah and Lund, 1994)* | *Vignette* | *Ethical Issue* | *Moral Intensity Weighting*[**] |
|---|---|---|---|
| A. Personal use | | | |
| Doing personal business on company time | Pretest | Renewing driver's license on work time | 1 |
| B. Passing Blame | | | |
| Passing blame for errors to innocent co-worker | Posttest | Blaming co-worker for own failures | 3 |
| C. Bribery | | | |
| Accepting gifts/favors for preferential treatment | Posttest | Accepting gift certificate for product recommendation | 2 |
| D. Falsification | | | |
| Falsifying information | Posttest | Classifying experimental treatments as traditional | 3 |
| F. Deception | | | |
| Divulging confidential information | Pretest | Divulging information to professor | 3 |
| Not recording work above budget[***] | Pretest | Not billing all work hours ("eating time") | 1 |
| G. Other | | | |
| Accounting specific items | Posttest | Nonpayment of last year's audit fee | 1 |

***Notes:*** [*] Category E, Padding expenses, was not included in these vignettes.
[**] 1. Somewhat unethical behavior
2. Unethical behavior
3. Highly unethical behavior
[***] Similar to, but not explicitly identified by Newstrom and Ruch (1975) as "Taking longer than necessary to do a job".

***Table 2.*** Descriptive Statistics

| | *Issue* | *Treatment* | *Control* |
|---|---|---|---|
| Pre | Renewing driver's license on work time | .69 | .65 |
| | Divulging information to professor | .86 | 1.00 |
| | Not billing all work hours | .59 | .43 |
| | Mean ethical issues identified | 2.14 | 2.08 |
| | Mean ethical sensitivity (mean issues × ethics weighting) | 10.69 | 11.61 |
| | Mean ethical issue recognition (mean sensitivity × moral intensity weighting) | 18.90 | 22.04 |
| Post | Client failure to pay previous year audit fees | .79 | .87 |
| | Classifying experimental treatments as traditional | .76 | .65 |
| | Accepting gift certificate | .83 | .78 |
| | Blaming co-workers for own failure | .86 | .61 |
| | Mean ethical issues identified | 3.24 | 2.91 |
| | Mean ethical sensitivity (mean issues × ethics weighting) | 15.90 | 13.91 |
| | Mean ethical issue recognition (mean sensitivity × moral intensity weighting) | 37.00 | 30.78 |

clarity of questions and directions. Feedback from practitioners and students lead to minor modifications.

Pretest data were gathered early in the semester and followed by an ethical intervention in the treatment group. The intervention involved discussing three auditing cases over the next seven weeks that included ethical and technical auditing/accounting aspects. The treatment group discussed both the technical and ethical aspects of the case while the control group reviewed only the technical auditing/accounting issues. The seven-step framework for stakeholder theory (Langenderfer and Rockness 1989) was used to sensitize the treatment group to ethical considerations of the consequences to society from the auditor's behavior. Posttest data were gathered near the end of the semester, one week after completion of the intervention. Matched pretest and posttest measures were completed by 52 students.

Seven ethical issues were embedded in two vignettes (pretest and posttest) developed by the researchers along with several accounting/auditing technical issues. Six of the ethical issues were chosen from a list of 17 discrete and nonrepetitive issues identified by Newstrom and Ruch (1975). The seventh issue related to the AICPA Code of Professional Conduct. The ethical issues were weighted for moral intensity based on rank orderings provided by Newstrom and Ruch (1975). Ethical issues embedded in the vignettes were classified as: (1) somewhat unethical behavior, (2) unethical behavior, or (3) highly unethical behavior (see Table 1).

***Table 3.*** Analysis of Change in the Number of Ethical Issues Identified

| *Source* | *DF* | *Sum of Squares* | *Mean Square* | *F value* | *Pr > F* |
|---|---|---|---|---|---|
| Model | 3 | 2.398 | 0.800 | 0.60 | 0.6179 |
| Error | 47 | 62.582 | 1.331 | | |
| Corrected total | 50 | 64.980 | | | |
| *R*-square | 0.037 | | | | |
| *Source* | *DF* | *Type I SS* | *F value* | *Pr > F* | |
| Treatment/Control | 1 | 1.018 | 0.76 | 0.3864 | |
| Grade point average | 1 | 1.286 | 0.97 | 0.3307 | |
| Ethics course | 1 | 0.094 | 0.07 | 0.7912 | |
| *Source* | *DF* | *Type III SS* | *F value* | *Pr > F* | |
| Treatment/Control | 1 | 1.151 | 0.86 | 0.3573 | |
| Grade point average | 1 | 1.207 | 0.91 | 0.3460 | |
| Ethics course | 1 | 0.094 | 0.07 | 0.7912 | |

Students completed an issue identification score sheet similar to Shaub et al. (1993, 159) to identify issues and answered several questions about each issue after reading the vignettes. An ethical sensitivity weighting was based on the student's determination on a seven point Likert-type scale of whether Sarah (Monica) would "resolve this issue based on her accounting/auditing knowledge or on her ethical values?" Student assessment of the importance of the issue was also evaluated on a seven point likert scale ranging from "very important" to "very unimportant." Other questions regarding the issues identified were included to minimize the likelihood of sensitizing the students to the objective of the study (see Appendix B).

After completion of the ethical issue recognition instruments, students completed a three-scenario version of Rest's (1986) DIT to measure moral development. The same three cases were used for both the pretest and posttest measures to prevent changes in moral reasoning to be inferred from differences in the cases. Demographic information was collected at the posttest to provide control variables in the statistical analysis.

## Data Analysis

The data were analyzed in stages, following the model of ethical issue recognition. Stage 1 consisted of issue identification while Stage 2 weighted the ethical issues recognized by the degree to which the subjects considered the issues to be ethical, creating an ethical sensitivity value. In the third stage, ethical sensitivity was multiplied by the moral intensity of the issues to create the value for ethical

**Table 4.** Analysis of Change in Ethical Sensitivity

| *Source* | *DF* | *Sum of Squares* | *Mean Square* | *F value* | *Pr > F* |
|---|---|---|---|---|---|
| Model | 3 | 189.931 | 63.310 | 1.27 | 0.2959 |
| Error | 47 | 2345.049 | 49.895 | | |
| Corrected total | 50 | 2534.980 | | | |
| *R*-square | 0.075 | | | | |
| *Source* | *DF* | *Type I SS* | *F value* | *Pr > F* | |
| Treatment/Control | 1 | 101.131 | 2.03 | 0.1611 | |
| Grade point average | 1 | 73.711 | 1.48 | 0.2303 | |
| Ethics course | 1 | 15.089 | 0.30 | 0.5850 | |
| *Source* | *DF* | *Type III SS* | *F value* | *Pr > F* | |
| Treatment/Control | 1 | 88.061 | 1.76 | 0.1904 | |
| Grade point average | 1 | 79.660 | 1.60 | 0.2126 | |
| Ethics course | 1 | 15.089 | 0.30 | 0.5850 | |

**Table 5.** Analysis of Change in Ethical Recognition

| *Source* | *DF* | *Sum of Squares* | *Mean Square* | *F value* | *Pr > F* |
|---|---|---|---|---|---|
| Model | 3 | 1817.892 | 605.964 | 2.84 | 0.0494 |
| Error | 47 | 10125.441 | 215.435 | | |
| Corrected total | 50 | 11943.333 | | | |
| *R*-square | 0.152 | | | | |
| *Source* | *DF* | *Type I SS* | *F value* | *Pr > F* | |
| Treatment/Control | 1 | 955.553 | 4.44 | 0.0406 | |
| Grade point average | 1 | 757.267 | 3.52 | 0.0670 | |
| Ethics course | 1 | 105.072 | 0.49 | 0.4884 | |
| *Source* | *DF* | *Type III SS* | *F value* | *Pr > F* | |
| Treatment/Control | 1 | 851.129 | 3.95 | 0.0527 | |
| Grade point average | 1 | 806.075 | 3.74 | 0.0591 | |
| Ethics course | 1 | 105.072 | 0.49 | 0.4884 | |

issue recognition. Table 2 presents the scores for individual ethical issues and the mean ethical sensitivity and ethical issue recognition scores for the control and treatment groups for both the pretest and posttest.

The data were analyzed using the General Linear Models procedure and correlation analysis in SAS. Dependent variables in the general linear models

**Table 6.** Analysis of Change in Moral Reasoning

| *Source* | *DF* | *Sum of Squares* | *Mean Square* | *F value* | *Pr > F* |
|---|---|---|---|---|---|
| Model | 3 | 908.234 | 302.745 | 1.17 | 0.3331 |
| Error | 42 | 10882.346 | 259.103 | | |
| Corrected total | 45 | 11790.580 | | | |
| *R*-square | 0.077 | | | | |
| *Source* | *DF* | *Type I SS* | *F value* | *Pr > F* | |
| Treatment/Control | 1 | 730.452 | 2.82 | 0.1006 | |
| Grade point average | 1 | 134.324 | 0.52 | 0.4755 | |
| Ethics course | 1 | 43.458 | 0.17 | 0.6842 | |
| *Source* | *DF* | *Type III SS* | *F value* | *Pr > F* | |
| Treatment/Control | 1 | 752.793 | 2.91 | 0.0957 | |
| Grade point average | 1 | 125.434 | 0.48 | 0.4904 | |
| Ethics course | 1 | 43.458 | 0.17 | 0.6842 | |

were, for Hypotheses 1 to 4, respectively, the difference between the students' pretest and posttest measures for: (1) issue identification, (2) ethical sensitivity, (3) ethical issue recognition, and (4) moral development while the independent variable of interest was the student's membership in the control or treatment group. Other variables included in the model were overall college grade-point average and a dichotomous variable indicating whether the student had taken a course in ethics.

## RESULTS

The test of Hypothesis 1 investigated whether the intervention would change students' ability to identify issues that have an ethical dimension. Results of the test, presented in Table 3, indicate that changes in the number of issues identified are not significantly different between the control and treatment groups and none of the independent variables in the model (control versus treatment, grade-point average, and ethics course) are significant. The control versus treatment variable had an incremental $F$ value of 0.86 ($p$ = .36). As a result of these findings, Hypothesis 1 is not accepted.

Table 4 presents the results of testing Hypothesis 2. This hypothesis examined the impact of the ethics intervention on the students' ability to classify ethical issues identified. Taken alone, changes in ethical sensitivity (issues identified weighted by their ethical content) did not differ between the control and treatment groups ($p$ = .30). As with Hypothesis 1, the control versus treatment variable was not statistically significant ($p = .19$), and Hypothesis 2 is not accepted.

Hypothesis 3 is the test of ethical issue recognition, that is, combining ethical sensitivity with the moral intensity of the issues identified. Ethical sensitivity was weighted by moral intensity of the issues as suggested in the rank orderings of Newstrom and Ruch (1975).[1] The intervention, which emphasizes consequences across groups of stakeholders, significantly increased the ethical issue recognition ($p = .05$). In this analysis, the treatment versus control variable is a significant variable at approximately .05 and we can accept Hypothesis 3.

Results of the test of Hypothesis 4 are presented in Table 6. This hypothesis investigates the relationship between the ethical intervention and changes in moral reasoning as measured by the DIT. The changes in moral development did not differ significantly ($p < .33$) and results indicate that the ethical intervention did not have a significant impact on the level of moral reasoning for the control versus the treatment group ($p = .10$). As expected, an intervention which is limited in scope to ethical issues in business and auditing does not affect moral development in the broader sense, supporting Hypothesis 4.

Hypothesis 5, the relationship between changes in ethical issue recognition and changes in moral development, was tested using correlation analysis. The correlation coefficient was not significant ($r = .11$; $p = .47$). As a result of this test, we find no evidence of a relationship between changes in ethical issue recognition and changes in the level of moral development measured by the DIT, supporting Hypothesis 5.

## CONCLUSIONS

Ethical issue recognition, the first element in models which depict the moral reasoning process, has two aspects studied in prior works: ethical sensitivity (Shaub et al. 1993) and moral intensity (Jones 1991). The present study extends prior research by examining how they are linked in the recognition process and whether subjects can be sensitized to recognize the ethical dimensions of an issue. The relationship of recognition and moral reasoning was also explored.

The results indicate that recognizing an issue's ethical dimensions stems from the combination of an individual's sensitivity and the issue's moral intensity and that subjects can be sensitized to recognize ethical dimensions when issues are discussed in terms of the severity of their consequences to society. The intervention based on a stakeholder framework was effective in enabling subjects in the treatment group to recognize ethical issues. These findings have significance for improving behaviors in practice. According to Jones (1991), the greater the moral intensity of an issue, the greater the moral responsibility of the individual facing the dilemma; and, the greater the ability to recognize ethical issues, the more likely that individual will apply an ethical schemata to evaluate the issue.

While increasing ethical issue recognition is not linked to changes in the level of moral development in the present study, it follows that the ethical dimensions of an issue must be recognized to employ moral judgment or an ethical schemata to evaluate the issue. These findings are consistent with other studies (Ponemon 1993; St. Pierre et al. 1990) and support suggestions that subjects employ different levels of moral reasoning as appropriate (Elm and Weber 1994; Weber 1990). Thus, in terms of the sequencing of Rest's four-component model, improving ethical issue recognition may have positive effects by ensuring that an ethical dimension will be part of the analysis process that leads to behavior.

The stakeholder theory intervention focused participants' attention on the differing levels of moral intensity in issues. This included consequences of behavior to society, establishing social consensus, and evaluating the probability of effect. The result was an awareness of social concerns that leads to increased ethical issue recognition, making members of the treatment group more likely to recognize ethical issues with greater consequences to stakeholders than members of the control group. As a result, the individuals subjected to stakeholder theory interventions may be more likely to identify issues as ethical and apply an ethical schemata in the decision making process. The ability to identify and classify ethical issues as ethical did not differ between the control and treatment groups. When moral intensity and ethical sensitivity are joined in ethical issue recognition (Hypothesis 3) and subjects learn to consider the consequences of the auditor's actions, the stakeholder framework has a significant effect on the subject's ability to recognize ethical issues.

Elm and Weber (1994) indicated that individuals may employ reasoning at a stage above or below their dominant stage in different circumstances. It is possible that subjects would resolve issues at different levels of moral reasoning depending on the moral intensity of the issue involved and that subjects do not perceive the same moral intensity in dilemmas involving business as they would for socially sensitive issues such as those found in the DIT. Ethical interventions and pre- and posttest vignettes in this study were limited to those issues likely to occur to individuals in the practice of accounting and auditing. Instruction in the recognition of these types of issues may not be related to the moral reasoning levels required in situations such as the Heinz dilemma, as suggested by Elm and Weber (1994). As a result, educational interventions may achieve the goal of improving ethical issue recognition for business dilemmas while no change occurs in moral reasoning measured by the DIT.

The results of the current study must be interpreted in light of research limitations. The sample is a relatively small ($n = 52$) and homogeneous student group from a private college in the Northeast. A limited number of interventions occurred during a one-semester course in Auditing and follow-up studies to examine lasting effects have not been conducted. Additionally, behavioral research of this type is dependent on the participants taking the survey completion seriously.

There may be other factors that are related to ethical issue recognition and moral development. If different levels of moral reasoning are used in solving business-related ethical issues than are measured by the DIT as has been suggested, an instrument which measures levels of moral reasoning for business related issues should be developed and employed in future research to determine the effects of educational interventions on moral judgment for those situations.

Ponemon (1993) sought to answer the question, "Can ethics be taught in Accounting?" His conclusion was "no—at least with respect to moral development and free-riding behavior" and he also notes that "it may be necessary for educators to find pedagogical methods that are specifically designed to foster the moral development of accounting students" (p. 207). The current study further examined the effects of teaching ethics in the context of the first stage of Rest's model, the recognition of ethical issues. Our results indicate that measurement of ethical issue recognition needs to consider the moral intensity of the issue and the ethical sensitivity of the individual. We also found that reviewing cases using the stakeholder framework may help to sensitize students to ethical issues whose moral intensity is perceived to be stronger. More extensive use of cases involving ethical issues and applications of stakeholder theory may increase the ability to recognize issues of lower moral intensity.

## APPENDIX A

### Pretest Vignette

Below is a scenario you could encounter as an auditor. To the extent possible, place yourself in Sarah's shoes.

Different aspects of the scenario would vary in significance to you if actually encountered on the job. You will be asked to identify the issues found in each paragraph and answer questions following the scenario.

Sarah Hughes is a staff accountant with the CPA firm of Nobil, Tobias, and Thread. While auditing the revenue account of Greenway Grass Seed Co., she discovered that a material sales transaction had been recorded even though the product had not been shipped as of December 31, 1994, the end of Greenway's fiscal year. She discussed this matter with Tracy Morse, the controller of Greenway. Tracy indicated that the buyer, Leo's Landscaping, had been a Greenway client for a long time, and that they had agreed to purchase a fixed amount of seed each quarter for a two-year period. He further indicated that because there was a known buyer for the seed, the revenue could be recognized before shipment. When Sarah asked for documentation of this agreement, Tracy indicated the agreement was based on a handshake.

After lunch, Sarah went to Donna Monette, the Nobil, Tobias and Thread senior assigned to the Greenway audit, to discuss the transaction. Donna told

Sarah that she was too busy to discuss the issue because she would be meeting with Tracy Morse and Frank Sweeney, the audit partner, for most of the afternoon. Sarah was unsure of how to continue with her fieldwork, so she decided to return to the CPA firm office to work on other aspects of the Greenway audit. On her way to the office, she spent several hours at the Department of Motor Vehicles to renew her license.

In her MBA class that evening, Sarah mentioned that she was working on the Greenway audit and that Greenway had recorded revenue before the product was shipped for a particular order. Her professor told her this issue would be covered in a subsequent course in the MBA program, and there was not sufficient time to discuss that issue during the current class. As a result, Sarah was unable to get an authoritative answer as to the proper accounting for the transaction.

The following day, Sarah was still unable to decide whether the transaction was recorded in accordance with GAAP because she was unsure of the applicable rules. She again talked to Donna, who told her she could research the transaction further if she wanted to. To stay within the time budget, Sarah decided not to charge the firm for the hours she spent researching the matter.

## Posttest Vignette

Below is a scenario you could encounter during an audit. To the extent possible, place yourself in Monica's shoes.

Different aspects of the scenario would vary in significance to you if actually encountered on the job. You will be asked to identify the issues found in each paragraph and answer questions following the scenario.

Monica is the senior accountant for Moseky, Moseky and Cloot, CPAs. She is currently working on the audit of the Apple-a-Day Health Clinic. There are many doctors working on the staff who are involved in medical research, and this research attracts patients from across the country. Despite the profitability of the clinic, Apple-a-Day has always been slow in paying its audit fee. In fact, the clinic still owes the CPA firm last year's audit fee.

Health insurance companies often do not pay for experimental treatments. This has resulted in occasional cash flow difficulties for the clinic. Monica became aware of this situation and suggested to the client that some items could be classified as traditional treatments when billing the insurance company. Classifying experimental treatments as traditional would allow the clinic to collect from the insurance companies for treatments that would not otherwise be reimbursed. The client agreed that this approach would significantly improve its cash flow position and considered changes that could be made in the insurance reporting procedures.

To assist in implementing the new reporting procedure, Monica recommended that the client purchase a new software package called "InsurePay." The client purchased the software package from Bill Houseman, a college friend of Monica's. To encourage Monica to continue recommending "InsurePay," Bill gave Monica a gift certificate sufficient to buy dinner for two at Monica's favorite restaurant. The audit of accounts receivable began soon after the software was purchased. Monica knew the new system was going to be implemented soon, so she told the staff accountant on the engagement not to waste time testing the allowance for uncollectible accounts.

When the partner reviewed the workpaper files, he noted the lack of testing of the allowance account and indicated to Monica that this testing should have been done. He told Monica to return to the client and perform the testing. After testing the allowance for doubtful accounts, Monica found that the balance was significantly understated. The staff accountant received low evaluation marks from Monica as a result of failure to test the allowance account.

## APPENDIX B

### Pretest and Posttest Issue Recognition Sheet

Please write the last 4 digits of your Social Security Number on this line: ______

Please *Identify* any issues *Sarah* faces in this scenario. Any given paragraph may contain no issues, one issue, or more than one issue. *There is no need to fill in every issue line provided.* On the pages that follow, you will be asked a series of questions about each issue you identify.

*Issue #*

Paragraph 1

1 Issue ____________________________________________

____________________________________________

2 Issue ____________________________________________

____________________________________________

3 Issue ____________________________________________

____________________________________________

Paragraph 2

4 Issue ____________________________________________

____________________________________________

5 Issue ____________________________________________

____________________________________________

6 Issue ____________________________________________

____________________________________________

Paragraph 3

7 Issue ____________________________________________

____________________________________________

8 Issue ____________________________________________

____________________________________________

9 Issue ____________________________________________

____________________________________________

Paragraph 4

10 Issue ____________________________________________

____________________________________________

11 Issue ____________________________________________

____________________________________________

12 Issue ____________________________________________

____________________________________________

A. Issue Number

________

| | Unimportant | | | Moderately Important | | | Very Important |
|---|---|---|---|---|---|---|---|
| B. When considering the facts of this case, how important is the issue identified above? | ① | ② | ③ | ④ | ⑤ | ⑥ | ⑦ |

| | Highly Likely | | | Moderately Likely | | | Highly Unlikely |
|---|---|---|---|---|---|---|---|
| C. Is the identified above likely to have a significant impact on the financial statements? | ① | ② | ③ | ④ | ⑤ | ⑥ | ⑦ |

| | Entirely Ethical Values | | | Equally Ethical Values and Accounting/Auditing | | | Entirely Accounting/Auditing |
|---|---|---|---|---|---|---|---|
| D. Will Sarah resolve this issue based on her accounting/auditing knowledge or on her ethical values? | ① | ② | ③ | ④ | ⑤ | ⑥ | ⑦ |

| | No Impact | | | Little Impact | | | Great Impact |
|---|---|---|---|---|---|---|---|
| E. Evaluate the impact of this issue on the audit opinion? | ① | ② | ③ | ④ | ⑤ | ⑥ | ⑦ |
| F. What will be the likely impact of this issue on Sarah's job/career prospects? | ① | ② | ③ | ④ | ⑤ | ⑥ | ⑦ |

## NOTE

1. The weightings assigned using the Newstrom and Ruch rank orderings are supported by students' evaluation of the importance of the issues; the correlation between the two measures was significant at $p < .0001$.

## REFERENCES

Akaah, I.P., and D. Lund. 1994. The influence of personal and organizational values on marketing professionals' ethical behavior. *Journal of Business Ethics* 13: 417-430.

American Accounting Association Committee on the Future Structure, Content, and Scope of Accounting Education (AAA). 1986. Future accounting education: Preparing for the expanding profession. *Issues in Accounting Education* 1(Spring): 168-195.

American Assembly of Collegiate Schools of Business (AACSB). 1988. *Accreditation Council Policies, Procedures and Standards.* St. Louis, MO: AACSB.

Armstrong, M.B. 1987. Moral development and accounting education. *Journal of Accounting Education* 5(Spring): 27-43.
Armstrong, M.B. 1993. Ethics and professionalism in accounting education: A sample course. *Journal of Accounting Education* 11(Spring): 77-92.
Clarkson, M.B.E. 1995. A stakeholder framework for analyzing and evaluating corporate social performance. *Academy of Management Review* 20: 92-117.
Claypool, G.A., D.F. Fetyko, and M.A. Pearson. 1990. Reactions to ethical dilemmas: A study pertaining to certified public accountants. *Journal of Business Ethics* 9: 699-706.
Cohen, J. 1988. *Statistical Power Analysis for the Behavioral Sciences,* 2nd edition. Hillsdale, NJ: Lawrence Earlbaum.
Cohen, J.R., L.W. Pant, and D. Sharp. 1992. Methodological issues in cross-cultural ethics research. Working paper.
Cohen, J.R., and L.W. Pant. 1989. Accounting educators perceptions of ethics in the curriculum. *Issues in Accounting Education* 4(Spring): 70-81.
Colby, A., L. Kohlberg, J. Gibbs, and M. Lieberman. 1983. A longitudinal study of moral development. *Monographs of the Society for Research in Child Development*, Series 200, 48(1&2): 1-107.
Donaldson, T., and L.E. Preston. 1995. The stakeholder theory of the corporation: Concepts, evidence, and implications. *Academy of Management Review* 20: 65-91.
Elm, D.R., and J. Weber. 1994. Measuring moral judgment: The moral judgment interview or the defining issues test? *Journal of Business Ethics* 13: 341-355.
Ferrell, O.C., and L.G. Gresham. 1985. A contingency framework for understanding ethical decision making in marketing. *Journal of Marketing* 49(Summer): 87-96.
Ferrell, O.C., L.G. Gresham, and J. Fraedrich. 1989. A synthesis of ethical decision models for marketing. *Journal of Macromarketing* (Fall): 55-64.
Freeman, R.E. 1984. *Strategic Management: A Stakeholder Approach.* Boston: Pitman Press.
Hebert, P., E.M. Meslin, E.V. Dunn, N. Byrne, and S.R. Reid. 1990. Evaluating ethical sensitivity in medical students: Using vignettes as an instrument. *Journal of Medical Ethics* 16: 141-145.
Hebert, P., E.M. Meslin, and E. V. Dunn. 1992. Measuring the ethical sensitivity of medical students: A study at the University of Toronto. *Journal of Medical Ethics* 18: 142-147.
Hiltebeitel, K.M., and S.K. Jones. 1991. Initial evidence on the impact of integrating ethics into accounting education. *Issues in Accounting Education* 6(Fall): 262-275.
Hunt, S.D., and S. Vitell. 1986. A general theory of marketing ethics. *Journal of Macromarketing* (Spring): 5-16.
Jones, T.M. 1991. Ethical decision making by individuals in organizations: An issue-contingent model. *Academy of Management Review* 16(April): 366-395.
Jones, T.M. 1995. Instrumental stakeholder theory: A synthesis of ethics and economics. *Academy of Management Review* 20: 404-437.
Kohlberg, L. 1969. Stage and sequence: The cognitive-developmental approach to socialization. In *Handbook of Socialization Theory and Research*, ed. D.A. Goslin, 347-480. Chicago, IL: Rand McNally.
Kohlberg, L. 1976. Moral stages and moralization: The cognitive development approach. In *Moral Development and Behavior: Theory Research and Social Issues*, ed. T. Lickona, 29-53. New York: Holt, Rinehart, & Winston.
Langenderfer, H.Q., and J.W. Rockness. 1989. Integrating ethics into the accounting curriculum: Issues, problems, and solutions. *Issues in Accounting Education* 4(Spring): 58-69.
Loeb, S.E. 1988. Teaching Students Accounting Ethics: Some Crucial Issues." *Issues in Accounting Education* 3(Fall): 316-329.
National Commission on Fraudulent Financial Reporting (The Treadway Commission). 1987. *Report of the National Commission on Fraudulent Financial Reporting.*
Newstrom, J.W., and W.A. Ruch. 1975. The ethics of management and the management of ethics. *MSU Business Topics* (Winter): 29-37.

Newstrom, J.W., and W.A. Ruch. 1976. Managerial values underlying intraorganizational ethics. *Atlanta Economic Reivew* 26(May/June): 12-15.

Nisbett, R., and L. Ross 1980. *Human Inference: Strategies and Shortcomings of Social Judgment.* Englewood Cliffs, NJ: Prentice-Hall.

Piaget, J. 1932. *The Moral Judgment of the Child.* New York: Free Press.

Ponemon, L. 1992. Ethical reasoning and selection-socialization in accounting. *Accounting Organizations and Society* 17(April/May): 239-258.

Ponemon, L. 1993. Can ethics be taught in accounting? *Journal of Accounting Education* (Fall): 185-209.

Ponemon, L., and D. Gabhart. 1990. Auditor independence judgments: A cognitive developmental model and experimental evidence. *Contemporary Accounting Research* 7: 227-251.

Ponemon, L., and A. Glazer. 1990. Accounting education and ethical development: The influence of liberal learning on students and alumni in accounting practice. *Issues in Accounting Education* 5(Fall): 195-208.

Rest, J. 1986. *Moral Development: Advances in Research and Theory.* New York: Praeger.

St. Pierre, K.E., E.S. Nelson, and A.L. Gabbin. (1990). A study of the ethical development of accounting majors in relation to other business and nonbusiness disciplines. *The Accounting Educators' Journal* 3(Summer): 23-35.

Shaub, M.K. 1994. An analysis of the association of traditional demographic variables with the moral reasoning of auditing students and auditors. *Journal of Accounting Education* 12(Winter): 1-26.

Shaub, M.K., D.W. Finn, and P. Munter. 1993. The effects of auditors' ethical orientation on commitment and ethical sensitivity. *Behavioral Research in Accounting* 5: 145-169.

Trevino, L.K. 1986. Ethical decision making in organizations: A person-situation interactionist model. *Academy of Management Review* 11(October): 601-617.

Weber, J. 1990. Managers' moral reasoning: Assessing their responses to three moral dilemmas. *Human Relations* 43: 687-702.

# ORGANIZATIONAL MORAL DEVELOPMENT:

## LESSONS FROM MORAL REASONING FRAMEWORKS

Kristi Yuthas


## ABSTRACT

This paper uses models of individual moral development proposed by Kohlberg and Gilligan as metaphors for understanding the morality of organizations. The paper discusses the two models and examines how their perspectives on morality in humans might be applied to the study of the morality of organizations. Using insights gained from this analysis, a model of organizational moral development (OMD) is constructed. Just as moral development in individuals is evidenced by the modes of reasoning they adopt, this paper suggests that organizational moral development is evidenced by the use of management control mechanisms and reward systems and that accounting system characteristics will vary across the three stages of the OMD model. The model provides a framework for thinking about how morality in organizations develops, and for exploring the conceptual linkages between management control/accounting systems and organizational morality.


**Research on Accounting Ethics, Volume 4, pages 53-72.**

**ISBN: 0-7623-0339-5**

## INTRODUCTION

It is widely believed that as individuals undergo cognitive development, they also develop morally. Piaget was an early proponent of the view that as humans progress from childhood to adulthood, they move through various stages of both cognitive and moral development. Each stage in this progression is thought to include those previous to it, and movement from one stage to the next is sequential. Once individuals have attained a particular stage of development, they either progress or stagnate but, in general, do not regress. Various models of how moral development takes place in individuals have been proposed and two have been widely accepted.

In recent years, we have come to recognize that organizations, like individuals, learn. In individuals, learning is accompanied by moral development. This paper suggests that as organizations learn and mature, they also move through stages of moral development. Individuals at lower levels of development are self-centered and concerned with their own needs. As they mature, they begin to balance the needs and rights of the self and others. The same may be true for organizations. Less developed organizations may focus on their own survival and prosperity without regard for the needs and rights of various constituents affected by their actions. Morally mature organizations may attempt to consider the mutual interests of all stakeholders as they select and implement strategies.

The purpose of this paper is to explore the concept of moral development in organizations by relating it to the concept of moral development in individuals. The paper relies on insights from two theoretical models of human moral development and from models of corporate social performance to construct a model of organizational moral development. The model outlines the stages of moral development in organizations, and identifies the phenomenal forms or symbols of morality that organizations might exhibit at various stages.

The discussion of organizational moral development is presented in three sections. In the first, literature on organizational morality is reviewed. The reciprocal relationship between individual and organizational morality is examined, and extant models of corporate social performance and morality are reviewed. In the second section, two competing models of individual moral development, proposed by Kohlberg and Gilligan, are presented. Each model is discussed from an individual perspective, and applied to organizations by translating human characteristics into organizational ones. The third section develops the three-stage model of organizational moral development, and discusses the management control strategies and accounting system characteristics of organizations at each stage.

# ORGANIZATIONAL MORALITY

The model of organizational morality proposed here translates individual morality into organizational terms. The translation is informed by and builds upon models of corporate social responsibility. To justify the appropriateness of the mapping, the following discussion explores the relationship between the morality of an organization and the morality of its individual participants, and discusses a number of models from the corporate social responsibility literature that divide organizations into categories or levels based upon their social performance or morality.

## The Relationship Between Individual and Organizational Morality

The link between organizational and individual perceptions, values, and beliefs has long been recognized. Williamson (1985) has suggested that managers will engage in opportunistic behavior despite the pressures of their fiduciary roles. Many others believe that managerial behavior is strongly influenced by organizational culture and norms. These environmental forces affect the manner in which managers understand and act upon the organization (Victor and Cullen 1988). Frederick and Weber (1987) have argued that organizational values may supersede individual values when an agent is acting in an organizational context. Levitt and March (1988) propose that, through time, organizations develop routines that guide behavior, and that individuals within those organizations adopt these behavioral norms. Morgan (1991) has argued that these routines can continue as individual members of the organization come and go. For Sridhar and Camburn (1993), "organizations, like individuals, develop into collectivities of shared cognitions and rationale, over a period of time" (p. 727). In so doing, the individuals concerned develop a shared sense of right and wrong.

But where do these organizational values arise, and how do they evolve? Quinn and Jones (1995) have examined inconsistencies in common perceptions of the morality of organizational agents. Their findings suggest that the morality of organizations is strongly influenced by the morality of their individual members, and that both are structured by societal forces. They argue that managers have multiple duties, and that it is unlikely that their fiduciary duty to shareholders takes precedence over all other duties. Further, Quinn and Jones suggest that individual moral characteristics such as trust necessarily underlie many transactions in the marketplace. Organizational learning, including learning new shared values and ethical norms, is influenced by organization members. For example, leadership changes can result in new behavior norms or new organizational paradigms (Argyris 1990). Similarly, organizational culture is strongly linked with the characteristics of top management. When a change in management occurs, behavior norms and values are likely to follow.

Both managerial and nonmanagerial organization members can influence the moral character of an organization. When a decision is made or a strategy pursued by an organization, the choice is really made by an individual or group. In their roles as organizational participants, individuals may use the same modes of moral reasoning and exhibit the same characteristics of moral development as they do in their personal lives. The parameters of the strategies they feel free to pursue are affected by personal, organizational, and societal factors. Thus, individuals, both privately and in their roles as members of organizations, take actions that are fitting in the light of their own values as well as those of others surrounding them.

These arguments suggest that the morality of an organization is a reflection of the morality of its agents, and vice versa. Organizations are not free to pursue actions that are unacceptable to their agents or to society in general; individuals are not free to pursue actions proscribed by their organizations. It is for these reasons that models of individual morality provide an appropriate foundation for the development of organizational morality—morality that is created and enacted by individuals.

## Corporate Social Performance

In the past two decades, numerous researchers have studied the relationship between corporate social performance and economic performance. Because of the important societal roles played by organizations, these papers were premised on the assumption that social performance would have a positive effect on economic performance. Although some studies have found positive relationships, others have found negative, weak, or no relationships (Cochran and Wood 1984). Taken together, the papers suggest that socially responsible activities do not necessarily lead to superior profitability. Despite this finding, increasing attention has been paid to the social impact of firms' activities. Beginning with Freeman's (1984) seminal work on the subject, numerous stakeholder models have been developed. These models take into account not only the economic well-being of a small group of stockholders but also the social benefits created by a firm for other stakeholders, such as employees, customers, managers, and local community members. Research on corporate social performance focuses on the social rather than the economic benefits created by the firm's activities. Performance on the social dimension may be enhanced by increasing social benefits, such as employment, employee morale, or charitable donations, or by reducing social costs, such as pollution, unsafe working conditions, or use of nonrenewable resources. Stakeholder and social responsibility models suggest that most firms focus on both economic and social costs and benefits.

A number of models have been developed in the business and society literature that identify categories for corporate social responsibility or organizational morality. These models are generally normative, rather than descriptive, and seek to differentiate between firms that are highly responsible or moral and those that are not. Freeman (1984) and Freeman and Gilbert (1988) provide classification schemes for firm strategies. The categories cover a spectrum that begins with a focus on maximizing the interests of stockholders or a narrow set of stakeholders and ends with a focus on maximizing the interests of all stakeholders or creating social harmony. Several earlier models have also categorized firms on the basis of corporate social performance. Sethi (1979) categorized firms as reactive, defensive, or responsive; Carroll (1979) used the categories of reactive, defensive, accommodative, and proactive.

Meznar, Chrisman, and Carroll (1991) seek to go beyond these early models by proposing a scheme that takes into account both the types of stakeholders to which a firm attends, and the values and benefits offered to these stakeholders. They develop a model that categorizes firms as defensive, offensive, and accommodative. Defensive firms seek to reduce social costs for one or more groups of stakeholders, attempting to avoid crises before they occur. These firms may take such actions as postponing use of potentially dangerous chemicals while they are under government investigation, before laws banning or allowing their use are established. Offensive firms are concerned with increasing social goods. These firms may become involved in social or charitable causes in an effort to promote the public welfare. Accommodative firms are very concerned with the interests of their stakeholders and seek to promote the interests of various stakeholder groups. These firms seek to both reduce social costs and increase social benefits as necessary to accommodate stakeholder interests.

Wood (1991) has developed perhaps one of the best known taxonomies of corporate social performance, building on the work of Wartick and Cochran (1985). Wartick and Cochran developed a social responsiveness model that addresses principles which motivate firm behavior, processes of social responsiveness, and policies developed to address social issues. The processes used by Wartick and Cochran were the same as those used by Carroll (1979)—reactive, defensive, accommodative, and proactive. Wood (1991) expands and clarifies the model proposed by Wartick and Cochran. Wood's model defines the processes through which social responsiveness can be addressed: environmental assessment, stakeholder management, and issues management. A firm pursuing an environmental assessment strategy focuses on survival through understanding and adapting to conditions in a continually changing environment. A firm pursuing a stakeholder management strategy focuses on the links between company functions and external stakeholders and attempts to manage these stakeholder relationships in order to perform successfully. A firm pursuing an issues management strategy develops policies designed to address social issues directly. Such firms devise and manage processes, both internally and externally, to deal effectively with these issues.

Although the stakeholder models mention values and socially responsible action, they do not discuss organizational morality per se. Victor and Cullen made an early link between Kohlberg's stages and organizational concerns. They suggest that each level of development is based on a different type of social perspective, ranging from micro to macro concerns, and they attempt to measure the ethical dimensions of the organizational work climate. Another recent paper more directly addresses organizational morality (Sridhar and Camburn 1993). It suggests that, over time, organizations develop into collectivities of individuals with shared meanings and values. Sridhar and Camburn further argue that the moral development of organizations can be inferred by examining the rationales they use in explaining their behavior. The study develops an instrument based on the Defining Issues Test (Rest 1986), which in turn draws from Kohlberg's model of individual moral development. Sridhar and Camburn assign firms to Kohlberg's stages of moral development by analyzing statements made by organization spokespersons to defend their actions during periods of ethical crisis. Redenbach and Robin (1991) also suggest that an organization's stage of moral development is signaled by corporate behaviors. They present case scenarios typifying organizations at each of Kohlberg's developmental stages.

The model presented in the current paper goes beyond previous research in a number of ways. First, it develops an organizational model based not on a linear transformation of Kohlberg's stages but on a combination of Kohlberg's and Gilligan's models reinterpreted to apply more directly to organizations. Second, it uses research on corporate social responsibility and moral philosophy to inform development of the model. Finally, it links organizational morality to management control and accounting systems, providing mechanisms through which the morality of organizations can be appraised.

## INDIVIDUAL MORAL DEVELOPMENT

Just as notions of organizational learning are patterned after principles of individual learning, the model of organizational moral development is based on theories of individual moral development. Two theories of moral development provide the foundation from which an organizational moral theory can be developed: Kohlberg's justice-based theory and Gilligan's responsibility-based theory. Both theories are empirically grounded and combine moral philosophy with cognitive psychology. They suggest that individual learning (cognitive development) is a necessary prerequisite for moral reasoning (Selman 1971; Kuhn, Langer, and Kohlberg 1971).

Kohlberg's theory (1969, 1976, 1981) has at its center the notion that moral development culminates with the fair and just application of abstract universal principles. The theory is that humans have rights, and that individual choices are moral if they uphold those rights. Gilligan's theory (1977, 1979, 1982) suggests that individuals follow different paths of moral development. Some follow Kohlberg's path of justice reasoning while others follow a path that focuses more heavily on the responsibilities of individuals within specific social circumstances. For Gilligan, moral development culminates with responsible choices that recognize the need for compassion and fairness for one's self and others.

***Table 1.*** Theories of Moral Development

| *Rights-Based Ethics* | | *Responsibility-Based Ethics* | |
|---|---|---|---|
| KOHLBERG | | GILLIGAN | |
| *Individual* | *Organizational* | *Individual* | *Organizational* |
| | | **Preconventional** | |
| Punishment and obedience | Compliance | Individual survival | Solvency |
| Individual interest | Self-interested contracts | From selfishness to responsibility | Stakeholder focus |
| | | **Conventional** | |
| Interpersonal conformity | Conforming to industry norms | Self-sacrifice and social conformity | Industry norms |
| Social system and conscience main tenance | Enlightened self-interest | From goodness to truth | Balanced scorecard |
| | | **Postconventional** | |
| Prior rights and social contracts | Respect for constituent rights | Morality of nonviolence | Strategic alliances |
| Universal ethical principles | Rights supersede profits | Ethic of responsibility | Social mission |

The two theories have often been viewed as competing or incompatible, and their relative merits have been widely debated (cf. Blum 1988; Larrabee 1993; Meyers and Kittay 1987; Walker 1984).[1] A growing body of literature has attempted to synthesize these perspectives (cf. Brabeck 1983; Nunner-Winkler 1984). This work suggests that the theories are not inconsistent and that common ground can be established. The organizational model presented in this paper is based on the synthesis of the two perspectives.

In the remainder of this section, the stages of individual moral development put forth in the two theories are briefly described. The stages are then examined from an organizational perspective, and examples are presented regarding how each stage might be interpreted if the entity of interest is an organization rather than an individual. Individual moralities and the values of organizational agents are bound to be reflected in the actions these agents take on behalf of the organization (Quinn and Jones 1995). Thus, this paper takes an anthropomorphic view of organizations and imbues them with attributes of morality. Using individual moral development as a metaphor for organizational moral development allows for exploration of the possibility of and predicates to moral development in organizations. The Kohlberg and Gilligan stage models, along with their organizational interpretations, are shown in Table 1.

## KOHLBERG'S RIGHTS-BASED THEORY OF MORAL DEVELOPMENT

The notion of justice forms the philosophical foundation for Kohlberg's work. The justice perspective is grounded in the tradition of social liberalism; it is committed to personal liberty within a social contract model. The social contract model suggests that it is possible to specify principles of conduct that can be mutually agreed upon by rational members of society. In this model, individuals could be expected to cooperate to develop a set of universal principles that promote the interests of all. In this sense, Kohlberg's model is related to Mill's utilitarian perspective in which moral action promotes the greatest happiness for all. The notion of personal liberty recognizes that individuals have differing views of `the good life' and should be allowed the autonomy to pursue differing goals. From this perspective, human rights, such as Locke's life, liberty, and property, are promoted by eliminating unnecessary restrictions on human activity.

Kohlberg's model (Table 1) posits six stages of moral development that must be attained by individuals in order to achieve full morality. The first two stages comprise the preconventional stage of moral development in which behavioral norms are viewed as external to the individual. In the first stage, the goodness of an action is determined by its consequences. The self-centered individual defers to power and avoids punishment. In the second, the individual seeks actions which will satisfy needs. Individuals may also seek to satisfy the needs of others who will ultimately benefit from the action.

For organizations, the preconventional stage is again one in which right behavior is determined exogenously. As in the traditional market model, a company seeks behaviors that maximize the interests of its shareholders. At this stage, firms may focus on complying with legal and contractual obligations, attempting to avoid negative consequences such as loss of business and legal action.

Stages 3 and 4 make up the conventional stage, in which norms are internalized by an individual. In Stage 3, the goodness of behavior is determined by external forces; the individual seeks to please and receive approval from others. In Stage 4, the individual begins to use external rules and social norms to determine right action.

In organizational terms, the conventional stage can be represented by an expanded view of the organization. In this stage, the organization becomes aware of its relationships with and responsibilities to external constituents. In Stage 3, a company may seek to please a variety of individual and external stakeholders. The organization acts in a manner consistent with industry norms, regardless of whether formal regulations are in place to monitor this behavior.

The postconventional stage consists of the final two stages of development. In these stages, an individual complies with group norms only when they are consistent with individual values and principles. The fifth stage represents consideration of social welfare when determining which rules to obey. Personal values are used as means to evaluate rules or laws. In the final stage, the individual evaluates behavior on the basis of universal ethical principles. Right behavior is behavior that would be accepted by consensus among rational individuals.

Postconventional development applied to organizations implies the adoption of behaviors that conform to an organization's values or mission, even when they exceed the standards adopted by the industry. An organization in Stage 5 would comply with industry standards only when they coincided with the moral principles adopted by the organization. Stage 6 organizations focus on exhibiting behaviors consistent with their principles. Principles would be derived by anticipating the rules for organizational conduct that would be agreed upon by a group of rational organizational and constituent groups' representatives.

## GILLIGAN'S THEORY OF MORAL DEVELOPMENT

Gilligan's theory is based on the notion of mutual responsibility among individuals. Gilligan's model suggests that moral decisions are made within a specific contextual frame. When individuals make decisions, they can and should consider their relationships with and responsibilities to others. The right course of action is determined through discourse that takes seriously the interests of both one's self and others.

The roots of Gilligan's theory can be traced back to Aristotle and Hume (Meyers and Kittay 1987; Baier 1987). These philosophers stress the importance of virtue rather than justice in moral deliberation. Aristotle viewed humans as social beings and suggested that moral decisions are a result of the careful cultivation of character, rather than of mere application of abstract principles. Hume similarly argued that reason alone cannot inform moral decisions, but that moral sentiments guide ethical conduct. From the responsibility perspective, circumstantial and contextual factors circumscribe the application of moral principles. A moral individual is, therefore, one who expands the notion of self by assimilating responsibilities to self and other. The preconventional stage in Gilligan's model suggests that individuals focus on survival, first through individual means and later through instrumental exchanges with others. In the first stage, morality is a matter of self-imposed sanctions that promote survival of the self. The individual is concerned only with the self. In the second stage, the individual begins to recognize interdependence between the self and others. In this stage, the individual begins to perceive a conflict between self-interested behavior and responsibilities to others.

For organizations, Gilligan's Stage 1 may represent a focus on continued organizational existence. In this stage, moral action is that which promotes or maintains solvency. In Stage 2, organizations begin to consider the ramifications of their actions on constituent groups other than shareholders and managers. Companies at this stage begin to weigh profitability against good business practices from the perspective of other stakeholders.

Stages 3 and 4 represent a shift in focus from the self-interested needs of an individual to the needs of others with whom the individual comes into contact. In Stage 3, an individual begins to develop a strong concern for others. In this stage, goodness is equated with sacrificing the needs of self for those of others. In Stage 4, the individual becomes aware of the responsibilities to oneself. The individual in this stage begins to balance the needs of others with the needs of self.

For organizations, the conventional stage is characterized by a shift in concern from the individual organization and its stakeholders to industry and social norms and consensus. In Stage 3, corporations shift their efforts to an external focus, attempting to mimic the business practices accepted by other firms. In so doing, they shift attention from organizational survival to satisfying the needs of various constituent groups, consistent with industry norms. In Stage 4, they begin to develop strategies in which the interests of the organization are weighed against those of constituents to develop a more balanced approach to determining action.

In Gilligan's postconventional stage, the conflict between the needs of the self and others is resolved through the notion of responsibility. The individual accepts responsibility for avoiding actions that harm the self or others. In Stage 5, the individual develops and modifies relationships with others in a manner that incorporates the particular interests of the self and others. This consolidation provides stability in the conflicts that arise in the conventional stage. In Stage 6, the individual adopts a principle of universal condemnation of exploitation and harm to others. The individual in this stage understands the particularities of human relationships and attempts to maintain integrity without neglecting the needs of others.

In the organizational model, the postconventional stage is characterized by adherence to ethical business practices independent of industry or societal norms. In Stage 5, the organization reformulates its mission and goals to encompass the interests of stakeholders, thereby mitigating the conflicts associated with attempting to balance the conflicting needs of various stakeholder groups. In Stage 6, the organization develops its own individual morality. This morality considers the broader welfare of society in relationship to that of traditional stakeholders.

## Differing Perspectives on Moral Development

The theories of Kohlberg and Gilligan differ in several important respects. Before presenting a unified model of organizational moral development, it is useful to briefly examine the two theories of moral development individually and in contrast. First, Kohlberg's work is characterized by an "impartialist" perspective of morality (Blum 1988) in which the rights of all individuals are seen as equally important. Gilligan views morality as inseparable from the particular interests and values of the self and others, as well as from the particular contextual setting in which a moral dilemma arises (Brabeck 1983). Kohlberg focuses on independent individuals, whereas Gilligan views the individual as a self connected to others. Kohlberg focuses on the development of universal principles developed by rational individuals; Gilligan focuses on the development of context-specific actions determined by emotional, caring individuals.

Lyons (1983) summarizes the two perspectives in terms of a separate/objective self versus a connected self. The separate/objective self is concerned with reciprocity, treating others as one would wish to be treated. Relationships are mediated through rules that maintain fairness and are grounded in roles arising from duties and commitments. The connected self is concerned with responsiveness, treating others as one believes the others would wish to be treated. Relationships are mediated through caring relationships and grounded in interdependence arising from recognizing human interconnectedness.

Exploring these two perspectives in organizational terms leads to two quite distinct prototypes of a "moral" organization. The separate/objective organization would tend to use a morality of justice or fairness. This organization would develop stakeholder relationships that rest on duty and obligation. The organization would make commitments and live up to them. Conflicting claims between various stakeholder groups would be mediated by invoking impartial rules or principles. For example, this firm might have a set of rules for determining choice of supplier, incorporating such criteria as lowest bid, highest quality, and least environmental impact. Prespecified rules would be applied impartially, and the supplier receiving the highest rating would be selected. The firm might likewise have policies regarding compensation and promotion, perhaps basing them on a combination of productivity measures. An employee who achieved high productivity would be rewarded. In both cases, the firm would make decisions in such a way that stakeholders are treated impartially, consistently, and fairly.

The connected organization would tend to use a morality of care or responsibility. The organization would develop relationships with stakeholders that rest on understanding and promoting the unique needs of individual stakeholder groups. The organization would create alliances with stakeholders and maintain them. The organization would resolve conflicting claims through an analysis of the particular circumstances surrounding the conflict. The firm would make decisions in such a way that the welfare of stakeholders would be promoted and relationships maintained. For example, in selecting a supplier, such an organization would consider the particular circumstances. It might choose to go with a small supplier, with which it has had previous dealings, that is being driven from the market through the predatory pricing policies of a larger firm. It might choose to reward a long-term employee who has worked particularly hard rather than the highest-producing employee. In these examples, the firm makes decisions attempting to promote the well-being of individuals.

## A MODEL OF ORGANIZATIONAL MORAL DEVELOPMENT

Viewing each model separately creates the impression that the Kohlberg and Gilligan models, as well as firms meeting these ideal types, would differ considerably. However, the distinctions separating the two quickly dissolve under scrutiny. Responsibilities and personal relationships often come into conflict with the just application of universal principles, and it is difficult to visualize a real-life situation in which either approach alone could adequately resolve the dilemma. As organizations take seriously the rights of individuals in various groups holding stakes in the organization, they also accept the responsibility of taking seriously the rights and interests of these groups. Organizations and stakeholder groups do not experience personal relationships or merit natural human rights (life, liberty,

***Table 2.*** A Stage Model of Organizational Moral Developmen

| *Stage of Moral Development* | *Moral Characteristics* | *Accounting System Characteristics* |
|---|---|---|
| | **Stage 1: Preconventional Level** | |
| Individual organization | Focus on organizational survival, maximization of shareholder wealth, avoidance of penalties and regulations | Traditional financial and operational performance-based measures, such as ROI, stock price, EPS, efficiency indicators, and evidence of compliance with regulations |
| | **Stage 2: Conventional Level** | |
| Stakeholder managing organization | Focus on survival via instrumental approach to stakeholder management, comply with industry norms, adopt progressive policies for long term gain | Traditional measures plus direct measures of stakeholder interests such as quality, supplier relationships, customer satisfaction, corporate image, employee compensation and rewards, employment equity, product safety |
| | **Stage 3: Postconventional Level** | |
| Organization in a community | Focus on intrinsic value of stakeholder interests, perceive role as member of community, incorporate stakeholder and community interests into corporate mission | Traditional and stakeholder measures plus measures of the impact of the organization on its industry and community such as environmental impact, unemployment rates, quality of work life, social investment and donations, community relations, public policy involvement, public health and safety |

and property). Their moral decisions, therefore, are those that consider the responsibilities, rights, and interests of the organization, stakeholders, and society.

A model of organizational moral development (OMD) is shown in Table 2. The model proposes three stages of development in which the organization: (1) focuses on its own self-interests; (2) focuses on the interests of constituents as an instrumental means to promote its own; and (3) expands its strategic mission to encompass the interests of all members of the community. The model as specified is not intended to be descriptive or testable. Rather, it provides a framework for thinking about moral issues from an organizational perspective. The Kohlberg and Gilligan models have been criticized for their failures in terms of their descriptive accuracy and normative content. Despite these problems, they have served a valuable role in the study of human morality. Similarly, the OMD model is intended to provide a mechanism for explicating and evaluating the concept of organizational morality.

Following the individual models, the OMD model is presented in three basic stages representing preconventional, conventional, and postconventional forms of ethical development. Unlike the individual models, the OMD model does not suggest that there is a natural progression from one stage to the next over time, or that once an organization reaches a particular stage, it will not regress. However, there is evidence that, as a group, organizations in the United States have shifted from lower to higher stages over time. The "ethicization of business" attests to the fact that many organizations have moved to Stage 2 or beyond in recent years. In addition, as presented, the OMD model proposes that the three stages represent progressively higher forms of moral development, and that the postconventional stage can serve as an ideal for moral conduct.

In discussing the three stages, prototypical organizations representative of each stage of moral development are described, along with their activities. These are not intended to be representative of real organizations. Rather, they are meant to illustrate the contours of each stage of morality identified in the model. Thus, the organizations discussed are caricatures developed to promote consideration of possibilities for moral action.

## Stage 1: Preconventional Morality

The first stage of moral development represents the standard neoclassical-economic approach to ethics. The ethical organization is self-centered, pursuing those activities that will ensure its own survival and prosperity. The organization aggressively seeks opportunities for maximizing shareholder wealth, and all activities that promote this goal are seen to be ethical. This view is characteristic of the minimalist ethics of Adam Smith and Milton Friedman (1970). As it has generally been interpreted, this perspective suggests that it is up to the market, not individual organizations, to determine what is in the best interests of society.

With the massive push to promote business ethics, this perspective is no longer in vogue, and may be superseded by the strategic use of ethical practices in Stage

2. At Stage 1, however, concern for others is purely instrumental; organizations engage in relationships to promote their own interests, and any benefits accruing to other parties are incidental. Organizations in this stage define ethical activities generally as those that promote maximization of their own long-term wealth. They believe this approach to be the best course for pursuing societal goals.

Stage 1 organizations also usually comply with laws and regulations in order to avoid financial penalties and other negative consequences that would hamper their profitability objectives. Likewise, they may choose to ignore laws when the expected value of probable outcomes is positive. In other cases, they may exceed the requirements of the law and comply with ethical norms of corporate behavior. They may, for example, forgo opportunities for predatory pricing if they believe that their image, and opportunities for future gain, could be damaged. Further, these companies may, in some circumstances, pursue socially beneficial policies. If they perceive a likelihood of new regulations or other restrictions on behavior, they may take proactive steps to avoid such measures. Again, they seek to promote organizational interests and avoid punishment.

Management control structures and reward systems in these organizations are profit-oriented, and focus on traditional financial and operational definitions of performance. Measures gathered and reported by the accounting systems of these organizations have an internal focus. Performance indicators, such as return on investments, earnings per share, and increases in stock prices, are likely to be the most important measures. Other operational measures, such as efficiency or throughput measures expected to lead to future profitability, may also be reported. Like the reporting systems, the reward structures of these organizations are expected to be highly focused toward long-term profitability.

## Stage 2: Conventional Morality

The second stage of organizational moral development can be characterized by the conventional approach to stakeholder management. Companies legitimize stakeholder concerns on behalf of "enlightened self-interest." Stage 2 morality is based on the notion that ethical practices ensure profitability. Moral intentions are thought to lead to the same practices as pursuit of long-term profitability. Stage 2 is an intermediate stage and takes on characteristics of Stages 1 and 3.

Like the organizations in Stage 1, Stage 2 firms seek their own interests; however, in this stage, they come to recognize the interdependence between themselves and their stakeholders. With this understanding, they begin to actively pursue courses of action intended to serve the needs of multiple stakeholder groups (not for the benefit of the stakeholders, as in Stage 3, but because the firm's own interests will also be promoted).

Evidence suggests that today, in general, organizations are taking ethical concerns seriously, and have reached the conventional stage of moral development. According to Stark (1993), "surveys suggest that over three-quarters of America's major corporations are actively trying to build ethics into their organizations." At the conventional stage, however, these efforts are geared toward long-term self-interest.

Clarkson and Deck (1993) studied 70 large corporations and found that they actively manage relationships with their stakeholders, but not with society. Such companies often engage in what has been termed "instrumental" or "strategic" ethics—using ethics as a means of pursuing other goals (Quinn and Jones 1995). One study of corporate codes of ethics suggests that these codes are often used defensively as a control mechanism over employee behavior and are designed to protect the organization and its assets from employees (Clarkson and Deck 1993).

As with Stage 1, firms at the conventional level may also pursue ethical courses of action to avoid legal difficulties. Donaldson and Preston (1995) discuss a recent American Law Institute document that states that although corporations are under no legal obligations to pursue ethical actions, they are still accountable for the same moral standards as are individuals. Thus, companies in Stage 2 may take a progressive role in pursuing policies beneficial to constituents to reduce the likelihood that future activities will be restricted. For example, firms may provide generous employee benefits packages to reduce the likelihood of unionization movements.

## Stage 3: Postconventional Morality

Stage 3 organizations recognize interests of stakeholders as intrinsically valuable and they consider the broader interests of society in their actions. Although there is skepticism about the very idea of "business ethics," there is evidence that at least some companies seek morality for its own sake. As the rapid growth of mutual funds targeted toward environmentally or socially responsible corporations attests, individuals may forgo some personal monetary gain for social gain. Corporations likewise may pursue policies that consider broader societal interests.

Organizations in Stage 3 recognize that ethical policies do not necessarily promote self-interest, and that moral decision making can be a serious detriment to financial performance in many cases (Hoffman, 1989). However, they do not believe that human welfare is best promoted through the blind pursuit of profits. Movement to Stage 3 does not imply that organizations are no longer allowed to pursue their interests, or that they should shed their own goals to pursue some notion of social good. Contrarily, the perspective supported by the OMD model is one in which the organization expands its mission to encompass the interests of stakeholders and society. Rather than attempting to balance the competing interests of the organization and its constituents, Stage 3 organizations integrate these interests by viewing the organization and its constituents as mutually responsible members of a community.

The differences between Stage 2 and 3 organizations may be difficult to detect. Stage 2 organizations may, for example, pursue quality not because quality is intrinsically good but because customers will pay more for it; they may engage in employee empowerment initiatives, not to improve the lot of employees but to motivate them to perform better. The outward appearance of these organizations may be the same as those in Stage 3, but their motives for pursuing ethical policies differ. The moral status of these organizations is precarious; the companies may revert to planned obsolescence and Taylorism if they come to believe this best promotes corporate goals.

The possibility that organizations can attain postconventional morality has been explored from a variety of perspectives. Although the terminology varies, researchers suggest that corporations can move from Stage 2 to Stage 3. Donaldson and Preston (1995) refer to instrumental stakeholder theory, which links stakeholder management to economic performance, and normative stakeholder theory, which recognizes the intrinsic value of shareholder interests. They argue that all stakeholder models have a normative component, suggesting that organizations that adopt them will necessarily move toward Stage 3. Quinn and Jones (1995) refer to the two stages using the terms "instrumental ethics" and "noninstrumental ethics." These authors suggest that instrumental ethics is logically problematic and cannot be morally binding on managers. They suggest that a noninstrumental approach can be used to produce policy recommendations that are morally binding. They further show that, even within the logic of the market, a number of moral principles necessarily have precedence over firm profits.

Literature on corporate social responsibility has long argued that organizations can and should consider societal effects of their actions. Preston and Post (1975), for example, suggest that businesses are responsible for the social problems they cause and for helping solve problems and issues related to their activities. Wood (1991) states that according to the principle of public responsibility, organizations have the responsibility to act affirmatively to promote social welfare.

Organizations that expand their notions of self-interest to incorporate those of stakeholders and society would reflect this expanded mission by incorporating social impact measures into their accounting systems. Postconventional organizations would likely report the financial and operational measures of Stage 1 as well as the stakeholder measures of Stage 2. As discussed previously, managers in these organizations would attend to the stakeholder measures not only as instruments to profits, but because providing for interests such as product quality, employee attitudes, and customer satisfaction are valued for their own sake. In addition, organizations at this level would look beyond their immediate stakeholders to the communities and societies within which they operate. Accounting systems would report socially oriented internal measures such as employment equity, quality of work life, health and safety, and product safety. In addition, they would report measures of community impact such as employment levels, social investment, community relations, and public policy involvement. Although not every measure applies to each Stage 3 organization, these organizations would be sensitive to the situational and contextual circumstances surrounding moral decisions and would exhibit responsibility to stakeholders and society in their actions.

## SUMMARY AND IMPLICATIONS

This paper has examined the well-known competing moral development theories of Kohlberg and Gilligan. The theories are summarized and examined in light of the lessons they provide to the organizational domain. From these theories, a three-stage theory of organizational morality is constructed. The model suggests that ethical activity in organizations is exhibited by: (1) organizations that pursue their own self-interests, (2) organizations that balance their own and stakeholder interests, and (3) organizations that consider societal interests. These forms of morality are objectified through management control and reward systems and result in distinct accounting system characteristics.

Like the individual subjects in Kohlberg's and Gilligan's research, organizations are likely to exhibit characteristics of multiple stages. There is a great deal of doubt about whether humans can be classified at all, and the same concern is apparent with the organizational model. To further complicate the issue, it would be difficult to determine empirically whether the stages actually exist and whether a given organization falls into a particular stage. What defines the stages most clearly is the *reasons* for which organizations pursue the paths they select. So, while accounting system characteristics may provide a reasonably good indication of an organization's motivation, they are only appearances. Evaluating the "true morality," if one exists, would rely on extensive interviews and assessments of top management and perhaps all other individuals in the organization to attempt to identify the reasons that decisions are made and strategies are pursued.

The research presented in this paper has a number of implications for researchers, managers and accountants. The model of organizational moral development presented in the paper is based on the notion that individual and organizational morality are closely linked. The relationship between organizational culture and individual moral development has long been recognized. Corporate employees, particularly in top management, help steer organizational culture. Their values, beliefs, and goals help shape those of the organization. Likewise, shared organizational beliefs and norms of conduct influence the actions and beliefs of individuals associated with the organization. The OMD model uses well-known models of individual moral development to provide an ethical foundation for understanding corporate social performance and, thus, provides a link between organizational and individual morality. It suggests that, like individuals, organizations learn and develop over time, in both technical and moral capabilities.

The paper also posits a link between management control/accounting systems and organizational morality. Management and accounting reports have long been thought to provide a clear indication of an organization's interests, objectives, and goals. This paper argues that these reports also provide information about the organization's ethical concerns and moral development. For example, an organization that takes seriously its impact on the environment would be expected to work environmental considerations into its capital budgeting system, its cost accounting system, and its reward systems. All of these actions would result in accounting measures and reports that differ in content from those of a similar organization that is not actively interested in environmental impacts. Because reports reflect organizational concerns, they also drive them. To organization members, the information contained on accounting reports may be a more concrete indicator of corporate objectives than high-level strategy and policy. Accounting reports may therefore serve as an important conduit for managing or changing the ethical attitudes and behaviors of individuals.

The model may have additional prescriptive value for organizations seeking to attain postconventional morality. When an individual's morality differs substantially from that of an organization, it is unlikely that the relationship will continue in the long run. However, it is also unlikely that all members of an organization share its ethical perspectives or that they are all at the same stage of individual moral development. The OMD model suggests that the level of morality exhibited by individuals can be brought up to the level expected by the organization, just as the level of moral development for the organization as a whole can increase. Through devices such as socialization programs and employee newsletters, the organization's culture and behavioral norms can be made known to individual employees. Through reward structures, behavior complying with organization policy can be encouraged in a manner more evident to the individual. Accounting reports can make corporate policies and reward structures known to employees in addition to providing measures of the degree to which they are achieved through time. Through this range of management control structures, organizations can influence individual behavior in a variety of ways. Thus, even if specific employees are at a preconventional level of individual moral development and act only to increase personal self-interest, management controls can encourage them to act at a level of morality acceptable to the organization. Employees need not necessarily share the organization's values in order to act as if they do, and an employee's behavior in personal life may differ substantially from that exhibited in an organizational capacity. Of course, this can also work in a negative direction; individuals may feel institutional pressure to take actions inconsistent with their personal ethical beliefs. Thus, for organizations seeking high levels of morality, the paper suggests that management control structures, including accounting reporting systems, provide a mechanism through which organizational values can be communicated and institutionalized.

Finally, the model provides a foundation for future research. Grounded in models of individual organizational morality and corporate social performance, the OMD model provides a normative theoretical foundation for research efforts. In accounting, for example, researchers seeking empirical links between accounting practices and organizational morality or corporate social performance may find lessons from the OMD model useful. The paper may likewise be of interest to those interested in the effects of changes of accounting practices or reporting policies on organizational morality. As posited here, accounting may play an important role in forming the moral character of the organization, however little attention has been paid to this possibility in either the accounting or the business and society literature. By providing an ethical foundation for the corporate social performance model, the paper takes a step toward development of a theoretical basis for linking accounting to organizational morality.

## NOTE

1. In particular, Kohlberg's model has been criticized because it is seen to be more representative of the development of males than of females. However, there is evidence suggesting that both males and females can adopt justice-based and responsibility- or care-based reasoning (Brabeck 1983; Larrabee 1993), and a meta-analysis of studies of moral development failed to find evidence supporting sex differences in moral reasoning (Walker 1984). However, Gilligan's model does seem to correspond with our *stereotypes* of how females behave, and may be considered a mythic truth if it is not an empirical truth. The myth provides with insights for understanding and comparing the two alternative perspectives on moral development.

## REFERENCES

Argyris, C. 1990. *Overcoming Organizational Defenses*. Boston, MA: Allyn and Bacon.

Baier, A.C. 1987. Hume, the women's moral theorist? In *Women and Moral Issues*, eds. D.T. Meyers and E.F. Kittay, 37-55. New Jersey: Rowman and Littlefield.

Blum, L. A. 1988. Gilligan and Kohlberg: Implications for moral theory. *Ethics* 98: 472-491.

Brabeck, M. 1983. Moral judgment: Theory and research on differences between males and females. *Development Review* 3: 274-291.

Carroll, A.B. 1979. A three-dimensional conceptual model of corporate social performance. *Academy of Management Review* 4: 497-505.

Clarkson, M.B.E., and M.C. Deck. 1993. Applying stakeholder management to the analysis and evaluation of corporate codes. In *Business and Society in a Changing World Order*, ed. D.C. Ludwig, 55-76. New York: Mellen Press.

Cochran, P.L., and R.A. Wood. 1984. Corporate social responsibility and financial performance. *Academy of Management Journal* 27: 42-56.

Donaldson, T., and L.E. Preston. 1995. The stakeholder theory of the corporation: Concepts, evidence, and implications. *Academy of Management Review* 20: 65-91.

Frederick, W.C., and J. Weber. 1987. The values of corporate managers and their critics: An empirical description and normative implications. In *Research in Corporate Social Performance and Policy: Empirical Studies of Business Ethics and Values*, eds. W.C. Frederick and L.E. Preston, 131-152. Greenwich, CT: JAI Press.

Freeman, R.E. 1984. *Strategic Management: A Stakeholder Approach*. Boston, MA: Pitman.

Freeman, R.E., and D. Gilbert. 1988. *Corporate Strategy and the Search for Ethics*. Englewood Cliffs NJ: Prentice Hall.

Friedman, M. 1970. The social responsibility of business is to increase its profits. *New York Times Magazine* (September 13): 32-33, 122, 126.

Gilligan, C. 1977. Concepts of the self and of morality. *Harvard Educational Review* 47: 481-517.

Gilligan, C. 1979. Women's place in man's life cycle. *Harvard Educational Review* 49: 431-446.

Gilligan, C. 1982. *In a different voice: Psychological theory and women's development.* Cambridge, MA: Harvard University Press.

Hoffman, W.M. 1989. The cost of a corporate conscience. *Business and Society Review* (Spring): 46-47.

Kohlberg, L. 1969. Stage and sequence. In *Handbook of Socialization Theory and Research,* ed. D. Goslin, 347-480. Chicago, IL: Rand McNally.

Kohlberg, L. 1979. Moral stages and moralization. In *Moral Development and Behavior*, ed. T. Likona, 31-53. Chicago, IL: Rand McNally.

Kohlberg, L. 1982. *Essays on Moral Development,* Vol. 1: *The Philosophy of Moral Development.* San Francisco, CA: Harper and Row.

Larrabee, M.J. 1993. Gender and moral development: A challenge for feminist theory. In *An Ethic of Care*, ed. M.J. Larrabee, 3-16. London, UK: Routledge Chapman and Hall.

Levitt, B., and J. March. 1988. Organizational learning. *Annual Review of Sociology* 14: 319-340.

Lyons, N.P. 1983. Two perspectives: On self, relationships, and morality. *Harvard Educational Review* 53: 125-145.

Meyers, D.T., and E.F. Kittay. 1987. Introduction. In *Women and Moral Issues*, eds. D.T. Meyers and E.F. Kittay, 3-16. New Jersey: Rowman and Littlefield.

Meznar, M.B., J.J. Chrisman, and A.B. Carroll. 1991. Social responsibility and strategic management: Toward an enterprise strategy classification. *Business and Professional Ethics Journal* 10: 49-66.

Newton, L. 1992. Virtue and role: Reflections on the social nature of morality. *Business Ethics Quarterly* (July): 357-365.

Nunner-Winkler, G. 1984. Two moralities? A critical discussion of an ethic of care and responsibility versus an ethic of rights and justice. In *Morality, Moral Behavior and Moral Development*, eds. W. Kurtines and J. Gerwitz, 348-361. New York: John Wiley and Sons.

Preston, L.E., and J.E. Post. 1975. *Private Management and Public Policy.* Englewood Cliffs: Prentice-Hall.

Quinn, D.P., and T.M. Jones. 1995. An agent moral view of business policy. *Academy of Management Review* 20: 22-42.

Redenback, R.E., and D.P. Robin. 1991. A conceptual model of corporate moral development. *Journal of Business Ethics* 10: 273-284.

Rest, J.R. 1986. Moral research methodology. In *Lawrence Kohlberg: Consensus and Controversy,* eds. S. Modgil and C. Modgil, 455-470. London: Falmer Press.

Selman. R. 1971. The importance of reciprocal role-taking for the development of conventional moral thought. In *Recent Research in Moral Development*, eds. L. Kohlberg and E. Turiel. New York: Holt.

Sethi, S.P. 1979. A conceptual framework for environmental analysis of social issues and evaluation of business response patterns. *Academy of Management Review* 4: 63-74.

Sridhar, B.S., and A. Camburn. 1993. Stages of moral development of corporations. *Journal of Business Ethics* 12: 727-739.

Stark, A.S. 1993. What's the matter with business ethics? *Harvard Business Review* (May-June): 38-48.

Victor, B., and J.B. Cullen. 1988. The organizational bases of ethical work climates. *Administrative Science Quarterly* 33: 101-125.

Wartick, S.L., and P.L. Cochran. 1985. The evolution of the corporate social performance model. *Academy of Management Review* 13: 16-22.

Williamson, O.E. 1985. *The Economic Institutions of Capitalism.* New York: Free Press.

Wood, D.J. 1991. Corporate social performance revisited. *Academy of Management Review* 16: 691-718.

# EMPHASIS ON ETHICS IN TAX EDUCATION

Lawrence P. Grasso and Steven E. Kaplan

## ABSTRACT

Accountants engaged in tax practice frequently face moral and ethical dilemmas arising from their dual responsibilities to their clients and to the public at large. A review of the research on ethics in the accounting curriculum and textbooks used in introductory accounting courses suggests that there has been little attempt to include coverage of ethics in tax courses. The academic tax community has failed to increase the coverage of ethics despite numerous calls for coverage of ethics throughout the accounting curriculum, and empirical evidence that suggesting that context-specific coverage of moral and ethical issues may have beneficial effects. The standard arguments advanced to explain the lack of ethics coverage in tax courses are examined and are found wanting. We contend that there are economic incentives within the current institutional structure that discourage increased coverage of ethics in tax courses and encourage continuation of the status quo. The implications of this contention are discussed.

**Research on Accounting Ethics, Volume 4, pages 73-87.**

**ISBN: 0-7623-0339-5**

## INTRODUCTION

Tax preparers making compliance decisions and recommendations face moral conflict. The conflict arises because tax preparers have professional responsibilities to the taxpayer and to the public. In the United States, the American Institute of Certified Public Accountants has explicitly acknowledged the profession's dual responsibilities in the *Code of Professional Conduct* (AICPA 1991a). With rare exception, the *Code of Professional Conduct* applies to all services accountants provide, including tax services.

By an overwhelming margin, members of the AICPA indicate that tax issues represent the job situation that poses the most difficult ethical or moral problems (Finn et al. 1988). This may not be surprising, given that the application of tax law to the facts and circumstances of a particular situation is often ambiguous. The AICPA's *Statements on Responsibilities in Tax Practice* (SRTP), a set of "advisory opinions" and recommendations specifically directed towards CPAs engaged in tax practice, are evidence of a recognition within the profession that there are significant moral and ethical dilemmas unique to tax practice (AICPA 1991b).

In this paper, we contend that because of the moral conflict inherent in resolving tax issues the academic tax community has a large responsibility to incorporate ethics in the tax curriculum, and that this community has been shirking this responsibility, especially at the introductory level. This latter contention is supported by a review of research on ethics in the accounting curriculum. This review indicates that the academic tax community has only minimally attempted to incorporate ethics into courses on taxation. Several explanations have been offered by the academic tax community as to why tax courses have little emphasis on ethics. In this paper, we critically evaluate these explanations. Our analysis suggests that the standard explanations represent red herrings. We contend that the academic tax community's unwillingness to emphasize ethics may represent an attempt to "serve" the short-term economic interests of the professional tax community, possibly at the expense of the public.

The rest of the paper is organized as follows. The next two sections present a discussion on moral and ethical issues related to taxation from the taxpayers' and tax preparers' perspectives. These sections provide the relevant background used to discuss the academic tax community's responsibility to incorporate ethics in their curriculum. A review of the current status of ethics in accounting curriculum and arguments in support of greater coverage of ethics in introductory tax courses are presented in the fourth section. The "standard" explanations for lack of ethics coverage in tax course are presented and discussed in the fifth section. We propose an alternative explanation for the lack of ethics coverage in the sixth section of the paper. The last section offers a summary and conclusions.

## TAX COMPLIANCE AS A MORAL ISSUE

The federal tax system in the United States is grounded in voluntary compliance. Taxpayers are encouraged to comply on their own volition. In this section, we characterize taxpayer's voluntary tax compliance choices as moral decision making. We begin by introducing four definitions, the first three from Jones (1991). First, a *moral issue* is present where a person's actions, when freely performed, may harm or benefit others. As Jones (1991, 367) states, "The action or decision must have consequences for others and must involve choice, or volition, on the part of the actor or decision maker." Second, a *moral agent* is a person who makes a moral decision, regardless of whether he or she is aware that moral issues are at stake. Third, an *ethical decision* is one that is both legal and morally acceptable to the larger community, whereas an *unethical decision* is either illegal or morally unacceptable to the larger community.

Etzioni (1988, 41-43), in contrast, maintains that, among other things, a *moral act* reflects an imperative on the part of the agent. The agent must act out of some sense of obligation or duty. Making the right decision unaware that moral issues are at stake, or making the right decision on the basis expected pleasure or extrinsic gain would be considered *amoral.*

A combination of the slightly different perspectives offered by Etzioni (1988) and Jones (1991) yields a stricter view of what constitutes ethical behavior and some observations relevant to the issue of ethics in tax education. Common to both Etzioni and Jones is the idea that the decision and the action taken are what is relevant from an ethical standpoint, not the outcome or consequence of the action. Jones's definitions suggest that ignorance is no excuse. A decision can be unethical if a moral agent made it without thought of its effects on others and, in fact, the decision does have a negative effect on others. On the other hand, an action that led to beneficial consequences for others that were not considered by the moral agent when deciding to take the action would not necessarily be considered a moral act by Etzioni. An action is not a moral act unless the decision to act was made for the right reasons.

With these definitions in mind, consider U.S. taxpayers' federal income tax payment decisions. In this situation, taxpayers are the *moral agents* and the amount of tax to pay (or even whether or not to pay any tax at all) represents a *moral issue.* Clearly, a taxpayer's decision to pay his/her taxes may benefit others, and a decision not to pay taxes may harm others. For example, the benefits funded by federal income taxes include education, roads and bridge construction, health-related research, aid to impoverished citizens, and defense. Alternatively, nonpayment creates burdens for others. In the United States, the Internal Revenue Service (1990) estimated that in 1992, the total legal-sector tax gap from noncompliance with federal income tax laws by individual taxpayers was expected to be between $99 and $114 billion. The absence of such funds translates into a substantially reduced level of service or a substantially increased level of national debt with its attendant transfer of wealth. Further, these aggregate consequences are non-trivial and indicate that many taxpayers are noncompliant.

According to Jones (1991), noncompliance represents an *unethical decision* since it is illegal, whether or not the taxpayer took the potential harm to other taxpayers and government program beneficiaries into consideration when making the decision. From Etzioni's (1988) perspective, it is possible to view noncompliance as a moral act under certain limited circumstances. For example, a taxpayer may use noncompliance as a protest against a provision in the tax code viewed as immoral, or in protest of government policies or actions viewed as immoral. However, to maintain noncompliance as an ethical act, noncompliance should not be conducted without disclosure to tax authorities. To mask the protest from tax authorities while deriving monetary benefit from noncompliance would fall short of being a moral act from Etzioni's perspective.

Tax researchers have explored both beliefs about the ethics and morals of noncompliance as well as the role such beliefs have on compliance. Regarding beliefs about the ethics of noncompliance, studies by Westat, Inc. (1980) and Song and Yarbrough (1978) have found that a significant number of taxpayers do not consider tax evasion unethical under many circumstances. Additionally, ethics (or moral beliefs) has been found to be one of the factors to be most consistently associated with compliance (Roth et al. 1989). Jackson and Milliron (1986) report that 15 of 16 studies they reviewed found a positive relationship between ethics and compliance. The results from one study were indeterminate. More recently, Kaplan et al. (1988) and Porcano (1988) have also found a positive relationship between ethics and compliance. Additionally, the relationship between ethics and compliance has been found to be robust to research method (e.g., survey and experimental) and operationalization of ethics. These results are important because they establish that many taxpayers do not recognize the moral dimension of noncompliance, and second, moral and ethical tax beliefs make a difference in taxpayers' compliance decisions.

## ETHICAL DILEMMAS IN TAX PRACTICE

CPAs acting as tax preparers also confront ethical and moral issues. CPAs are moral agents in the compliance-related decisions of their clients. Tax preparers influence decisions made by their clients and make decisions on behalf of their clients that may harm or benefit others. In highlighting this point, Fisher (1994, 3) states, "tax compliance choices should not be viewed as a simple gamble, but rather as a classic social dilemma involving conflict between financial self-interest and social cooperation." In this environment, CPAs may face or believe they face noncompliance pressure from their clients, particularly if they believe some or all of their clients do not view tax evasion as unethical. For example, the results of a survey prepared for the IRS indicate that a majority of returns are prepared by practitioners who experience problems with at least one client in a tax season (IRS 1987).[1]

The findings of Finn et al. (1988) are useful to demonstrate the scope of ethical problems arising from tax work. The authors conducted a study to examine the nature of ethical problems confronting senior level AICPA members. The authors analyzed responses to the following open-ended question: "In all professions, managers are exposed to at least some situations that pose a moral or ethical problem. Would you please briefly describe the job situation that poses the most difficult ethical or moral problem for you personally?" (Finn et al. 1988, 609). By far the most common problem, cited by 47% of senior level AICPA member respondents, concerned client proposals of tax alteration and tax fraud.[2] At the core of this problem is the difficulty tax preparers have sorting out their dual responsibilities to the client and an outside party (e.g., the IRS). One of Finn et al.'s (1988, 609) respondents articulated this well in the following response:

> Preparing income tax returns whereby the basis of taking a deduction on the tax return is not 100 percent supportable (in other words a "grey area"). In this situation, the preparer is compensated by the client yet the Internal Revenue Service can subject the preparer to penalties unless the preparer can show he relied on "**substantial**" authority. It seems the regulations are enacted to pressure the preparer into making tax decisions which favor the IRS. This can, at times, be a moral dilemma.

The IRS (1987) survey documents specific types of questionable behaviors common among tax preparers, as follows:

- Not probing for secondary sources of income,
- Not cautious about criminal violations when there is a suspicion that income is intentionally understated,
- Sign return when there is a strong suspicion it understates income,
- Sign return that has a large, undocumented deduction,
- Show deduction in such a way as to minimize the chances of being selected for an audit.

Results from research exploring the behavior of CPAs as tax preparers also raises concern over the extent to which responsibilities to interested parties other than their client are being discounted. Klepper and Nagin (1989) report that tax preparers function as exploiters of the tax code in ambiguous situations. Their study was based on a line-by-line analysis of audited tax returns. Two studies (Jackson, Milliron, and Toy 1988; Ayres, Jackson, and Hite 1989) examine the conditions under which preparers tend to be more taxpayer oriented. Jackson et al. (1988) further analyzed the IRS (1987) survey data. The authors report that more expert tax professionals tended to be more taxpayer oriented (e.g., more likely to engage in the questionable behaviors listed above). The Ayres et al. (1989) study examined whether the tax preparers' professional status (e.g., CPA versus non-CPA) was associated with how ambiguous tax reporting issues were resolved. The authors found that CPAs took significantly more aggressive positions than did unlicensed tax preparers.[3] In considering the above studies, it should be noted that they were not designed to examine whether tax preparers' judgments represented departures from professional responsibilities.

Studies by McGill (1988), Kaplan et al. (1988), and Helleloid (1989), however, provide more direct evidence that CPAs engaged in tax work may be falling short of their professional responsibilities. The first two studies examined the role of audit risk on tax preparer recommendations. Recommending a tax return position that exploits the Internal Revenue Service audit selection process is clearly considered improper under the professional standards established to guide CPAs (SRTP §112.08). McGill (1988) found that CPAs were more aggressive to clients when audit risk was low. Kaplan et al. (1988) manipulated audit risk by describing whether the district commissioner put a high or low priority on the tax issue. Based on SRTP, CPAs' recommendations to clients should not be influenced by this particular type of information. However, the results did show that subjects responses were affected by this type of evidence and suggests that tax preparers play the "tax lottery." These findings also indicate that CPAs practicing in taxation either are not fully aware of their ethical and professional responsibilities or do not care.

Helleloid (1989) examined whether tax professionals' judgments were sensitive to the quality of the client's documentation. The issue examined, business transportation expense, was selected largely because stringent documentation is required to support a deduction and preparers were expected to be aware of the documentation requirement. Based upon two experiments using tax professionals, Helleloid found that many tax professionals were aware of the documentation standards of the tax issue but did not believe that their role included rigorous enforcement of the documentation standards.

## COVERAGE OF ETHICS IN THE TAX CURRICULUM

Over the past decade, there has been a widespread effort to encourage greater emphasis on ethics in accounting education. For example, formal recommendations for increasing ethics instruction have been made by the American Assembly of Collegiate Schools of Business (AACSB 1988), the American Accounting Association (Bedford Committee 1986), the Federated Schools of Accountancy (Kiger et al. 1988), and the Accounting Education Change Commission (AECC 1990). These calls would seem to suggest a consensus among accounting educators that ethics should be an integral part of accounting education.

Within an educational environment placing greater value on the importance on ethics, and given the fact that compliance and other tax work is inherently a moral issue, one might expect tax courses would devote a great deal of attention to discussing professional standards and to the process of resolving ethical dilemmas involving tax issues. This is not the case. Introductory tax textbooks, with rare exceptions, offer neither a chapter covering ethical and professional responsibilities nor even a listing of the topic in the index. The scant attention to ethics and professional responsibilities in introductory tax textbooks is an important indicator of the academic tax community's level of interest and concern. That is, textbook authors respond to the demands of course instructors. Apparently there is no demand among the academic tax community for ethics materials in the introductory tax course.

It seems implausible that the academic tax community could claim that it is unaware that ethics and morals play a significant role in compliance decisions. It is this very group that has determined in study after study that it does. Tax academics have played an active role in research and in providing valuable insights into the understanding factors associated with tax compliance decisions (for a summary, see Roth 1989). For example, Jackson and Milliron (1986) report that 15 of 16 studies they reviewed found a positive relationship between ethics and compliance.

Several surveys have examined the extent of ethics coverage in the accounting curriculum (Armstrong and Mintz 1989; Cohen and Pant 1989; Ingram and Petersen 1989; Karnes and Sterner 1988; Thompson et al. 1992). While these surveys have generally found that attention to ethics is increasing, they also report that minimal coverage in accounting courses is still common. Armstrong and Mintz (1989) report that ethics coverage in tax courses averaged one session per semester. The results of a survey by of department chairpersons by Thompson et al. (1992) also raise concern over the attention given ethics by tax educators. Whereas only 4% of their respondents indicated that ethics was not addressed in auditing courses, an alarming 42% of their respondents indicated that ethics was not addressed in the individual tax course. The authors also report that the percentage not covering ethics in corporate tax courses was approximately the same (40%).

Thus, department chairpersons believe that a large minority of their tax faculty are not addressing ethics in their individual and corporate tax courses. Further, the authors report that chairpersons generally believe that ethics should be addressed in both individual and corporate tax courses. On a five-point scale in which one is anchored by "strongly agree" and five was anchored by "strongly disagree" the mean of individual and corporate tax courses was 1.86 and 1.89, respectively, in response to the statement that ethics should be addressed in this course. Finally, the study reports that while 49% of the respondents believed the accounting curriculum can have a significant impact on a student's further ethical conduct, only 16% believed that the accounting curriculum would have no significant impact on student's future ethical conduct.

Cohen and Pant (1989) also surveyed department chairpersons regarding their perceptions of ethics coverage in accounting courses on a scale from one (no coverage) to seven (great deal of coverage). They report a mean of 3.3 for tax courses. Auditing courses again had a significantly higher mean of 5.3.

Overall, we believe the results from research on the extent of ethics coverage in accounting courses, in conjunction with our previous discussion, support our contention that ethics are not being sufficiently emphasized in tax courses. While an argument can be made that ethics coverage throughout the accounting curriculum is far from ideal, a comparison of the level of ethics coverage in tax courses with that in auditing courses suggests that tax educators are far behind their auditing counterparts in their efforts to address ethical issues. We believe that auditing is an appropriate benchmark for taxation because both auditors and tax preparers have dual responsibilities to the client and outside parties even though their fees are always paid by the client, and because the consequences to outside parties and to society at large of actions taken by either auditors or tax preparers can be enormous.

## "STANDARD" EXPLANATIONS FOR LACK OF ETHICS COVERAGE IN TAX COURSES

Why have tax academics largely failed to incorporate coverage of ethics and professional responsibilities in tax courses? We identify and critically evaluate three possible explanations that have been advanced in the literature. First, the large amount of technical material that must be covered in the accounting curriculum, especially tax courses, is cited as a factor contributing to the low level of ethics coverage (Armstrong and Mintz 1989; Langenderfer and Rockness 1989; Loeb 1988). Traditionally, the core of introductory tax courses has been teaching various provisions of the Internal Revenue Code. Perhaps, tax academics falsely justify this approach by claiming that the Internal Revenue Code is relatively more detailed and technical than other areas of accounting.

Whether the Code is or is not more technical than the FASB or the GASB is irrelevant when considering the nature of tax laws. Tax laws, like other laws, are socially constructed through a specific political process involving the legislative branch, the executive branch, and interested outside parties (e.g., lobbyists). While other accounting rule-making bodies also produce socially constructed outcomes, our strong speculation is that tax laws are the most dynamic and subject to change. We are unconvinced of the need to teach tax laws that change with great regularity at the expense of integrating other material. Holding on to teaching technical tax material is in complete opposition to the general consensus of practitioners, educators, and standard-setters that accounting educators have disproportionately and inappropriately taught technical material to the exclusion of other material more important to their long-term success as a business professional and a member of society. Students learning a specific set of technical rules will be outdated as soon as those rules change. In tax, the rules change often. Thus, we believe that the "need" to teach technical tax material represents a red herring.

Second, the low level of ethics coverage has been attributed, in part, to tax educators' lack of ethics training and ethics teaching materials (Cohen and Pant 1989).

This argument is hollow in that it is incapable of explaining why academic auditors are able to include coverage of ethics and professional responsibilities in their courses and have developed sufficient teaching materials. Academics who teach auditing do not have an obvious comparative advantage in teaching ethics. They have, however, made a collective judgment that it is important to teach ethics and professional responsibilities. Consequently, auditing educators have made the collective commitment to teach ethics and professional responsibilities and have developed teaching materials. For example, if the academic tax community valued the teaching of ethics and professional responsibilities, they would demand coverage in their introductory textbooks and supplementary materials. Clearly, this demand does not exist.

Finally, concerns regarding the effectiveness of ethics coverage within the accounting and tax curriculum are raised as a factor explaining the low level of ethics coverage found (Armstrong and Mintz 1989; Lampe 1994; Langenderfer and Rockness 1989; Ponemon 1993). While we acknowledge that little empirical evidence on the outcomes of ethics coverage exists, we also believe this is a red herring. Assessing effectiveness is an important step in the process of continually improving course content and pedagogy. However, we believe any assessment of ethics coverage in tax education at this time would be premature. We might be a little old-fashioned on this point, but we do not think that an empirical assessment of the effectiveness of ethical coverage within tax courses can take place before ethical coverage is introduced into tax courses. Those questioning the effectiveness of ethics coverage apparently want a guarantee of success before any attempt is made to incorporate ethics into tax courses. It is doubtful that tax courses at all would exist if pedagogical techniques and all elements of course content, including technical tax material, were held to this standard.

In addition, the limited evidence that is available on the effects of introducing ethics coverage in accounting courses is at least promising. Hiltebeitel and Jones (1991) asked students to rank-order the importance of six considerations when resolving each of six ethical dilemmas. They found that exposure to an ethics module in an accounting course changed the level of moral reasoning used by accounting students (as evidenced by their rankings of the considerations) in responding to hypothetical professional ethical dilemmas. No change in the level of moral reasoning was found for students' responses to hypothetical personal dilemmas. In a second study, Hiltebeitel and Jones (1992) had students evaluate a set of 14 principals of ethical conduct that may guide moral choices. Critics may argue that the measures of moral reasoning are inappropriate or that the measured effect may not last long enough to influence ethical decisions as the students enter practice. The results are nevertheless promising, and they at least hint that there may be a benefit to exposing students to context-specific ethical training.

The results from a study by Fulmer and Cargile (1987) are also encouraging. The authors examined whether senior accounting students who have been exposed to the *Code of Professional Conduct* are more likely to view business practices in ethical terms than non-accounting business students. Students were given a case descibing a ethical dilemma and asked to indicate how they might have acted and felt had they actually faced the dillema and whether the behaviors of the actors in the case were ethical. Fulmer and Cargile found that the judgments of accounting majors were more ethical than non-accounting business students.

## AN ALTERNATIVE EXPLANATION FOR THE LACK OF ETHICS IN THE TAX CURRICULUM

Since the standard explanations for the lack of attention given to ethics by the academic tax community appear to be inadequate, we offer an alternative explanation. The academic tax community's unwillingness to give greater attention to ethics may represent an attempt to serve the short-term economic interests of the professional tax community, possibly at the expense of the public. That is, the academic tax community is responding as if it can serve the economic interests of the professional tax community by ignoring or deemphasizing ethics. To consider this alternative explanation, we briefly characterize the market for tax services and contrast it with the market for auditing services.

The market for tax services is similar to auditing services in that the client pays the fees. The markets are also similar in that CPAs performing either service must adhere to professional standards. However, CPAs providing tax services compete with other groups that offer tax services, while CPAs have a monopoly on providing audit services. In addition, the taxpayer's use of a paid tax preparer is entirely discretionary, while audits are a prerequisite for a business wanting access to the capital markets and many private sources of capital.

The relationships of tax preparers to their clients and to third parties are also somewhat different from those of auditors. Tax preparers and auditors are required to be objective, but tax preparers must also be advocates for their clients while auditors must be independent. Further, the consequences associated with failing to fulfill responsibilities to third parties are very different in magnitude for auditors and tax preparers. Auditor settlements to third parties have reached the hundreds of millions of dollars. While large in aggregate, penalties and settlements for individual cases are much smaller for tax preparers.

These features suggest that the market for tax services is likely to be highly competitive. The CPAs may believe that their ability to charge for tax services depends in large part on the taxpayer's belief that engaging a CPA will result in a lower tax liability than would filing without the assistance of a paid preparer. Preparers may therefore perceive strong pressures to resolve questionable issues in the taxpayer's favor. This perceived pressure may be fueled by interactions with taxpayers that do not consider tax evasion unethical.

Within this marketplace for tax services, Ayers et al. (1989) contend that the comparative advantage of CPA firms is in taking aggressive, questionable positions on behalf of their clients. The authors base their contention on the economic theory of regulation. Under this theory, regulation allows CPAs and other enrolled agents, relative to unenrolled paid preparers, "to interpret the law more to the benefit of the taxpayer, further increasing the value of the regulated practitioner's services to the client, and thus, the total industry profitability" (Ayres et al. 1989, 301). As indicated earlier, the empirical results supported their contention. This theory and evidence suggests that CPA firms have a culture that extends the limits of advocacy beyond those of non-CPAs. Even though professional standards require adherence to both objectivity as well as advocacy, the economic theory of regulation would suggest that a CPA firm's culture would place little emphasis on adhering to the standard of objectivity.

In this context, consider the results of a recent study by Johnson (1993). Johnson found that CPAs' advocacy role interferes with their ability to evaluate tax information objectively. While Johnson does not speculate on the origins or consequences of this bias, one can argue that the existence of this type of bias facilitates CPA firms' tax departments ability to take aggressive positions and minimize reported tax liabilities for their clients in order to justify premium fees for their tax services. Such a bias may represent a manifestation of an organizational culture that does not value the public's interests.

Now consider the demand for ethics education in the professional tax community. The above analysis suggests that this community would not demand the integration of ethics into the tax curriculum and, in fact, may view such an integration as dysfunctional. The socialization of new tax employees into a culture encouraging the adoption of aggressive tax positions for clients could be made more difficult if the employees had been exposed to a broad ethical foundation highlighting both the client's and public's interests in tax compliance. Beyond the issue of avoiding IRS penalties, coverage of ethics may simply be considered an investment that does not maximize income and, therefore, has no value.

Finally, consider the relationship between the academic tax community and the professional tax community. Often, members of the academic tax community also act as tax preparers. Further, many members of academic tax community supplement their academic salaries by teaching in continuing education programs for tax preparers. We believe that these relationships encourage many members of the academic tax community to align their beliefs with and/or cater to the preferences of the professionals engaged in tax practice.

The academic tax community has a significant responsibility for educating students who are specializing in taxation and who aspire to careers as tax professionals. Since only a small percentage of these students will be hired by government agencies, the academic tax community also has incentives to adopt the view that they are serving their students best by satisfying the preferences of more students' more likely employer—firms performing tax services. In this regard, the academic tax community appears to have made a collective judgment to limit ethics education and to allow individual firms to decide both the extent and kind of coverage given to ethics and professional responsibilities.

## SUMMARY AND CONCLUSIONS

In this paper, we have examined the moral and ethical conflicts that confront accounting professionals engaged in tax practice and the extent to which the academic tax community has responded to the need to prepare students to deal with these conflicts. Our examination led us to the following conclusions.

First, due to the economic environment in which tax practice occurs, the dual responsibilities of the tax professional, the characteristics of the tax professional's relationship with clients and the public at large, and the magnitude of the economic consequences involved, tax practice gives rise to greater moral and ethical conflict than any other type of professional accounting services. Second, CPAs individually and as a professional group are aware of the unique character of tax practice and the moral and ethical conflict associated with tax practice.

Third, there has been a general call for increased attention to ethics throughout the accounting curriculum. Fourth, the limited empirical evidence that exists on the effectiveness of incorporating ethics into accounting education, while hardly conclusive, holds the promise that there are real benefits to be gained. Fifth, to date there has been very little integration of ethics coverage in tax courses, especially introductory courses. Sixth, the standard arguments advanced in the literature to explain the lack of ethics coverage appear to be unfounded.

Finally, we suggest that the standard explanations are obscuring what may be the real reason why there has not been a greater effort to incorporate ethics coverage in tax courses. That is, the economic incentives currently in place in tax practice and in academia discourage the incorporation of ethics into tax courses. These conclusions have implications for accounting and tax education and, ultimately, for tax practice as well.

Regarding tax education, our analysis indicates that the obstacle to increased coverage of ethics in tax courses is institutional in nature and is more fundamental than a lack of training in teaching ethics or a lack of instructional material. The academic tax community is acting as if there is little or no demand for (increased) coverage of ethics and, therefore, no reason to invest human capital in developing ethics-based education and related instructional materials. The obstacle is a set of currently existing incentives which provide no reason to place more emphasis on integrating ethics into tax courses.

Our analysis suggests that the incentive system has to change before the academic tax community will place greater emphasis on ethics. It is unlikely that the professional tax community will of its own volition change its behavior toward the academic tax community. Thus, any changes in the incentive system must come from the broader educational environment in which the academic tax community operates. The academic tax community has felt that it could cater to the preferences of the professional tax community because there was no cost. The academic accounting community has a responsibility to create a cost for members who choose not to integrate ethics into their cirriculum.

As students graduate and enter tax practice, they will invariably encounter moral and ethical dilemmas. Given the current level of coverage of ethics in tax courses, it is almost certain that these students will be ill-equipped to deal with the dilemmas they encounter. It appears, however, that the profession prefers such ill-equipped employees because they may be more likely to resolve ethical dillemmas in favor of clients.

We believe it is clear that moral and ethical dilemmas in tax practice represent significant problems for the profession and for society. It is also clear that the prospect of beneficial effects from coverage of ethics in the tax curriculum is large enough to warrant immediate implementation of "field trials" of a variety of materials and pedagogical techniques. The question that remains unanswered is whether the academic accounting community has the collective will to overcome the pressures created by the economic incentives that support the status quo.

## NOTES

1. Research by Hite and McGill (1992), however, suggests that much of this pressure may be illusory. In an experimental study using a national sample of U.S. taxpayers, they found that their subjects did not have a preference for aggressive tax advice. Instead, they found a tendency toward conservative tax reporting positions. This suggests that tax preparers may erroneously project their own aggressive tendencies, or the aggressive tendencies of a vocal minority of taxpayers, onto all of their clients.

2. To get a sense of the dominance of this problem, the second most cited problem, conflict of interest and independence, was cited by only 16% of the respondents.

3. The unlicensed tax preparers in the Ayres et al. (1989) sample also were not enrolled agents. Enrolled agents may represent their clients in all matters related to the IRS.

## REFERENCES


Accounting Education Change Commission (AECC). 1990. Objectives of education for accountants: Position statement number one. *Issues in Accounting Education* 5(Fall): 307-312.

American Institute of Certified Public Accountants (AICPA). 1991a. *Code of Professional Conduct.* New York: AICPA.

American Institute of Certified Public Accountants. 1991b. *Statement on Responsibilities in Tax Practice*. New York: AICPA Federal Tax Division.

American Assembly of Collegiate Schools of Business (AACSB). 1988. *1987-88 Accreditation Council Policies Procedures and Standards.* St. Louis, MO: AACSB.

Armstrong, M.B., and S.M. Mintz. 1989. Ethics education in accounting: Present status and policy implications. *Government Accountants Journal* 38(2, Summer): 70-76.

Ayres, F.L., B.R. Jackson, and P.A. Hite. 1989. The economic benefits of regulation: Evidence from professional tax preparers. *The Accounting Review* 64(2, April): 300-312.

Cohen, J.R., and L.W. Pant. 1989. Accounting educators' perceptions of ethics in the curriculum. *Issues in Accounting Education* 4(1, Spring): 70-81.

Committee on the Future Structure, Content, and Scope of Accounting Education (The Bedford Committee). 1986. Future accounting education: Preparing for the expanding profession. *Issues in Accounting Education* 1(Spring): 168-195.

Etzioni, A. 1988. *The Moral Dimension: Toward a New Economics.* New York: The Free Press.

Finn, D.W., L.B. Chonko, and S.D. Hunt. 1988. Ethical problems in public accounting: The view from the top. *Journal of Business Ethics* 7(August): 605-615.

Fisher, D.G. 1994. The use of the four-component model of moral reasoning as a framework for tax compliance research. Paper presented at the Accounting, Behavior, and Organizations Research Conference, San Antonio, TX, March.

Fulmer, W., and B. Cargile. 1987. Ethical perceptions of accounting students: Does exposure to a code of professional ethics help? *Issues in Accounting Education* 2(2, Fall): 207-219.

Helleloid, R.T. 1989. Ambiguity and the evaluation of client documentation by tax professionals. *Journal of the American Taxation Association* 11(1, Fall): 22-36.

Hiltebeitel, K.M., and S.K. Jones. 1991. Initial evidence on the impact of integrating ethics into accounting education. *Issues in Accounting Education* 6(1, Fall): 262-275.

Hiltebeitel, K.M., and S.K. Jones. 1992. An assessment of ethics instruction in accounting education. *Journal of Business Ethics* 11(1): 37-46.

Hite, P.A., and G.A. McGill. 1992. An examination of taxpayer preference for aggressive tax advice. *National Tax Journal* 45(4, December): 389-403.

Ingram, R.W., and R.J. Petersen. 1989. Ethics education in accounting: Rhetoric or reality? Paper presented at the Ohio Regional Meeting of the American Accounting Association, May.

Internal Revenue Service (IRS). 1987. *Survey of Tax Practitioners and Advisors: Summary of Results by Occupation.* Washington, DC: Office of the Assistant Commissioner, Research Division, IRS, June.

Internal Revenue Service (IRS). 1990. *Income Tax Compliance Research: Net Tax Gap and Remittance Gap Estimates.* Publication 1415, Supplement to Publication 7285. Washington, DC: Internal Revenue Service, April.

Jackson, B.R., and V.C. Milliron. 1986. Tax compliance research: Findings, problems, and prospects. *Journal of Accounting Literature* 5: 125-166.

Jackson, B.R., V.C. Milliron, and D. Toy. 1988. Tax practitioners and the government. *Tax Notes* (October 17): 333-341.

Johnson, L.M. 1993. An empirical investigation of the effects of advocacy on preparers' evaluations of judicial evidence. *Journal of the American Taxation Association* 15(1, Spring): 1-22.

Jones, T.M. 1991. Ethical decision making by individuals in organizations: An issue-contingent model. *Academy of Management Review* 16(2, April): 366-395.

Kaplan, S.E., P.M.J. Reckers, S.G. West, and J.C. Boyd. 1988. An examination of tax reporting recommendations of professional tax preparers. *Journal of Economic Psychology* 9(4): 423-443.

Karnes, A., and J. Sterner. 1988. The role of ethics in accounting education. *The Accounting Educators' Journal* 1(2, Fall): 21-29.

Kiger, J., R. Sterling, R. Hermanson, and A. Mitchell. 1988. Report of the General Education Guidelines Committee. In *Proceedings of the Eleventh Annual Meeting of the Federation of Schools of Accountancy,* 37-40. Urbana-Champaign, IL: Federation of Schools of Accountancy.

Klepper, S., and D.S. Nagin. 1989. The role of tax preparers in tax compliance. *Policy Sciences* 22: 167-194

Lampe, J.C. 1994. The impact of ethics education in accounting curricula. In *Proceedings of the Ernst & Young Research on Accounting Ethics Symposium,* 220-236. Binghamton, NY: SUNY—Binghamton.

Langenderfer, H.Q., and J.W. Rockness. 1989. Integrating ethics into the accounting curriculum: Issues, problems, and solutions. *Issues in Accounting Education* 4(Spring): 58-69.

Loeb, S.E. 1988. Teaching students accounting ethics: Some crucial issues. *Issues in Accounting Education* 3(Fall): 316-329.

McGill, G.A. 1988. The CPA's role in income tax compliance: An empirical study of variability in recommending aggressive tax positions. Ph.D. dissertation, Texas Tech University.

National Commission on Fraudulent Financial Reporting (The Treadway Commission). 1987. *Report of the National Commission on Fraudulent Financial Reporting.* New York: AICPA.

Ponemon, L.A. 1993. Can ethics be taught in accounting? *Journal of Accounting Education* 11: 185-209.

Porcano, T.M. 1988. Correlates of tax evasion. *Journal of Economic Psychology* 9: 47-67.

Roth, J.A., J.T. Scholz, and A.D. Witte, eds. 1989. *Taxpayer Compliance: An Agenda for Research*, Vol. 1. (Philadelphia, PA: University of Pennsylvania Press.

Song, Y., and T.E. Yarbrough. 1988. Tax ethics and taxpayer attitudes: A survey. *Public Administration Review* (September-October): 442-452.

Thompson, J.H., T.L. McCoy, and D.A. Wallestad. 1992. The incorporation of ethics into the accounting curriculum. In *Advances in Accounting*, Vol. 10, ed. B.N. Schwartz, 91-103. Greenwich, CT: JAI Press:

Westat, Inc. 1980. *Industrial income Tax Compliance Factors Study: Qualitative Research Results.* Prepared for the Internal Revenue Service, February 4, 1980, by Westat, Inc., Rockville, MD.

# A LONGITUDINAL STUDY OF THE EFFECT OF TEACHING ETHICS THROUGHOUT THE ACCOUNTING CURRICULUM

Patricia Casey Douglas and Bill N. Schwartz

## ABSTRACT

This paper describes a study which examined the effects over time of an instruction program incorporating ethics components throughout the accounting curriculum. The study extends the ethics literature by examining evidence of the effect of such instruction on individual students' moral development over a two-year period. Such longitudinal studies have long been suggested in the accounting literature. Using Rest's Defining Issues Test, the authors analyzed both group and individual changes over the test period as well as changes across all developmental stages, not only those at the most advanced or "principled" reasoning stage. As predicted by moral development theory, results indicated an increase in students' principled reasoning across groups, although no differences between experimental and control groups were observed. These findings have important implications for both accounting education

**Research on Accounting Ethics, Volume 4, pages 89-112.**

**ISBN: 0-7623-0339-5**

and practice. Educators would be wise to widen their pedagogical approach, clarify goals, and seek other assessment techniques. The objectives of accounting ethics education should be both broader than "moral development" in that they address the wider considerations of a pluralistic profession, and narrower in that they address issues specific to the accounting profession.

## INTRODUCTION

Accounting instruction in ethics has reached a crossroads. Practitioner and academic groups call for more emphasis on ethics in the classroom, but previous research has been unable to persuasively demonstrate its effectiveness. Lampe (1994) points out that pressures placed on accounting curricula by an expanding code, budget constraints, and requirements of 150 hour programs lead to the real possibility of reducing or eliminating "ethical education interventions that appear to lack effectiveness" (234). That course of action would not meet the expressed needs and wants of the profession, which *needs* accountants prepared to face "even more complex ethical dilemmas with increasing frequency" (234) and which *wants* that preparation to be accomplished by preprofessional education.

A number of respected individuals and groups have called for more emphasis on ethics. Some have written articles discussing the need for ethics instruction. Others have conducted studies examining the effects of ethics interventions in classroom settings. However, as yet no one has assessed the longitudinal effects on one group of students of an ethics program that incorporates ethics components throughout the accounting curriculum.[1] The current study makes such an attempt.

The discussion of this study proceeds along the following lines. First, pertinent literature is reviewed. Calls for ethics instruction from organizations, scholars, and practitioners are presented. The current status of ethics instruction and previous empirical studies are depicted. Second, the research methodology is discussed, including its theoretical underpinnings and applications in accounting of the research instrument and treatment. The research setting also is described. Third, results are presented. Finally, conclusions and implications for education and practice are considered, and future research topics are suggested.

## BACKGROUND

The call for increased ethics instruction in accounting comes from a number of quarters. Numerous organizations as well as individual academics and practitioners have discussed increased ethics instruction in the classroom. These groups and individuals may define "ethics" differently and may have varying views on what, how and when ethics should be taught, but it is clear that they want more attention paid to ethics.

Such organizations as the American Accounting Association (The Bedford Committee) (1986), the National Commission on Fraudulent Financial Reporting (The Treadway Commission) (1987), and the American Assembly of Collegiate Schools of Business (1993) all recommend increased emphasis on ethics. The American Institute of CPAs, discussing education requirements for entry into the profession, stresses "the importance of developing ethical values in the accounting student and integrating the coverage of ethical issues and considerations with the coverage of technical knowledge throughout the curriculum" (1988, 23). The managing partners of the then-Big Eight CPA firms, in their white paper on accounting education, emphasize ethics instruction, too. They say that "practitioners must . . . be able to identify ethical issues and apply a value-based reasoning system to ethical questions" (AICPA 1989, 6).

A number of academics have called for increased ethics instruction. As early as 1964, Grimstad commented in the *Journal of Accountancy*:

> no profession can continue to be effective unless adherence to a high standard of ethical conduct is maintained. This in turn requires that young [people] entering the profession be imbued with a sense of professional responsibility and ethical propriety (p. 82).

Kunitake and White (1986) urge ethics instruction in auditing classes to instill professional ethics in young accountants. Pamental (1989) stresses the need for ethics instruction in introductory accounting so non-accounting, as well as accounting students, will get exposure. Huss and Patterson (1993) provide suggestions for implementing values education in accounting ethics while Beets (1993) suggests role-playing techniques. Lantry (1993) describes a more pragmatic approach, stressing real problems in current practice in his "Professional Ethics and Legal Responsibilities" course. Others (e.g., Cooke et al. 1987-88; Hosmer 1988; Lehman 1988; Loeb 1988; Mintz 1990; Murray 1987; Smith 1993a, b) also encourage more ethics instruction. Ahadiat and Mackie (1993) find that students' ethical behavior is a very strong factor in recruiting decisions by public accounting firms. They conclude that "our results strongly suggest the need for much greater emphasis on the teaching of ethics to college students" (255).

Individuals with strong practitioner backgrounds agree with the academics. John Burton, former SEC Chief Accountant, and Robert Sack, former SEC Chief Accountant in the Enforcement Division and Big Eight firm partner, strongly support ethics instruction (Burton and Sack 1989). Another former Big Eight firm partner, Robert Fess (1987), expresses his views when he says:

> The most feasible approach [to teaching ethics] . . . is to include more ethics in existing courses so as to reach all students in a minimal way. The ideal solution would include both the infusion of ethics material into all accounting courses and a separate course in ethics (p. 60).

## Current Status of Ethics Instruction

Ethics instruction is not pervasive in the accounting curriculum. George (1987) determined that 21% of respondents to a survey of deans from AACSB-accredited schools require an ethics course for all business students. Pamental (1989) noted that very little ethics instruction is included in introductory accounting texts. Karnes and Sterner (1989) learned from a survey of accounting chairs that only 8.5% offer a separate accounting ethics course. A majority offer a module in auditing or some other accounting course. Cohen and Pant (1989) and Armstrong and Mintz (1989) similarly found that most ethics instruction occurs in auditing courses.

## Previous Empirical Studies

A few studies have considered the ethics of accounting practitioners (e.g., Armstrong 1987; Flory et al. 1992; Shaub et al. 1993; Ward et al. 1993) and academics (e.g., Engle and Smith 1992; McNair and Milam 1993; Carver et al. 1993). Several have studied ethics and the effectiveness of ethics instruction in the accounting classroom. A review of their findings follows.

Arlow and Ulrich (1985) conducted a longitudinal study of 73 undergraduate business (including accounting) students enrolled in an upper-division Business and Society course. The course included coverage of business ethics both as a specific topic area and integrated throughout discussions of the business-society relationship. Students were surveyed at the beginning and end of the course and surveyed again by mail four years later using a questionnaire "designed to appraise personal ethical standards rather than behavior in business" (p. 14). Results showed no significant differences in ethics scores over the three time periods, suggesting that business ethics instruction has virtually no impact over time.

Fulmer and Cargile (1987) examined whether senior accounting students who have been exposed to the AICPA Code of Professional Ethics were more or less likely to view business practices in ethical terms than senior business non-accounting majors. Sixty-eight accounting and 132 business non-accounting students in five sections of a business policy course at one university were given a case to read and asked to complete a short questionnaire. Accounting students demonstrated more ethical perceptions but not more ethical actions.

Davis and Welton (1991) examined the ethical level of 391 business and accounting students (131 lower division, 184 upper division, and 76 graduate students) at one university. Some of the students had been exposed to ethics, some had not. A set of 17 statements representing various situations involving "gray" areas in terms of ethical behavior was presented. The objective was to evaluate similar and dissimilar ethical responses. Lower division students differed from upper division students, but the latter did not differ from the graduate students. Prior ethics instruction did not seem to have any effect.

Hiltebeitel and Jones (1991) tested whether an ethics module taught in an accounting course affected the way students resolve ethical dilemmas. A pretest/posttest design was used with experimental and control groups. The authors developed six ethical dilemmas for use in the study. A total of 171 students from two introductory accounting classes at one university and two cost-accounting classes at a second university participated. The module did not change the students' ability to resolve ethical dilemmas.

Hiltebeitel and Jones (1992) conducted a different study with a larger sample. A total of 354 students (four cost accounting classes at one university; five auditing, two advanced accounting, and four introductory accounting classes at a second university) participated. The individual instructors designed their own ethics modules. A pretest/posttest design was employed. A questionnaire was administered which was based on Lewis' (1989) principles of ethical conduct. In this situation, the principles on which students rely when making ethical decisions were affected by the ethics module.

Giacomino (1992) used the same case as Fulmer and Cargile (1987). Giacomino studied responses from 363 freshmen and 80 seniors (132 accounting and 311 business non-accounting) at one university. Accounting students' ethical perceptions did not differ from those of business non-accounting students. In addition, very few differences existed between the ethical standards of freshmen and seniors.

Borkowski and Ugras (1992) examined ethical positions of 130 students (51 freshmen, 39 juniors, and 40 graduate students) at one university. They were exposed to two ethical dilemmas and completed a questionnaire afterward. Undergraduate students were suggested to be relatively more justice-oriented while graduate students were relatively more utilitarian.

Several studies have examined levels of students' moral reasoning and the effect of classroom ethical interventions using Rest's (1979b) Defining Issues Test (DIT). Ponemon and Glazer (1990) examined the ethical development of students from two universities, 44 from a private liberal arts college offering a concentration in accounting through its business administration department and 56 from an AACSB-accredited accounting program. Forty-six were freshmen and 54 were seniors. Seniors at each school demonstrated higher levels of moral reasoning than freshmen. Students at the liberal arts school scored higher than the students at the school with the accounting-accredited program.

St. Pierre et al. (1990) used one application of the DIT to test the ethical development of 479 seniors in seven business and accounting and three non-business disciplines at one university. The three non-business majors showed higher levels of moral reasoning than the seven business and accounting disciplines.

Jeffrey (1993) compared levels of moral development across different majors and between lower division and senior students at one university. She used a total of 503 students from six groups: 76 senior accounting majors, 135 senior business non-accounting majors, 41 senior liberal-arts majors, 57 lower division accounting majors, 165 lower-division business non-accounting majors, and 29 lower division liberal arts majors. Contrary to the findings of other studies (i.e., Ponemon and Glazer 1990; St. Pierre et al. 1990), lower division accounting students had higher moral reasoning scores than the other two lower division groups. Senior accounting students had higher scores than the other two senior groups, and the size of the differences was constant across groups.

Ponemon (1993) assessed the influence of ethical interventions integrated into a one-semester auditing course at both the undergraduate and graduate levels. Sixty-five students participated in the experiment. Ponemon tested the effectiveness using: (1) the DIT with a pretest/posttest design, and (2) an economic-choice experiment based on the Prisoners' Dilemma. The ethics intervention did not increase the level of ethical reasoning and did not curtail students' free-riding behavior.

Armstrong (1993) tested the impact of a separate "Ethics and Professionalism" (E&P) course offered to senior accounting students on an elective basis. Rest's DIT was administered to the E&P students on the first and last days of class. Intermediate Accounting students were selected to act as a control and were similarly tested. Data from 21 E&P students and 33 control group students indicate a significant increase in ethical development level of the treatment group over that of the control group.

Shaub (1994) evaluated whether demographic variables traditionally associated with higher levels of moral reasoning in other populations are associated with auditing students. Shaub administered the DIT to 91 senior auditing students at one university. Results indicated that auditing students' moral reasoning is associated with academic success, gender (women scored higher than men), and ethics education.

Lampe (1994) surveyed 328 auditing students over a four-year period (1990-1994) and 144 accounting information systems students over a two-year period (1991-1993) at a single university. During the test period, students were subjected to ethical decision cases added to existing courses throughout the accounting curriculum. Lampe reports substantial overlap in the responses of the two groups but was not able to match specific student responses because they were collected anonymously. Results indicate that students' moral development (measured by Rest's DIT), reactions to ethical dilemmas, and attitudes toward ethical behavior remained largely unchanged and "strongly oriented to code-implied rules" (p. 231).

To summarize, previous research has been unable to persuasively demonstrate the effectiveness of ethics interventions in accounting classes. A number of studies have demonstrated that accounting students' moral development differs from that of non-accounting students (Ponemon and Glazer 1990; St. Pierre et al. 1990; Jeffrey 1993; Lampe 1994). Some report a positive effect of a generalized "exposure" to ethics instruction (e.g., Shaub 1994); others report no effect (e.g., Davis and Welton 1991).

Results of specific efforts to affect student ethics through interventions in the accounting curriculum appear to be mixed: Hiltebeitel and Jones (1992) and Armstrong (1993) report positive results; Fulmer and Cargile (1987), Hiltebeitel and Jones (1991), Ponemon (1993), and Lampe (1994) report no such success. Four of these intervention studies assessed ethics components introduced in a single accounting class: one with apparent success (Hiltebeitel and Jones 1992), the others with no effect (Fulmer and Cargile 1987; Hiltebeitel and Jones 1991; Ponemon 1993). Armstrong (1993) reports positive effects of a separate ethics class; Lampe (1994), no effect of ethics components integrated throughout the accounting curriculum.

From a methodological perspective, research has been conducted at either one or two schools using a number of research designs and treatments. Some (e.g., Davis and Welton 1991; Giacomino 1992) simply used an instrument on a one-time basis. Others (e.g., Fulmer and Cargile 1987) used a treatment and posttest. Five studies (Arlow and Ulrich 1985; Hiltebeitel and Jones 1991, 1992; Armstrong 1993; Ponemon 1993) used a pretest/posttest design; however, both pretest and posttest were conducted during one semester. Lampe (1994) conducted a longitudinal study of changes in group response variables but did not track changes in individual students over time. Only Armstrong (1993) and Ponemon (1993) employed control groups.

Measurements also varied, although Rest's (1979b) Defining Issues Test (DIT) is the majority's instrument of choice. Seven of the studies reviewed here, including three of the six ethics intervention studies used the DIT either alone or in combination with other measures.

## RESEARCH METHODOLOGY

For the current study, the authors used a pretest/posttest experimental design on a two-year longitudinal basis with experimental and control groups. Davis and Welton (1991, 463) suggested "more research . . . undertaken to study a specific group of students through a longitudinal study." Ahadiat and Mackie (1993, 256) recommended that the effectiveness of different approaches for teaching ethics "be evaluated at different levels of exposure, as well as studying the effects over a longer period of time."

### Treatment

Loeb (1988, 1990), Langenderfer and Rockness (1989), and Scribner and Dillaway (1989) discuss the issues related to teaching accounting ethics and offer recommendations on how to resolve them. The issues are: (1) who should teach ethics; (2) how instructors would be trained; (3) should ethics be taught in one class and/or throughout the accounting curriculum; (4) finding materials such as cases, books, and so forth; and (5) selecting a method(s) of instruction (e.g., lecture, case, discussion, etc.). The investigators in the current study chose to investigate the effect of curriculum-wide ethics instruction. The experimental treatment employed ethics components included in all nine accounting courses above the introductory level, which was excluded because a majority of the students in that class are not accounting majors, and a significant portion of those who major in accounting transfer to the test institution from community colleges.

Most instructors applied the case method using one or two cases obtained either from the American Accounting Association (1992), Steven Mintz's *Cases in Accounting Ethics and Professionalism*,[2] The Robert M. Trueblood Accounting and Auditing Case Study Series, the American Accounting Association or Institute of Management Accounting videotapes. Auditing instructors discussed the AICPA Code of Professional Conduct and pertinent sections from the required auditing textbook. No specific text on ethics was used[3].

Accounting faculty required that students prepare a written analysis of an ethics case. Most allocated approximately 15 minutes to explaining how cases were to be analyzed and used a portion of one class period to discuss the case when the students turned in their papers.

Several of the accounting faculty attended the AAA Professionalism and Ethics Committee's seminars, where they received training in ethics instruction. Instructional materials received at the seminars were made available to the other faculty members.

## Instrument

Rest's (1979a, 1979b) Defining Issues Test (DIT) was used to pretest and posttest students. The theoretical basis for Rest's instrument is Kohlberg's (Colby and Kohlberg 1987) transformational stage model of moral cognition. Within this paradigm, stages of moral development are qualitatively different structures which serve the same basic function (moral judgment) at various points in human development. Higher stages of the model displace the structures found at lower stages. These stages form a "universal invariant sequence" (p. 9). Environmental factors may increase or attenuate the rate of stage development, but they cannot change its sequence.

Kohlberg presents a six-stage hierarchy of moral judgment which describes individuals' "sociomoral" perspective. This perspective is the characteristic point of view from which the individual formulates judgments of moral issues. The six stages are grouped into three levels which describe different relationships between the self and society's moral rules and expectations. Gilligan (1993) characterizes these stages as reflecting an expansion in moral understanding from an individual to a societal to a universal point of view.

In Level 1, the preconventional stage, rules and social expectations are external to the self. The preconventional level is that of most children under age nine, some adolescents, and many adolescent and adult criminal offenders. In Level 2, the conventional stage, the self has internalized the rules and expectations of others. This is the level of most adolescents and adults. In Level 3, the postconventional stage, self-chosen principles are differentiated from the rules and expectations of others. This stage is reached by only a minority of adults.

The DIT presents respondents with a standardized set of hypothetical scenarios depicting moral dilemmas. For example, the "Heinz Dilemma" presents the case of a very poor man whose wife is dying of a rare disease, the only cure for which is a drug developed and owned by a local pharmacist. Heinz cannot afford to purchase the drug and must decide whether to steal the drug or abide by the law. Respondents to the DIT are asked to state whether Heinz should steal the drug or not and to identify those situational factors of importance to them in reaching their decisions.

Individual scores on the DIT include measures of functioning at each developmental level and several summary measures. The measure of moral development cited in most studies using the instrument is Rest's P value, which measures "Principled" reasoning at Stages 5 and 6. Rest (1986) provides a summary and discussion of the findings of hundreds of studies using the DIT which confirm its reliability and external validity.

## Setting

One experimental and two control groups participated in this study. The experimental group consisted of all students enrolled in the first semester of Intermediate Accounting at a large, urban public university in the Southeast (University A, or U-A). One control group consisted of all students enrolled in the first semester of Intermediate Accounting at a large, rural public university in the Southeast (University B, or U-B). The second control group consisted of all business but non-accounting students enrolled in a Principles of Management course at U-A. All three groups consisted primarily of students who could be expected to graduate in spring of 1994.

The first control group was selected to represented an accounting group from another university, the second a business but non-accounting group. Neither University B's accounting department nor University A's business (non-accounting) departments supported a policy of providing ethics instruction at the time of this study. In the investigators' judgment, students enrolled in those programs were not exposed to an organized plan of ethics instruction during the test period.

Rest's DIT was administered initially to experimental and control group students during the second class meeting of the fall 1992 semester. The follow-up DIT was administered during the last month of the spring 1994 semester. As necessitated by student availability, it was either: (1) distributed and collected in classes but completed at home, or (2) mailed and completed at home.

A total of 562 students were tested originally. Of these, 150 responses contained incomplete data or failed to pass the DIT's internal consistency checks,

**Table 1.** Data Collection Summary

| | *Experimental Group* | *Control Groups* | | |
|---|---|---|---|---|
| | *U-A: Accounting* | *U-B: Accounting* | *U-A: Business* | *Total* |
| Tested originally | 187 | 164 | 211 | 562 |
| Unusable pretest responses | 28 | 25 | 97 | 150 |
| Unsable pretest responses | 159 | 139 | 114 | 412 |
| Unable to locate for posttest | 32 | 15 | 11 | 58 |
| Contacted for posttest | 127 | 124 | 103 | 354 |
| Failed to respond | 58 | 50 | 59 | 167 |
| Responded | 69 | 74 | 44 | 187 |
| Unusable posttest responses | 9 | 8 | 8 | 25 |
| Usable posttest responses | 60 | 66 | 36 | 162 |

resulting in usable data from 412 subjects. Fifty-eight of these students were not enrolled in classes at the retest date and could not be located. Retest materials were distributed to the other 354 students and 187 (45.4% of the original 412) responded. Twenty-five responses were eliminated for incomplete data or failure to pass the DIT's internal consistency checks, resulting in usable retest data from 162 subjects. Table 1 summarizes the data collected by experimental and control groups.

While the retest response rate was lower than anticipated, tests of response variables and demographic data indicate that a significant nonresponse bias can be ruled out. The mean pretest P score of those students in the pretest-only group ($n = 225$) versus those students retested ($n = 187$) is not significantly different: 37.5 versus 38.0, respectively ($p = .71$). Demographic data compared across the pretest-only and retested groups using the chi-square test show retest participation to be independent of demographic characteristics ($p = .85$).

## RESULTS

Rest (1975) suggests that longitudinal studies such as the current one, which a priori anticipate developmental change, utilize a strategy for analyzing that change.

**Table 2.** Comparisons of Pretest P Scores

| ***1. Overall*** | *Mean* | *Number of Students* |
|---|---|---|
| Overall | 37.7 | 412 |
| 1. U-A: Accounting | 38.3 | 159 |
| 2. U-B: Accounting | 39.0 | 139 |
| 3. U-A: Business | 35.4 | 114 |

***2. Whether students had an ethics course previously***

| | *Mean* | | *Number of Students* | |
|---|---|---|---|---|
| | *Yes* | *No* | *Yes* | *No* |
| Overall | 42.5 | 37.3** | 35 | 377 |
| 1. U-A: Accounting | 39.9 | 3.2 | 12 | 147 |
| 2. U-B: Accounting | 43.8 | 38.5 | 13 | 126 |
| 3. U-A: Business | 43.9 | 34.6*** | 10 | 104 |

***3. Gender***

| | *Mean* | | *Number of Students* | |
|---|---|---|---|---|
| | *Men* | *Women* | *Men* | *Women* |
| Overall | 35.7 | 39.4* | 184 | 228 |
| 1. U-A: Accounting | 37.4 | 39.0 | 67 | 92 |
| 2. U-B: Accounting | 36.4 | 41.8* | 66 | 73 |
| 3. U-A: Business | 31.9 | 37.6** | 44 | 70 |

***4. Marital Status***

| | *Mean* | | *Number of Students* | |
|---|---|---|---|---|
| | *Married* | *Not Married* | *Married* | *Not Married* |
| Overall | 39.1 | 37.4 | 80 | 332 |
| 1. U-A: Accounting | 40.6 | 37.5 | 42 | 117 |
| 2. U-B: Accounting | 42.0 | 38.8 | 9 | 130 |
| 3. U-A: Business | 36.0 | 35.2 | 29 | 85 |

***5. Race***

| | *Mean* | | *Number of Students* | |
|---|---|---|---|---|
| | *Caucasian* | *Non-Caucasian* | *Caucasian* | *Non-Caucasian* |
| Overall | 38.2 | 35.5 | 338 | 74 |
| 1. U-A: Accounting | 39.1 | 35.5 | 122 | 37 |
| 2. U-B: Accounting | 39.0 | 38.2 | 126 | 13 |
| 3. U-A: Business | 35.7 | 34.1 | 90 | 24 |

***Notes:*** $^{*}p < .0125$.
$^{**}p < .025$.
$^{***}p = .028$.

That strategy involves analysis of change in both groups and individuals and changes in all stage scores, not only those included in the P score.

## Group Analysis

Table 2 presents pretest results of two-tailed *t*-tests comparing students' mean pretest P scores by group and demographic characteristic. To control for the inflated experiment-wide error rate caused by the use of multiple *t*-tests, the "definition" of significance must be adjusted. For the four separate *t*-tests used here, a Bonferroni-type adjustment produces planned significance levels (alphas) for each separate test of .0125 or .025. These yield an overall experimental alpha level of approximately .05 or .10, respectively.[4] Table 2 reveals several pretest P score differences which are statistically significant at the prescribed levels.

Pretest P score differences among the experimental and control groups are not significant. Pretest P scores are different overall and for the U-A:Business control group, depending on whether or not students had an ethics course previously. As

***Table 3.*** P Scores and Sample Demographics: Experimental and Control Groups

| | *Experimental Group* | | *Control Groups* | | | |
|---|---|---|---|---|---|---|
| | *U-A: Accounting (n = 60)* | | *U-B: Accounting (n = 66)* | | *U-A: Business (n = 36)* | |
| | *Mean* | *Std. Dev.* | *Mean* | *Std. Dev.* | *Mean* | *Std. Dev.* |
| Mean Pretest P Score | 37.2 | 14.3 | 40.4 | 13.1 | 36.2 | 14.4 |
| Mean Posttest P Score | 37.4[a] | 10.9 | 42.3 | 11.9 | 41.5 | 14.7 |
| Mean Change in P Score | + 0.2 | 11.7 | + 2.0 | 12.0 | + 5.3 | 16.0 |
| Mean Age | 25.8 | 6.4 | 23.4 | 4.2 | 29.2 | 8.4 |
| Previous Ethics Course | 6 Yes (10.0%) | 54 No (90.0%) | 6 Yes (9.1%) | 60 No (90.1%) | 1 Yes (2.8%) | 35 No (97.2%) |
| Gender | 26 Male (43.3%) | 34 Female (56.7%) | 32 Male (48.5%) | 34 Female (51.5%) | 12 (Male (33.3%) | 24 Female (66.7%) |
| Marital Status | 11 Married (18.3%) | 49 Married (81.7%) | 6 Married (9.1%) | 60 Not Married (90.1%) | 11 Married (30.6%) | 25 Not Married (69.4%) |
| Race | 45 Caucasian (75.0%) | 15 Non-Caucasian (25.0%) | 61 Caucasian (92.4%) | 5 Non-Caucasian (7.6%) | 30 Caucasian (83.3%) | 6 Non-Caucasian (16.7%) |

***Note:*** [a]Experimental Group's posttest P score is significantly different from U-B: Accounting Control Group (p = .018) but not from U-A: Business Control Group. No other between-group differences are significant.

might be expected (and consistent with Shaub's 1994 findings), students who had ethics courses previously have higher P scores, but the number of such students in any of the groups is very small.

Pretest P scores are different overall and for both the U-B:Accounting and U-A:Business control groups, depending on gender. Again consistent with Shaub's (1994) findings, women have higher P scores than men. Rest (1979b) notes that correlation of gender with P scores is *usually* nonsignificant or very low, suggesting that the current finding is not unique. No significant pretest P score differences are evident by marital status or race.

Table 3 reports retest P scores and sample demographic characteristics. The experimental group's mean posttest P score is lower and significantly different from that of the U-B:Accounting control group (two-tailed *t*-test, $p = .018$), and lower but not significantly different from that of U-A:Business group ($p = .16$).[5] U-A:Accounting's score is also lower and significantly different from that currently reported by the Center for the Study of Ethical Development at the University of Minnesota (the scoring service for the DIT) as the norm for college students (43.19, *s.d.* 14.32, $p \leq .001$). The control groups' posttest P scores are not significantly different from each other ($p = .76$), nor from the college student norm (U-B:Accounting, $p = .56$; U-A:Business, $p = .49$).

As predicted by moral development theory, the mean change in P score is positive and significantly different from zero ($p = .047$) across groups. However, it is not significantly different between experimental and control groups ($p = .17$), implying no treatment effect.

Age differences are relatively small but significant across experimental and control groups ($p \leq .001$). Neither age nor other demographic variables are significantly correleted with pretest or posttest P scores or with change in P score. Ponemon (1993) likewise observed no correlation of DIT P scores with age, and explained this seeming contradiction to Rest's (1986) findings as "probably caused by the homogeneity of participating . . . students in terms of . . . ages" (p. 194). Other demographic data compared across the experimental and control groups using the chi-square test show demographic characteristics to be independent of group membership ($p = .89$).

Table 4 presents the results of t-tests comparing students' mean posttest P scores by group and demographic characteristic. As already noted, U-A:Accounting's score is significantly lower than both U-B:Accounting's and the norm for college students. The below-norm finding is consistent with that of other studies (e.g., Ponemon and Glazer 1990) comparing accounting students and practitioners with those in other disciplines.

Accountants' self-selection into the rules orientation of the profession and university training for accountants that is less conducive to moral development have been advanced as explanations for accountants' lower P scores (Lampe 1994). The findings of the current study would appear to refute at least the first of these explanations: pretest P scores for accounting students at both universities are not

**Table 4.** Comparisons of Posttest P Scores

**1. Overall**

| | Mean | Number of Students |
|---|---|---|
| Overall | 40.3 | 162 |
| 1. U-A: Accounting | 37.4** | 60 |
| 2. U-B: Accounting | 42.3 | 66 |
| 3. U-A: Business | 41.5 | 36 |

**2. Whether students had an ethics course previously**

| | Mean | | Number of Students | |
|---|---|---|---|---|
| | *Yes* | *No* | *Yes* | *No* |
| Overall | 35.5 | 40.7 | 13 | 149 |
| 1. U-A: Accounting | 33.9 | 37.8 | 6 | 54 |
| 2. U-B: Accounting | 35.0 | 43.1 | 6 | 60 |
| 3. U-A: Business | 48.3 | 41.3 | 1 | 35 |

**3. Gender**

| | Mean | | Number of Students | |
|---|---|---|---|---|
| | *Men* | *Women* | *Men* | *Women* |
| Overall | 39.4 | 41.0 | 70 | 92 |
| 1. U-A: Accounting | 36.6 | 38.0 | 26 | 34 |
| 2. U-B: Accounting | 41.0 | 43.6 | 32 | 34 |
| 3. U-A: Business | 41.1 | 41.7 | 12 | 24 |

**4. Marital Status**

| | Mean | | Number of Students | |
|---|---|---|---|---|
| | *Married* | *Not Married* | *Married* | *Not Married* |
| Overall | 41.6 | 40.1 | 28 | 134 |
| 1. U-A: Accounting | 34.6 | 38.1 | 11 | 49 |
| 2. U-B: Accounting | 41.5 | 42.4 | 6 | 60 |
| 3. U-A: Business | 48.6 | 38.3 | 11 | 25 |

**5. Race**

| | Mean | | Number of Students | |
|---|---|---|---|---|
| | *Caucasian* | *Non-Caucasian* | *Caucasian* | *Non-Caucasian* |
| Overall | 40.8 | 38.1 | 136 | 26 |
| 1. U-A: Accounting | 36.4 | 40.4 | 45 | 15 |
| 2. U-B: Accounting | 43.0 | 34.0 | 61 | 5 |
| 3. U-A: Business | 42.7 | 35.5 | 30 | 6 |

***Notes:*** *$p < .0125$.
**$p < .025$.

different from those of the non-accounting control group, providing no evidence of self-selection.

U-B:Accounting's posttest P scores are not significantly different from those of the non-accounting group nor from the college student norm, providing no evidence of a *generalized* bias in university training for accountants. However, a survey conducted independent of this study (Douglas 1995) indicates that students at U-A may be trained differently than those at U-B. Senior accounting students at both universities A and B indicated that U-A faculty may be relatively more rules-oriented than U-B faculty, providing an environment that is less conducive to moral development. A statistically significant difference by institution was found on student responses to the question: "Accounting faculty generally stress conforming to rules rather than applying individual judgment." U-B's mean student response disagreed with that statement; U-A's was equivocal.

The P score differences by gender and previous ethics instruction observed in the pretest (Table 2) are not evident in posttest scores. Students indicating previous ethics instruction would have taken these classes in their first or second year of college, as students tested were primarily juniors at the time of the pretest. The effect of that instruction appears to have dissipated by the end of senior year. Arlow and Ulrich (1985) explain their similar finding by concluding that business ethics instruction may stimulate students' awareness of ethical issues, but the impact is only temporary. Once that stimulus is removed, "students revert to previously held values" (p. 16).

While analysis of P score differences provides interesting information, Rest (1975) cautions against a focus on "one step progressions" in the analysis of developmental change. He suggests that "it is more appropriate to talk about upward shifts in distributions of responses (increases in higher stages, decreases in lower stages)" (p. 740). Figure 1 presents a graphical representation of pretest and posttest mean stage scores for the experimental and control groups.

Figure 1 reveals virtually no shifts of the type Rest (1975) suggested and observed in his longitudinal study of junior and senior high school subjects,[6] and no noticeable differences between the experimental and control groups. Rest's senior high school subjects' apparent movement over two years was from conventional morality to principled morality (i.e., large decrease in Stage 3, increases in Stages 5A and 5B). Junior high school subjects showed shifts from preconventional to conventional morality (i.e., large decreases in Stages 2 and 3, large increases in Stages 4 and 5A). The current study's lack of finding may indicate a lower rate of change for the older college students in the current study than for their much younger counterparts in Rest's study, and again indicates no effect of the treatment.

## Individual Analysis

Analysis of individual subjects' patterns of change "address[es] the question of step-by-step movement" (Rest 1975, 743). Characterization of change for

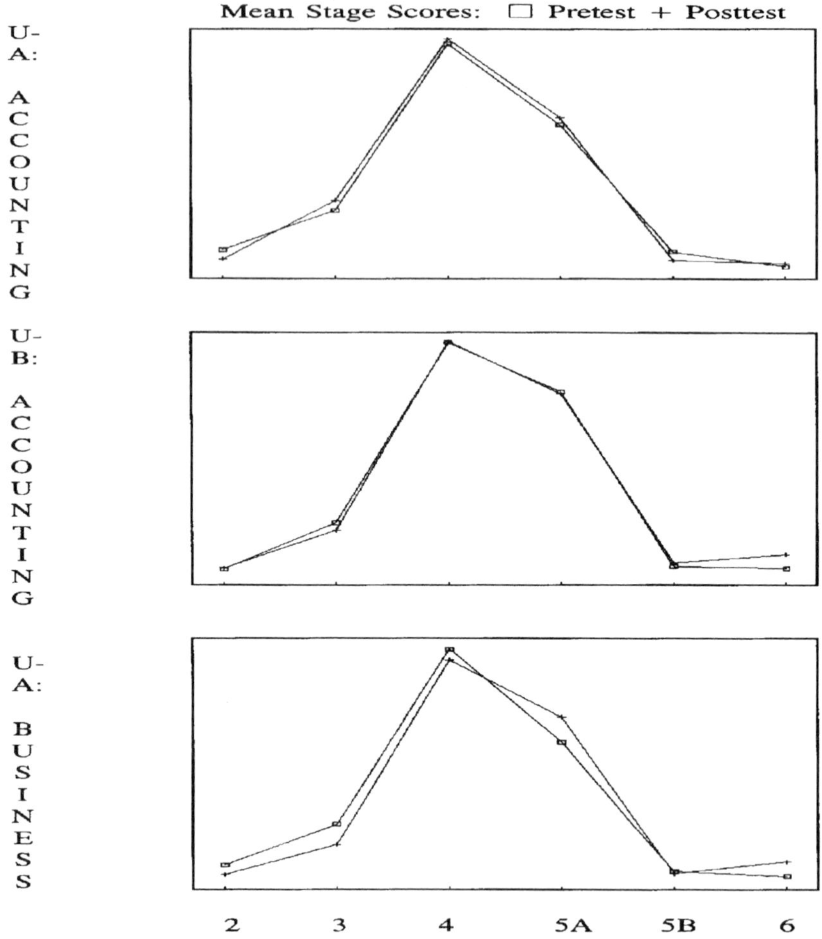


*Note:* Mean Stage Scores: □ Pretest + Posttest

***Figure 1.*** Group Stage Shifts

each subject must indicate which stage scores increased and which decreased. Rest cautions that only increases and decreases in excess of the standard error of the measure can be confidently said to represent real developmental change in the subject. For this study, the standard error of each stage score was calculated from test-retest correlations that Rest (1986) reported.[7] For Stage 2, the difference between pretest and posttest stage score had to exceed 2.6 points to be considered real change; for Stage 3, 2.8; Stage 4, 4.6; Stage 5A, 4.3; Stage 5B, 1.8; Stage 6, 1.8.

**Table 5.** Individual Stage Shifts[a]

| | *Upward Change Patterns* | *Downward Change Patterns* | *Ambiguous Change Patterns* | *Total* |
|---|---|---|---|---|
| U-A: Accounting | | | | |
| Number | 18 | 10 | 32 | 60 |
| Percent | 30.0% | 16.7% | 53.3% | 100.0% |
| U-B: Accounting | | | | |
| Number | 21 | 8 | 37 | 66 |
| Percent | 31.8% | 12.1% | 56.1% | 100.0% |
| U-A: Business | | | | |
| Number | 16 | 4 | 16 | 36 |
| Percent | 44.4% | 11.2% | 44.4% | 100.0% |
| Overall | | | | |
| Number | 55 | 22 | 85 | 162 |
| Percent | 33.9% | 13.6% | 52.5% | 100.0% |

***Note:*** [a]Positive (negative) changes in higher-stage scores accompanied by negative (positive) changes in lower-stage scores were classified as upward (downward) change patterns. All other stage score change patterns were classified as ambiguous. All positive and negative changes are in excess of the standard error of the measure.

Individual stage scores on both the pretest and posttest were compared, and changes exceeding the standard error were categorized. Positive (negative) changes in higher stage scores accompanied by negative (positive) changes in lower-stage scores were classified as upward (downward) change patterns. All other stage score change patterns were classified as ambiguous. Table 5 summarizes the individual stage shifts.

Table 5 shows the three classifications of individual changes: those that are consistent with developmental theory (i.e., upward change patterns), those that are inconsistent with developmental theory (i.e., downward change patterns), and those that cannot be regarded as confirming or disconfirming theory (i.e., ambiguous change patterns). Overall, observed change patterns in slightly more than one-third (33.9%) of this study's subjects are consistent with developmental theory. More than one-half (52.5%) of observed changes are ambiguous. Rest (1975) reported 65.4% of changes consistent with theory among his younger subjects. As with the group shifts previously discussed, this may indicate a lower rate of change for the older college students. A chi-square test of individual stage shifts by group indicates no differences in change patterns among experimental and control groups ($p = .59$), again implying no significant effect for the experimental treatment.

## CONCLUSIONS AND IMPLICATIONS

Results of this longitudinal controlled study of ethics instruction throughout the accounting curriculum do not provide evidence that such instruction affects students' ethical development. As predicted by moral development theory, the mean change in P score over the two-year test period is positive and significantly different from zero. This indicates an increase in students' principled reasoning across groups. However, no significant differences in mean change in P score are observed between experimental and control groups.

Analysis of the shifts in students' response distributions across the stages of moral reasoning fail to provide strong evidence in support of development theory among the age groups tested. Graphical representation of group mean stage scores reveals little change between pretest and posttest stage scores. Analysis of individual stage shifts show more than one-half of the shifts to be ambiguous. Neither group nor individual stage shift analysis reveals any difference between experimental and control groups.

The current study's findings imply that the program of ethics instruction employed had no effect on the sample of 60 accounting students included in the experimental group. These findings are consistent with previous studies which assessed efforts to affect student ethics through interventions in the accounting curriculum (e.g., Fulmer and Cargile 1987; Hiltebeitel and Jones 1991; Ponemon 1993; and Lampe 1994). Three of these intervention studies assessed ethics components introduced in a single accounting class; Lampe (1994) assessed ethics components integrated throughout the accounting curriculum. These findings have important implications for both accounting education and practice.

With respect to accounting education, Ponemon (1993) suggests that it may be premature to conclude that ethics instruction will not foster students' moral development. Rest (1986) suggests that the type of ethics instruction employed in the current study (i.e., classroom discussion of ethics cases) will foster moral development in college, graduate, and professional-school students. Both the Arthur Andersen & Co. Business Ethics Seminars and the AAA's Professionalism and Ethics Seminars recommend this approach. However, Ponemon (1993) concludes in discussion of his similar (negative) findings that "perhaps other teaching approaches and course materials . . . are needed to advance the moral development and ethical behavior of accounting students" (p. 206).

Kenny and Eining (1994) observe that ethics instruction in financial accounting classes generally uses pedagogies similar to those common in auditing courses. However, the reasoning processes underlying financial accounting activities are fundamentally different from those involved in auditing. Their results suggest the need for innovation in curriculum design consistent with the distinct reasoning processes of different accounting activities.

Armstrong (1993) advocates a "sandwich approach" to teaching ethics, and reports its positive effect on students' moral development. This approach consists of a general course in ethics taught outside of the business school, combined with a variety of accounting courses in which ethics instruction is included, followed by a capstone course in Ethics and Professionalism (E & P) taught by accounting faculty. In Armstrong's sample E&P course, specific treatment is given to discussion of theoretical foundations including moral development theory and the sociology of professions, to discussions of professional and governmental guidance and oversight, and to public opinion. Case studies are integrated throughout the course.

The importance of faculty support for the teaching of ethics also cannot be dismissed. Faculty participating in the current study were asked to assess the general level of student interest in ethics as well as the faculty member's own level of interest in allocating class time to ethics. A review of the responses to both questions showed the two responses were quite similar; that is, faculty appear to view the students' level of interest in ethics as quite similar to their own. In addition, when asked in which courses ethics should be covered, those faculty who had a high level of interest thought it should be taught in all accounting courses. Those faculty who indicated they had a low level of interest suggested that ethics should be covered only in a limited number of courses. Faculty need to enthusiastically support the program of ethics instruction if it is to be successful, as students are likely to be sensitive to faculty attitudes.

Arlow and Ulrich (1985) suggest that both the nature of the respondent group and curricular differences influence the effectiveness of business ethics instruction. Pretest P scores provide no evidence of the self-selection that has been suggested to explain accounting students' and practitioners' apparent rules orientation: students just beginning the major coursework in accounting are no more rules-oriented than non-accounting business students are. Posttest P scores of only the U-A:Accounting group are significantly lower than the norm for college students, and that difference is associated with perceived differences in faculty orientation.

To Ponemon's (1993) suggestion of other pedagogical approaches must be added suggestions for goal clarification and different assessment approaches. Huss and Patterson (1993) imply that moral development of students is the goal of ethics education in accounting when they characterize that education as "based on the premise that moral development can be enhanced through the educational process" (p. 235). Just as intelligence has come to be defined as that which is measured by the IQ test, much of the literature of accounting ethics serves to define "moral development" only within the context of Rest's (1979a) work—that is, as that which is measured by the Defining Issues Test, or P score.

Loeb (1988, 322) suggests a much different definition in his list of possible goals of accounting ethics education:

1. Relate accounting education to moral issues.
2. Recognize issues in accounting that have ethical implications.
3. Develop "a sense of moral obligation" or responsibility.
4. Develop the abilities needed to deal with ethical conflicts or dilemmas.

5. Learn to deal with the uncertainties of the accounting profession.
6. "Set the stage for" a change in ethical behavior.
7. Appreciate and understand the history and composition of all aspects of accounting ethics and their relationship to the general field of ethics.

Rest's Defining Issues Test cannot begin to assess these objectives, which are both broader than "moral development" in that they address the wider considerations of a pluralistic profession, and narrower in that they address issues specific to the accounting profession.

Future research should approach the question of what makes accountants and accounting students more or less ethical from different perspectives. For example, Huss and Patterson (1993) propose "*teaching* values" (236, emphasis added) as part of ethics education but do not explore the issue of whether values can be taught. Lampe (1994) indicates that accounting students' DIT scores, reasoning in ethical dilemmas, decisions, and attitudes toward ethical behaviors are "strongly oriented to code-implied rules" (p. 231) but does not explore environmental factors (such as U-A's faculty rules orientation) contributing to those student values.

Implications for accounting practice may include the necessity to reevaluate its expectations. To overcome what might be called a crisis of credibility and public confidence, "the profession is now striving to modify and improve the ethical behaviors of individual accounting professionals" (Ponemon 1992, 239). That "striving" has included efforts to encourage increased ethics instruction in preprofessional education, in the belief that accountants can be *educated* into being more ethical. Ethics interventions in the accounting curriculum may be largely ineffective or, at best, short-lived.

In Arlow and Ulrich's (1988) survey, business school graduates (including accountants) ranked school and university training last among six factors that influence ethical conduct. More important were family training, conduct of superiors, practices of industry, conduct of peers, and religious training. These imply a relatively larger role for accounting firms, and a relatively smaller role for educational institutions, in improving professional ethics.

Three of the five factors ranked in Arlow and Ulrich's survey as more important than school and university training in influencing ethical conduct have to do with firm ethical culture (i.e., conduct of superiors, practices of industry, conduct of peers). Its apparent importance has largely been ignored in accounting research to date, probably due to the difficulty of conducting research in this sensitive area. Future research should investigate the roles of firm ethical culture and of other environmental variables on ethical decision making and behavior in practice.

Despite the low ranking assigned to it, respondents to Arlow and Ulrich's (1988) survey endorsed business school ethics courses as a way to improve business ethics. The authors' explanation for this contradiction applies equally to academics' and practitioners' continued interest in ethics education: it is evidence of the complexity of attempts to improve business ethics. While ethics education may be the least effective way to influence ethical conduct, compared with other factors it is an area that can be most easily controlled and managed.

## ACKNOWLEDGMENT

The authors wish to thank Mary Beth Armstrong and Lawrence A. Ponemon for their helpful comments on earlier versions of this paper.

## NOTES

1. Lampe (1994) studied the effects of classroom ethical interventions over a four-year period but was not able to match students, that is, follow one group over time.
2. Permission to use the cases was obtained from the publisher.
3. Several excellent texts are available in addition to Mintz (1992), including Albrecht (1992), Armstrong (1993), and Windal (1991).
4. The calculations are:

$$1 - (1-.0125)^4 = .0491, \text{ or } 1 - (1-.025)^4 = .0963.$$

5. Appropriateness of the Bonferroni adjustment of *t*-test significance level is verified by analysis of variance applied to these data. The ANOVA model is significant at $p = .07$, consistent with *t*-test *p*-value .018. (Note that .018 X 4 = .072.) Tukey's test of pairwise comparisons, which likewise controls for experiment-wide error rate, indicates a significant difference between only U-A:Accounting and U-B:Accounting at a 90% confidence level.
6. Rest (1975) uses "raw weighted ranks" for his graphical analysis. Rank analysis applied to the current study data revealed no patterns substantively different from the mean stage scores in Figure 1.
7. Rest (1986) reports test-retest reliability generally to be in the high .70s or .80s. For this study, .75 was assumed. Following Rest (1975), the formula $S\sqrt{(1-r^2)}$ was used, where $r = .75$ and $S$ = the standard deviation of the pretest stage score.

## REFERENCES

Ahadiat, N., and J.J. Mackie. 1993. Ethics education in accounting: An investigation of the importance of ethics as a factor in the recruiting decisions of public accounting firms. *Journal of Accounting Education* 11(Fall): 243-257.

Albrecht, W.S. 1992. *Ethical Issues in the Practice of Accounting.* Cincinnati, OH: Southwestern Publishing (available from the AICPA).

American Accounting Association. 1992. *Ethics in the Accounting Curriculum: Cases & Readings* (William W. May, ed.) Sarasota, FL: AAA.

American Assembly of Collegiate Schools of Business (AACSB). 1993. *Achieving Quality and Continuous Improvement Through Self-Evaluation and Peer Review: Standards for Business and Accounting Accreditation.* St. Louis: AACSB.

American Institute of Certified Public Accountants (AICPA). 1988. *Education Requirements for Entry into the Accounting Profession: A Statement of AICPA Policies*, 2nd ed. (revised). New York: AICPA.

American Institute of Certified Public Accountant (AICPA). 1989. *Perspectives on Education: Capabilities for Success in the Accounting Profession.*

Armstrong, M.B. 1993. *Ethics and Professionalism for CPAs*. Cincinnati, OH: South-Western.

Arlow, P., and T.A. Ulrich. 1985. Business ethics and business school graduates: A longitudinal study. *Akron Business and Economic Review* 16 (Spring): 13-17.

Arlow, P., and T.A. Ulrich. 1988. A longitudinal study of business school graduates' assessments of business ethics. *Journal of Business Ethics* 7(April): 295-302.

Armstrong, M.B. 1993. Ethics and professionalism in accounting education: A sample course. *Journal of Accounting Education* 11(Spring): 77-92.

Armstrong, M.B., and S.M. Mintz. 1989. Ethics education in accounting: Present status and policy implications. *Association of Government Accountants' Journal* (Summer): 70-76.

Beets, S.D. 1993. Using the role-playing technique in accounting ethics education. *Accounting Educators' Journal* 5(Fall): 46-65.

Borkowski, S.C., and Y.J. Ugras. 1992. The ethical attitudes of students as a function of age, sex and experience. *Journal of Business Ethics* 11(December): 961-979.

Burton, J.C., and R.J. Sack. 1989. Editorial: Ethics and professionalism in accounting education. *Accounting Horizons* (December): 114-116.

Carver, M.R., M.L. Hirsch, Jr., and D.E. Strickland. 1993. The responses of accounting administrators to ethically ambiguous situations: The case of fund raising. *Issues in Accounting Education* (Fall): 300-319.

Cohen, J.R., and L.W. Pant. 1989. Accounting educators' perceptions of ethics in the curriculum. *Issues in Accounting Education* 4(Spring): 70-81.

Colby, A., and L. Kohlberg. 1987. *The Measurement of Moral Judgment*. Cambridge, UK: Cambridge University Press.

Committee on the Future Structure, Content and Scope of Accounting Education. 1986. Future accounting education: Preparing for the expanding profession. *Issues in Accounting Education* 1(Spring): 168-195.

Cooke, R., H. Kanter, and S. Martens. 1987-1988. The importance of ethical training in an accounting education. *The Government Accountants Journal* 36(Winter): 64-67.

Davis, J.R., and R.E. Welton. 1991. Professional ethics: Business students' perceptions. *Journal of Business Ethics* 10(June): 451-463.

Douglas, P.C. 1995. An empirical investigation of ethical ideology and the judgment of ethical dilemmas in accounting: A cognitive-contingency model approach. Unpublished Ph.D. dissertation, Virginia Commonwealth University, Richmond, VA.

Engle, T.J., and J.L. Smith. 1992. Accounting faculty involvement with activities of ethical concern. *Accounting Educators' Journal* 4(Spring): 1-21.

Fess, R.C. 1987. Ethics in accounting: Can it be taught? *Outlook* (Summer): 60.

Flory, S.M., T.J. Phillips, Jr., R.E. Reidenbach, and D.P. Robin. 1992. A multidimensional analysis of selected ethical issues in accounting. *The Accounting Review* 67(April): 284-302.

Fulmer, W.E., and B.R. Cargile. 1987. Ethical perceptions of accounting students: Does exposure to a code of professional ethics help? *Issues in Accounting Education* 2(Fall): 207-219.

George, R.J. 1987. Teaching business ethics: Is there a gap between rhetoric and reality? *Journal of Business Ethics* 6(October): 513-518.

Giacomino, D.E. 1992. Ethical perceptions of other business majors: An empirical study. *Accounting Educators' Journal* 4(Fall): 1-26.

Gilligan, C. 1993. *In A Different Voice*. Cambridge, MA: Harvard University Press.

Grimstad, C.R. 1964. Teaching the ethics of accountancy. *Journal of Accountancy* (July): 82-85.

Hiltebeitel, K.M., and S.K. Jones. 1991. Initial evidence on the impact of integrating ethics into accounting education. *Issues in Accounting Education* 6(Fall): 262-275.

Hiltebeitel, K.M., and S.K. Jones. 1992. An assessment of ethics instruction in accounting education. *Journal of Business Ethics* 11(January): 37-46.

Hosmer, L.T. 1988. Adding ethics to the business curriculum. *Business Horizons* 31(July-August): 9-15.

Huss, H.F., and D.M. Patterson. 1993. Ethics in accounting: Values education without indoctrination. *Journal of Business Ethics* 12(March): 235-243.

Jeffrey, C. 1993. Ethical development of accounting students, non-accounting business students, and liberal arts students. *Issues in Accounting Education* 8(Spring): 86-96.

Karnes, A., and J. Sterner. 1989. Ethics education in university accounting programs. *Journal of Education for Business* (April): 307-309.

Kenny, S.Y., and M.M. Eining. 1994. Integrating ethics exercises into intermediate accounting classes: Using an attribution theory framework. In *Proceedings of the Ernst & Young Research on Accounting Ethics Symposium,* 203-219.

Kunitake, W.K., and C.E. White, Jr. 1986. Ethics for independent auditors. *Journal of Accounting, Auditing & Finance* (Summer): 222-231.

Lampe, J.C. 1994. The impact of ethics education in accounting curricula. In *Proceedings of the Ernst & Young Research on Accounting Ethics Symposium*, 220-236..

Langenderfer, H.O., and J.W. Rockness. 1989. Integrating ethics into the accounting curriculum: Issues, problems and solutions. *Issues In Accounting Education* 4(Spring): 58-69.

Lantry, T.L. 1993. A pragmatic approach to teaching ethics. *Accounting: A Newsletter for Educators* 3: 1, 3.

Lehman, C.R. 1988. Accounting ethics: Surviving survival of the fittest. In *Advances in Public Interest Accounting*, Vol. 2, ed. M. Neimark, 71-82. Greenwich, CT: JAI Press.

Lewis, P.V. 1989. Ethical principles for decision makers: A longitudinal study. *Journal of Business Ethics* 8(April): 271-276.

Loeb, S.E. 1988. Teaching students accounting ethics: Some crucial issues. *Issues in Accounting Education* 3(Fall): 316-329.

Loeb, S.E. 1990. Whistleblowing and accounting education. *Issues In Accounting Education* 5(Fall): 281-294.

Loeb, S.E., and J. W. Rockness. 1992. Accounting ethics and education: A response. *Journal of Business Ethics* 11(July): 485-490.

McNair, F., and E. Milam. 1993. Ethics in accounting education: What is really being done. *Journal of Business Ethics* 12(October): 797-809.

Mintz, S.M. 1990. Ethics in the management accounting curriculum. *Management Accounting* (June): 51-54.

Mintz, S.M. 1992. *Cases in Accounting Ethics and Professionalism*, 2nd edition. New York: McGraw-Hill.

Murray, T.J. 1987. Can business schools teach ethics? *Business Monthly* (April): 24-26.

National Commission on Fraudulent Financial Reporting (The Treadway Commission). 1987. *Report of the National Commission on Fraudulent Reporting*. Washington, DC: National Commission on Fraudulent Financial Reporting.

Pamental, G.L. 1989. Ethics in introductory accounting. *Journal of Education for Business* (January): 179-182.

Ponemon, L. 1993. Can ethics be taught in accounting? *Journal of Accounting Education* 11(Fall): 185-209.

Ponemon, L., and A. Glazer. 1990. Accounting education and ethical development: The influence of liberal learning on students and alumni in accounting practice. *Issues in Accounting Education* 5(Fall): 195-208.

Rest, J.R. 1975. Longitudinal study of the Defining Issues Test of Moral Judgment: A strategy for analyzing developmental change. *Developmental Psychology* 11: 738-748.

Rest, J.R. 1979a. *Development in Judging Moral Issues*. Minneapolis, MN: University of Minnesota Press.

Rest, J.R. 1979b. *Manual for the Defining Issues Test*. Minneapolis, MN: University of Minnesota Press.

Rest, J.R. 1986. *Moral Development: Advances in Research and Theory*. New York: Praeger.

Scribner, E., and M.P. Dillaway. 1989. Strengthening the ethics content of accounting courses. *Journal of Accounting Education* 7(Spring): 41-55.

Shaub, M.K. 1994. An analysis of the association of traditional demographic variables with the moral reasoning of auditing students and auditors. *Journal of Accounting Education* 12(Winter): 1-26.

Shaub, M.K., D.W. Finn, and P. Munter. 1993. The effects of auditors' ethical orientation on commitment and ethical sensitivity. *Behavioral Research in Accounting* 5: 145-169.

Smith, L.M. 1993a. Teaching ethics: An update—Part I. *Management Accounting* (March): 18-19.

Smith, L.M. 1993b. Teaching ethics: An update—Part II. *Management Accounting* (April): 18.

St. Pierre, K.E., E.S. Nelson, and A.L. Gabbin. 1990. A study of the ethical development of accounting majors in relation to other business and nonbusiness disciplines. *Accounting Educators' Journal* 2(Summer): 23-34.

Ward, S.P., D.R. Ward, and A.B. Deck. 1993. Certified public accountants: Ethical perception skills and attitudes on ethics education. *Journal of Business Ethics* 12(August): 601-610.

Windal, F.W. 1991. *Ethics and the Accountant: Text and Cases.* Englewood Cliffs, NJ: Prentice-Hall.

# EARNINGS MANIPULATION AND DISCLOSURE OF THE EARLY ADOPTION OF SFAS 96 IN MANAGEMENT'S LETTER TO SHAREHOLDERS

Cynthia Firey Eakin and John P. Wendell

## ABSTRACT

When the Financial Accounting Standards Board (FASB) allows a choice of the year in which a firm may adopt a new accounting standard, it provides an opportunity for firms to manipulate earnings through the timing of the adoption of the new standard. The combined results of prior research suggest that managers are more likely to adopt a standard early when it will have an advantageous impact on earnings, but the methodologies employed by these studies were unable to determine if earnings manipulation was a factor in the decision of a particular firm to early adopt. This study examined the content of shareholder letters to determine for individual firms whether early adoption of Statement of Financial Accounting Standards 96 was motivated by earnings manipulation considerations. The disclosure of the effects of the adoption were found to be misleading for 22 of the 40 firms sampled and adequate for 18.

**Research on Accounting Ethics, Volume 4, pages 113-123.**

**ISBN: 0-7623-0339-5**

## INTRODUCTION

The Financial Accounting Standards Board (FASB) often allows a choice of the year in which a firm may adopt a new accounting standard. This multiyear adoption policy has been criticized because it provides a means for managers to artificially manipulate reported earnings (Langer and Lev 1993; Gujarathi and Hoskin 1992). Manipulation of earnings is clearly inconsistent with management's ethical responsibility to provide financial statements that fairly present the financial position and results of operations of the firm. Do some managers time the adoption of new FASB standards to manipulate earnings and deceive financial statement users? This paper provides direct evidence that some managers do.

A substantial body of prior research (Ayres 1986; Eakin 1996; Gujarathi and Hoskin 1992; Langer and Lev 1993; Sami and Welsh 1991; Scott 1991; Trombley 1989) has attempted to answer the earnings manipulation question by determining whether managers are likely to time the adoption of a new standard so that it has an advantageous effect on earnings. The combined results of these empirical studies provide statistically significant evidence that managers are more likely to time the adoption of a new standard to have an advantageous impact on earnings than to have a disadvantageous impact. These results are consistent with the assertion that managers are deliberately timing adoptions to manipulate earnings.

However, the evidence from these studies is only circumstantial, and the conclusions based on it must be viewed as tentative. For example, it is always possible that statistical significance arose because some legitimate factors affecting timing of the adoption (e.g., technical issues involved in implementing the new standard) were correlated with the earnings effect and not included in the statistical model. In that case, opportunistic earnings manipulation only appears to be the cause of the timing decision. The possibility of such an unidentified covariate explaining these results is supported on theoretical grounds because the mandated disclosures of the effects of the adoption of new standards should prevent financial statement users from being deceived by such manipulations and thereby eliminate the incentive for attempting earnings manipulation.

The other limitation of this type of statistical evidence is that it cannot show that an individual firm behaved unethically (it provides no "smoking gun"). A statistically significant result could have been achieved because only a portion of firms are deliberately manipulating earnings. It is always possible that an individual firm's decision to early adopt a standard was for legitimate reasons and the advantageous impact on earnings was merely coincidental.

This paper contributes to the previous literature by supplementing this circumstantial evidence of earnings manipulation with direct evidence of deliberate manipulation found in shareholder letters. The shareholder letter, along with other management disclosures, is expected to increase the usefulness of financial information by identifying certain transactions, events, and other circumstances that affect the firm, and by explaining their financial impact on the firm (FASB

1978, 54). If management is timing the adoption of a new financial reporting standard in order to manipulate earnings and deceive financial statement users then the shareholder letter will contain deceptive and misleading statements (either directly or through omission) about the effect of the new standard. On the other hand, if management is not trying to deceive financial statement users, and the timing of the adoption is driven by legitimate considerations, the shareholder letter should contain a frank and honest discussion of the impact of the adoption.

This paper examines the shareholder letters of firms opting for the early adoption of Statement of Financial Accounting Standards Number 96 (SFAS 96; FASB 1987) to determine if they are misleading about the effects of the early adoption. SFAS 96 is particularly well suited to the study of earnings manipulation because the financial statement effects were large and varied, and because all SFAS 96 adopting firms did so voluntarily.

## OVERVIEW OF SFAS 96

SFAS 96 was issued in December 1987 and could be adopted by firms beginning with fiscal years ending in 1987. Adoption was to be mandatory beginning with fiscal years ending in 1990. SFAS 96 required disclosure in the notes to the financial statements of the effects of adoption on income from continuing operations, income before extraordinary items, net income, and related per share amounts for the year of adoption. It used the liability method of comprehensive interperiod income tax allocation. Under the liability method deferred tax amounts are measured using the tax laws and rates enacted for the future periods when the deferred amounts are expected to become taxable or deductible.

For the majority of firms with previously recorded deferred tax credits, the largest SFAS 96 adoption effect was the reduction of deferred tax liabilities resulting from the decrease in corporate tax rates enacted in the Tax Reform Act of 1986 (Eakin 1996; Gujarathi and Hoskin 1992; Knutson 1988). For these firms, SFAS 96 adoption reduced the deferred tax liability balance, increased retained earnings, and increased net income. Reduction of the deferred tax liability balance related to the adoption year increases income before extraordinary items (IBE). Reductions related to prior years result in a cumulative effect that increases to net income, but may have no effect on IBE. For some firms, SFAS 96 adoption increased net income through both the current year effect and the cumulative effect.

Firms with previously recorded deferred tax debits and firms that entered into business combinations in prior periods may have experienced a reduction in retained earnings upon adoption of SFAS 96 (Eakin 1996, Gujarathi and Hoskin 1992, Martin et al. 1989). For these firms, the balance of the deferred tax liability increased, and retained earnings decreased. In addition, some firms reported positive cumulative effects and negative current year effects, while other firms reported negative cumulative effects and positive current year effects.

For firms with existing net operating loss carryovers (NOLs), SFAS 96 required NOLs to be disclosed as a reduction to current income tax expense. The previous standard required NOLs to be disclosed as extraordinary items. For many firms the change in NOL accounting was the only effect adoption of SFAS 96 had on the income statement (Eakin 1996). For these firms, SFAS 96 adoption increased income from continuing operations but had no effect on net income.

In addition to the varied financial statement effects resulting from adoption of SFAS 96, firms were also allowed to apply the Statement either prospectively or retroactively. If applied prospectively, a cumulative adjustment to income was made in the year of adoption. If applied retroactively, one or any number of prior years, including those that preceded the earliest year presented were restated. As a result, the effect of adoption on retained earnings is not always an indicator of the effect of adoption on net income. For example, if a firm reports a positive cumulative effect and a negative current year effect, the overall effect on retained earnings will be positive if the cumulative effect is larger than the current year effect. If this firm chooses the prospective method of adoption, the adoption year net income effect will also be positive. However, if the same firm were to choose the retroactive method of adoption, the overall retained earnings effect would not change, but the adoption year net income effect would become negative. Similarly, firms reporting negative cumulative effects and positive current year effects can use retroactive adoption to report an increase to net income in the year of adoption. For firms not reporting a current year effect, retroactive adoption eliminates all effects from adoption year net income.

The earnings manipulation opportunities afforded by SFAS 96 were not overlooked by the auditing profession. But, instead of discouraging the use of SFAS 96 as an earnings manipulation tool, auditors promoted it. For example, they encouraged firms to "determine which year of adoption and application is the most advantageous" (Coopers and Lybrand 1988, 19), and to consider such factors as reporting "good" news, and deemphasizing "bad" news (Price Waterhouse 1988, 120).

## DISCLOSURE IN THE SHAREHOLDER LETTER

The shareholder letter usually contains a description of the year's operating results along with a comparison of the results to those achieved in prior years. Year-to-year comparisons of financial results are important because they are often used to chart past results and predict future results. Unless the financial statement effects of an accounting change are fully disclosed, the change distorts the comparison of current results with those of prior periods (FASB 1978, ¶68). Failure to fully disclose these one-time financial statement effects is misleading in both the year of adoption and subsequent years.

There are many ways in which the disclosure of SFAS 96 adoption can be manipulated to portray the financial results and the year-to-year comparison of those results in a misleading manner. For example, firms that reported large increases to net income from SFAS 96 adoption can simply omit any reference to SFAS 96 adoption in the shareholder letter. By omitting reference to the effects of SFAS 96 adoption in the year of adoption, the false impression is given that the increase in net income over prior years arises from real economic events. Similarly, firms for which adoption of SFAS 96 resulted in large decreases to net income can mislead users by failing to disclose SFAS 96 adoption in the year subsequent to the year of adoption. This makes the increase in net income in the year subsequent to adoption appear to be a real economic event when it is not.

## EXPERIMENT

The research question that this study is attempting to answer is whether or not individual firms were using the early adoption of SFAS 96 to manipulate earnings. To answer this question a sample of firms was selected for inclusion in the study. For each firm selected the relevant shareholder letters were examined and it was determined if the letter was misleading about the effects of the adoption of SFAS 96. If the shareholder letters were misleading about the effects of SFAS 96, then it is concluded that one consideration for the timing of the adoption of SFAS 96 was earnings manipulation. If the firm's shareholder letter treatment of the SFAS 96 effects was not misleading, it is concluded that earnings management was not a consideration in the firm's decision to adopt SFAS 96.

### Sample Selection

Firms electing early adoption of SFAS 96 were identified by searching the 1989 Compustat files for firms making changes in accounting for income taxes in any of the three years beginning with 1987. The financial statements of firms making income tax accounting changes were examined to ensure that the change was adoption of SFAS 96 and to determine the income effects of adoption. The search resulted in identification of 612 adopting firms representing 238 4-digit Standard Industrial Classification (SIC) codes. Of the 612 adopters, 386 reported an increase to retained earnings, 99 reported a decrease, and 126 firms reported no material effects to retained earnings from SFAS 96 adoption. Of the 485 adopters that reported retained earnings effects from SFAS 96 adoption, 372 reported increases to net income, 42 reported decreases to net income, and 71 reported no effects to net income. The adoption effects on net income ranged from –291% of net income to 4,000% (the 4,000% increase was a result of a small firm that reported net income of $10,000 that included a cumulative effect from SFAS 96 adoption of $400,000). The adoption effects on income before extraordinary items (IBE) ranged from –77% to 150%.

To select a sample of shareholder letters from the population of SFAS 96 adopters, firms were ranked on the percentage effect of SFAS 96 adoption on net income. The FILING file of the COMPNY library of the Lexis database was used to examine the shareholder letter for the 50 firms that reported the largest increases net income from SFAS 96 adoption, and the 20 firms that reported the largest decreases to net income. Firms for which shareholder letters for the year of SFAS 96 adoption and the year subsequent to adoption were not available in Lexis were dropped. Next, firms were ranked on the percentage effect of SFAS 96 adoption on IBE. The shareholder letters of the 15 firms that reported the largest increases to IBE, and the fifteen firms that reported the largest decreases to IBE effect were selected for the sample. Again, if the requisite shareholder letters were not available, the firm was dropped. The final sample consists of 40 firms. The net income effects of adoption in the final sample ranged from −15% to 4000%. The IBE effects of adoption ranged from −77% to 134% (including 17 firms in the final sample that reported no effects to IBE but were included because of the large net income effects).

## Method

The shareholder letters in both the adoption year and the subsequent year of the 40 sample firms were examined to determine whether the disclosure in the shareholder's letter regarding SFAS 96 adoption was acceptable. If the adoption of SFAS 96 resulted in an increase in net income in the adoption year, the adoption year letter was considered unacceptable if it failed to disclose the effects of SFAS 96. If the adoption of SFAS 96 resulted in a decrease in net income in the adoption year, the subsequent year letter was considered unacceptable if it compared the subsequent year's net income to the adoption year net income and failed to disclose the effects of SFAS 96. These categories covered the situations where failure to disclose the effects of SFAS 96 would be misleading in a way that was advantageous to the firm. Failure to disclose the effects of SFAS 96 when the failure was considered disadvantageous to the firm was considered acceptable disclosure. For example, letters of firms where the effect of SFAS 96 adoption was to decrease net income, but that did not disclose this effect in the year of adoption, were considered acceptable. This is because, if shareholders were misled, it would be to think management did worse, not better, than real economic events indicated.

The above criteria for determining unacceptable reporting were selected to give firms the benefit of the doubt with regard to earnings manipulation. For example, if earnings were decreased by the effect of SFAS 96 in the adoption year, letters in the subsequent year that failed to disclose the effects of SFAS 96 were found to be unacceptable only if they explicitly referenced the adoption year's net income. This is because, even though a failure to remind the subsequent year users that the previous year's income had been artificially depressed by SFAS 96 adoption could be misleading, it is not a blatant enough failure to support a finding of deliberate earnings manipulation. On the other hand, failure to disclose the effects of SFAS 96 adoption in the adoption year letter was considered unacceptable when the effect was to increase net income even when adoption year net income was not compared to previous years. This is because adoption year net income was artificially inflated by the SFAS 96 effect and is, therefore, misleading even without being compared to prior years.

One firm, Valero Energy Corporation, made a disclosure that was clearly unacceptable but did not neatly fit the criteria for unacceptable given above. Overall, the disclosure made in Valero's shareholder letter appears to be complete. After comparing adoption year net income with that of the prior year, the firm made several disclosures regarding items that affected net income in the adoption year and in the prior year. Included in these disclosures was the $7 million cumulative effect of SFAS 96 adoption, but missing was a discussion of the $21.7 million increase in net income due to the effect of SFAS 96 on income from continuing operations. Consequently, Valero's shareholder letter was classified as unacceptable.

## RESULTS

Table 1 lists all of the firms selected in the study. For each firm, the table indicates whether the disclosure of the SFAS 96 data was acceptable or unacceptable. If a firm's shareholder letter was unacceptable, the reason is given. Eighteen firms were found to be acceptable, and 22 were found to be unacceptable. From this, it is concluded that for 22 of the 40 firms, earnings manipulation was a motivation for the decision to adopt SFAS 96, and for 18 out of 40 it was not.

## DISCUSSION

This study is subject to several limitations. First, it is possible that the firms found to be unacceptable failed to disclose the SFAS 96 adoption effects because the management of these firms misunderstood the SFAS 96 income effect and actually thought that it represented a real economic event and not a one-time adjustment due to a change in accounting principle. If this was the case, it would seem unlikely that these firms would disclose any other nonrecurring income items. In fact, six of the firms found unacceptable (Bay Meadows Operating Company, Bolt Beranek & Newman, Christiana Companies, Computer Horizons, Downey Savings and Loan Association, and Mosinee Paper Corporation) did report other nonrecurring income items. In each case, the disclosure of nonrecurring income items were for ones that reduced net income. This makes the misunderstanding explanation untenable for these six firms. For the other 16 unacceptable firms, it is difficult to believe that they could have the technical sophistication to choose to early adopt SFAS 96 and not understand the one-time nature of the adoption year effects.

**Table 1.** Shareholder Letter SFAS 96 Reporting Disclosures

| *Company Name* | *Year of SFAS 96 Adoption* | *Percent of Net Income from SFAS 96 Adoption* | *Acceptable Reporting* | *Reporting Code* |
|---|---|---|---|---|
| Alco Standard | 1988 | 19 | No | A |
| Amerada Hess Corporation | 1987 | 20 | Yes | |
| American Television & Communication | 1989 | −12 | Yes | |
| Baldor Electric | 1987 | 34 | No | A |
| Bay Meadows Operating Company | 1989 | −14 | No | B |
| Bird | 1989 | 16 | No | A |
| Boeing | 1989 | 31 | Yes | |
| Bolt Beranek & Newman | 1989 | 7 | No | A |
| Christiana Companies | 1988 | 114 | No | A |
| Computer Horizons | 1989 | 104 | No | A |
| Continental Materials | 1987 | 222 | Yes | |
| CTS Corporation | 1988 | −13 | No | B |
| Devon Energy Corporation | 1988 | 119 | Yes | |
| Downey Savings and Loan Association | 1989 | 107 | No | A |
| Everest & Jennings International Ltd. | 1987 | 227 | No | A |
| ESI Industries | 1988 | 8 | Yes | |
| Grubb & Ellis Company | 1987 | 1,573 | No | C |
| Grumman | 1987 | 104 | Yes | |
| Hecla Mining Company | 1988 | −10 | No | B |
| Hornbeck Offshore Services | 1989 | 62 | Yes | |
| Joule | 1988 | 4,000 | Yes | |
| Kaneb Services | 1989 | −16 | No | B |
| Kuhlman Corporation | 1988 | 13 | No | C |
| Lancer Corporation-Texas | 1988 | 1 | No | A |
| Leucadia National Corporation | 1988 | −11 | Yes | |
| Mosinee Paper Corporation | 1989 | 76 | No | A |
| National Convenience Stores | 1988 | 97 | Yes | |
| Oryx Energy Company | 1989 | 155 | No | A |
| Peerless Tube Company | 1988 | 207 | No | C |
| Pogo Producing Company | 1987 | 102 | Yes | |
| Sandy Corporation | 1988 | 110 | Yes | |
| Santa Fe Pacific | 1989 | 1,438 | Yes | |
| Sears Roebuck | 1988 | 25 | Yes | |
| Service Corp International | 1988 | −155 | Yes | |
| Sprague Technologies | 1989 | −11 | No | B |
| Sunshine-Jr. Stores | 1988 | 32 | Yes | |

*(continued)*

***Table 1*** (Continued)

| *Company Name* | *Year of SFAS 96 Adoption* | *Percent of Net Income from SFAS 96 Adoption* | *Acceptable Reporting* | *Reporting Code* |
|---|---|---|---|---|
| Union Valley Corporation | 1988 | 176 | Yes | |
| U.S. Home Corporation | 1988 | 48 | No | A |
| Valero Energy Corporation | 1987 | 226 | No | |
| Westmoreland Coal Company | 1988 | 109 | No | A |

***Notes:*** *Code* *Explanation*

A SFAS adoption increased net income. In the adoption year the shareholder letter compared net income with previous years but did not disclose SFAS 96 effects. This is misleading because not disclosing that income was increased by SFAS 96 adoption leaves shareholders with the false impression that the increase was due to real economic events.

B SFAS adoption decreased net income. In the year subsequent to adoption the shareholder letter compared net income with adoption year net income but did not disclose the effects of SFAS 96 adoption. This is misleading because it gives the false impression that real economic events caused net income to increase by the amount of the SFAS 96 adoption effect in the year subsequent to the adoption date.

C SFAS adoption increased net income. In the adoption year the shareholder letter did not disclose SFAS 96 effects. This is misleading because not disclosing that income was increased by SFAS 96 adoption leaves shareholders with the false impression that the increase was due to real economic events. This is not as blatantly misleading as A because the adoption year net income was not directly compared to previous years' net income, but it still creates the false impression that all of the reported net income reflects real economic events.

Other See text discussion of Valero Inc.'s shareholder letter.

Second, it is possible that the shareholder letters were only opportunistic attempts at earnings manipulation undertaken after the decision to adopt SFAS 96 had been made for legitimate reasons. This seems unlikely, given the magnitude of the SFAS effects of the firms included in the study and the previously mentioned promotion of the earnings manipulation potential of SFAS 96 by auditors. But even if it was only an afterthought, it is still unethical behavior.

Third, the evidence from the shareholder letters only indicates that some firms tried to deceive financial statement users by manipulating earnings, not whether they succceded. There is some empirical evidence that misleading disclosures in shareholder letters can influence financial statement users (Kaplan et al. 1990), but the question of whether financial statement users were influenced in this case is left to other researchers. From an ethics standpoint, it matters little if they were successful—they failed the ethics test just by trying to deceive.

Finally, the sample selection method was not designed to select a sample of firms that could be said to be representative of firms in general. Thus it would not be appropriate to say that this study provides evidence that 55% of firms (22/40) will use discretionary accruals to manipulate earning given the opportunity. If anything, the sample might have been biased toward firms disposed to earnings manipulation because it selected only those firms who elected to early adopt SFAS 96 and for whom it had a large effect. Nonetheless, conclusions can be reached about the 40 firms selected and they are important in their own right.

In spite of these limitations, these results should be of interest to ethics researchers because they provide strong evidence that some firms do deliberately manipulate earnings. It is also interesting because it provides evidence that other firms do not attempt to fool stockholders even when they are presented with the opportunity to do so. In this sense there is as much to learn from the firms that behaved ethically as those that did not. It is hoped that further research will be able to explain why some of these firms behaved ethically and some did not.

## CONCLUSION

Previous research has demonstrated that managers are likely to time the adoption of new FASB standards to have the most advantageous impact on earnings. This study examined the shareholder letters of a sample of firms that chose to early adopt SFAS 96 and found strong evidence that some of these firms were motivated by earnings manipulation considerations and some were not. This is an important finding for ethics researchers because it demonstrates that different firms will behave in different ways, both ethical and unethical, when presented with similar opportunities to manipulate earnings. By identifying some firms that behaved ethically and others that behaved unethically this study has given researchers an opportunity to examine the firms identified to determine what caused one firm to behave ethically and another to behave unethically. It is also hoped that the methodology developed in this paper for identifying whether a firm's decision to early adopt a new standard was motivated by earnings manipulation considerations or legitimate factors will aid future researchers investigating earnings manipulation.

## ACKNOWLEDGMENT

The authors thank Matthew Graves for his help with data collection.

## REFERENCES

Ayres, F.L. 1986. Characteristics of firms electing early adoption of SFAS 52. *Journal of Accounting and Economics* (May): 143-158.

Coopers and Lybrand. 1988. *Accounting for Income Taxes: Focusing on FASB Statement 96*. Coopers and Lybrand.

Eakin, C.F. 1996. A comprehensive analysis of the adoption of SFAS 96: Accounting for income taxes. In *Advances in Accounting*, Vol. 14, ed. P.M.J. Reckers, 107-133. Greenwich, CT: JAI Press.

Financial Accounting Standards Board. 1978. *Statement of Financial Accounting Concepts No. 1: Objectives of Financial Reporting by Business Enterprises.* Stamford, CT: FASB.

Financial Accounting Standards Board. 1987. *Statement of Financial Accounting Standards No. 96: Accounting for Income Taxes*. Stamford, CT: FASB.

Gujarathi, M.R., and R.E. Hoskin. 1992. Evidence of earnings management by the early adopters of SFAS 96. *Accounting Horizons* (December): 18-31.

Kaplan, S.E., S. Pourciau, and P.M.J. Reckers. 1990. An examination of the effect of the president's letter and stock advisory service information on financial decisions. *Behavioral Research in Accounting* 2: 63-92.

Knutson, P.H. 1988. FAS 96—Implications for analysts. *Financial Analysts Journal* (November/December): 117-118.

Langer, R., and B. Lev. 1993. The FASB's policy of extended adoption for new standards: An examination of FAS No. 87. *The Accounting Review* (July): 515-533.

Martin, D.R., H.I. Wolk, and D. Beets. 1989. Illustrations and critique. *Ohio CPA Journal* (Autumn): 24-31.

Price Waterhouse. 1988. *The New Accounting for Income Taxes: Implementing FAS 96*. Price Waterhouse.

Sami, H., and M.J. Welsh. 1992. Characteristics of early and late adopters of pension accounting standard SFAS No. 87. *Contemporary Accounting Research* (Fall): 212-236.

Scott, T.W. 1991. Pension disclosures under SFAS No. 87: Theory and evidence. *Contemporary Accounting Research* (Fall): 62-81.

Trombley, M.A. 1989. Firms electing early adoption of SFAS No. 86. *The Accounting Review* (July): 529-538.

# WHISTLE-BLOWING ON THE AUDIT TEAM:

## A STUDY OF AUDITOR PERCEPTIONS

P. Richard Williams and G. William Glezen

## ABSTRACT

During the performance of an audit, an auditor may disagree with the resolution of an accounting or auditing problem. The auditor is required by professional standards to bring his or her concerns to the attention of the person in charge of the audit. If still dissatisfied with the solution of the problem, the auditor can file a written disassociation from the resolution of the problem, thereby bringing the problem to the attention of members of the audit firm outside of the audit team. We compare auditors' situational perceptions of the existence of an audit problem with their willingness to document their disagreement with the resolution of the problem. We also investigate auditors' perceptions of audit firm reaction to a documented disassociation. We find that auditors are more willing to agree that a problem exists than they are to agree that they would be willing to document a disassociation. Additionally, we find a consensus of perceptions that audit firms respond positively to the subject of a documentation of a disassociation, that is, the problem is given thorough con-

**Research on Accounting Ethics, Volume 4, pages 125-143.**

**ISBN: 0-7623-0339-5**

sideration by the audit firm. While some consensus of perceptions exists that the audit firm will respond positively to the individual auditor who documented the disassociation, there is a lack of consensus as to whether that individual would be assigned to high profile clients in the future. The foregoing findings suggest that audit firms may wish to emphasize the importance of the disassociation procedure as a quality control mechanism.

## INTRODUCTION

The accounting profession is constantly striving to enhance professionalism and reduce the extent and effect of legal liability. While the profession is pursuing these goals, it is also important to investigate potential causes of audit failures. One such potential contributing factor is the failure of audit team members to report disagreements with audit work or audit findings to audit firm management.

Audit firms employ many procedures to provide reasonable assurance that unqualified audit reports are not issued on financial statements that contain material misstatements. If those procedures are ineffective, professional standards are compromised and large legal judgments or settlements may result. Generally accepted auditing standards (GAAS) provide that audit team members should bring to the attention of appropriate individuals in the firm significant disagreements with conclusions reached in the audit. Additionally, professional standards provide that audit team members should have the right to document the disagreement and disassociate themselves from such conclusions (AICPA 1995). Anecdotal evidence suggests that this procedure is used infrequently, presumably because unresolved disagreements seldom occur or because audit team members do not feel the need to document the disagreement. However, its infrequent use also may be associated with the fact that documentation of disagreements is, in effect, a form of whistle-blowing for which the audit team members may fear retaliation.[1] From an auditor's personal standpoint, the filing of a written disassociation could be a critical event in his or her career. Nevertheless, it is an important fail-safe audit quality mechanism that is assumed to function in all auditing firms. The purpose of this study is to provide empirical evidence regarding the potential effectiveness of this procedure.

Professional requirements and whistle-blowing behavior are discussed in the next sections followed by the research questions, the data description, and the analysis. The final sections present factors associated with the likelihood of whistle-blowing and the conclusions.

## PROFESSIONAL REQUIREMENTS

Generally accepted auditing standards "deal with measures of the quality of the performance of those acts (auditing procedures)" (AICPA 1995, enacted in 1978). AU Section 311 (Planning and Supervision) paragraph 14 states:

> The auditor with final responsibility for the audit and assistants should be aware of the procedures to be followed when differences of opinion concerning accounting and auditing issues exist among firm personnel involved in the audit. Such procedures should enable an assistant to document his disagreement with the conclusions reached if, after appropriate consultation, he believes it necessary to disassociate himself from the resolution of the matter. In this situation, the basis for the final resolution should also be documented.

This wording is virtually identical to an attestation standard of field work adopted by the Auditing Standards Board (AICPA 1995), and similar wording appears in the *Statement of Quality Control Standards* (AICPA 1995), which lists and discusses the elements of quality control. One of the objectives of the supervision element of quality control is to "provide procedures for resolving differences of professional judgment among members of an engagement team."

Some individual auditing firms have developed interpretations of this requirement that provide detailed guidance for resolving differences between members of the audit team at all levels. The ultimate arbitrator of unresolved differences is often the chairman of the firm. Thus, it is apparent that the auditing profession considers the ability to resolve dis-agreements among audit team members to be a critical component in the performance of a quality audit.

Two illustrations of the damage that can occur to the professional careers of auditors who fail to comply with the disassociation requirements can be seen in SEC Accounting and Auditing Enforcement Releases (AAERs) 81 (SEC 1985) and 455 (SEC 1993).

AAER 81 reports that a manager in a national CPA firm concluded that sufficient competent evidence had not been obtained to support the inclusion of certain revenues in the financial statements. The partner had concluded that the evidence was sufficient. The SEC found that the manager did not employ the standard firm practice, based on AU section 311, that provides that an auditor register his or her disagreement with the manner in which an audit is performed in a memorandum that is included in the working papers. The SEC disciplined the manager (and also the partner) and stated:

> Had (the manager) documented his disagreement in the audit work papers, (the partner) and (the firm) would have found it necessary to consider (the manager's) professional opinion and reconsider the evidentiary support. (The manager) did not avail himself of (the firm's) disagreement procedure because, in his view, the procedure was rarely used and (the partner) made all final audit decisions in any event. Additionally, (the manager)was *concerned that he would jeopardize his future with the firm* by preparing a disagreement memo (emphasis added).

According to AAER 455, a manager with a Big Six firm disagreed with the partner about the proper accounting for a client's obsolete inventory, past due receivables, unidentifiable fixed assets, and the classification of notes payable. Despite the manager's objections, the partner told him to sign off on the audit, which the manager did, but only after the audit report was released by the partner. The SEC made a special point of the fact that the manager was aware of firm procedures which provided that an accountant participating in an audit could disassociate himself from the audit conclusions by including a memorandum concerning his disagreement with the conclusions in the audit file. In addition to censuring the partner, the SEC censured the manager, stating:

> An independent accountant, including an audit manager, cannot excuse his failure to comply with GAAS because of a sense of futility after his proposed approaches to certain accounting issues are repeatedly rejected. Similarly, such failure cannot be excused by pointing to pressure, whether from the client or partners. . . . by signing off on the audit, the conclusions of which he disagreed with, [the manager] *abdicated his role as an independent accountant by not documenting in the work papers, or otherwise bringing to the attention of another [accounting firm] partner, his concerns and disagreements with the audit partner* regarding the sufficiency of the audit evidence and the audit conclusions [emphasis added].

The existence in practice of situations where the disagreements are *not* resolved compels the auditor to adhere to professional standards and to consider the whistle-blowing behavior discussed in the next section.

## WHISTLE-BLOWING BEHAVIOR

The key element in any definition of "whistle-blowing" is the idea that the complaint is made public by informing some person or agency external to the group to which the whistle-blower belongs. Such a definition includes complaints like an auditor's documented disassociation because "so long as a complaint is made to someone other than or in addition to the immediate supervisor . . . making the complaint through other than prescribed channels (i.e., the chain of command) represents going public, insofar as all groups outside the immediate work group are viewed as the public" (Near and Miceli 1985). In this study, the immediate work group is the audit team, and the documentation of a disassociation, which highlights the problem to firm members outside the audit team, is viewed as whistle-blowing behavior.[2]

Near and Miceli (1985) captured the complexity of whistle-blowing in a detailed model composed of four different decisions.[3] The following discussion adapts these decisions to the auditor disassociation process. A graphic summary of the process is presented in Figure 1.

- *Decision 1.* The observer must decide whether the situation is actually "wrongful," however that term is defined in context. In the audit context, an audit team member must decide that the subject of the disagreement is likely to cause the financial statements to be materially misstated, auditor independence to be impaired, and so forth.
- *Decision 2.* The observer must decide to report the activity to an "outsider." In the audit context, the audit team member has discussed his or her concerns with an immediate supervisor, and the problem has not been

resolved to the satisfaction of the audit team member, thus forcing him or her to consider documenting a disassociation. The problem is thus drawn to the attention of outsiders.

- *Decision 3.* If the activity is reported, the organization must decide to respond to the complaint in some way (e.g., halting an activity, changing an action, etc.). In the audit context, if a disagreement is documented, the audit firm management must consider a response to the subject of the disagreement: change or expand an audit procedure, propose a change to the client's financial statements, change an anticipated audit opinion, withdraw from the engagement, and so forth.
- *Decision 4.* The organization must decide how to react to the fact that the whistle-blower has taken action. This reaction can be to ignore, punish, or praise. In the audit context, the audit firm management can consider: termination, slow or rapid promotions, reducing or increasing pay, ignoring, and so forth.

Note that the occurrence of whistle-blowing is a conditional event—that is, the auditor must decide "yes" to both Decision 1 *and* Decision 2 for the event to occur. A decision of "no" to either results in termination of the whistle-blowing process.

In this study, we explore auditors' perceptions regarding their willingness to make the first two decisions, and their perceptions regarding their firms' likely reactions as described in the latter two decisions.

## RESEARCH QUESTIONS

A wide range of questions could have been explored in this area. For example, this study could have approached the auditing situation as an example of "peer reporting" (Trevino and Victor 1992) or "prosocial behavior" (Miceli and Near 1988). We chose a more direct approach and limited the research questions to those treating the filing of a written disassociation as whistle-blowing. The following questions are addressed in this study:

1. Are individual auditors who perceive the existence of an unresolved accounting or auditing problem (Near and Miceli Decision 1) willing to document a written disassociation (Near and Miceli Decision 2)?
2. What are individual auditors' perceptions of audit firm management reaction to the subject of the disassociation (Near and Miceli Decision 3)?
3. What are individual auditors' perceptions of audit firm management reaction to the auditor who filed the written disassociation (Near and Miceli Decision 4)?

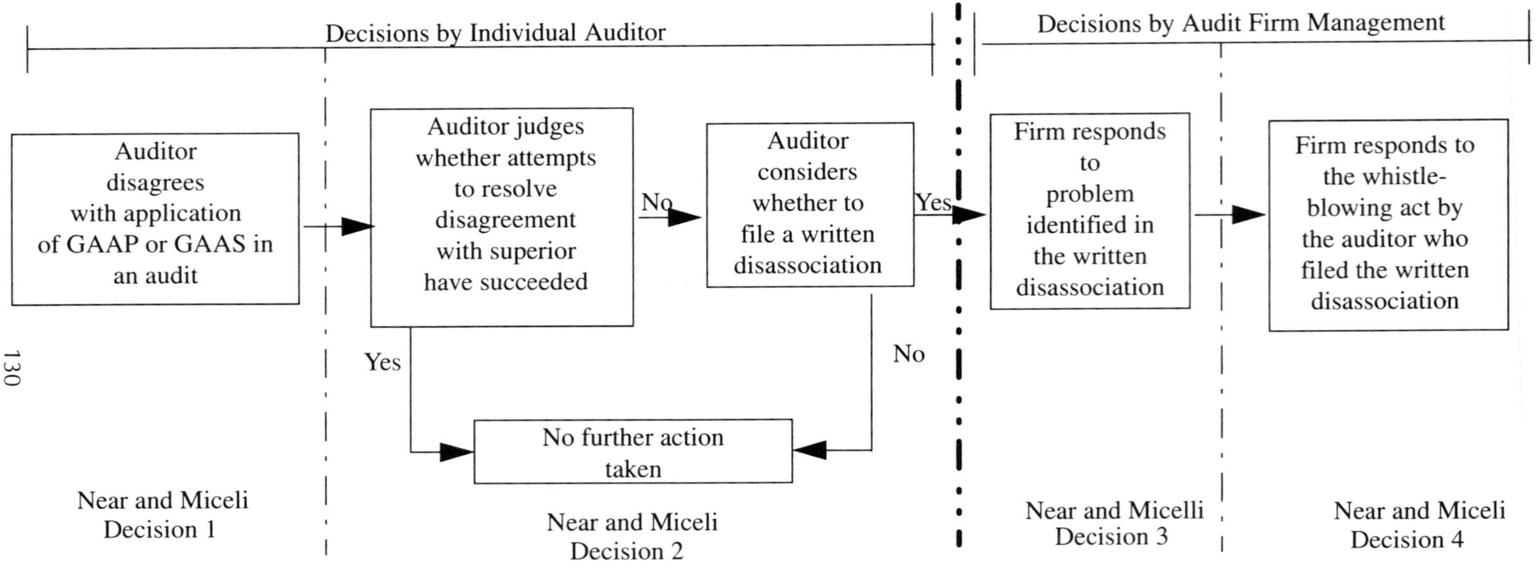


***Figure 1.*** Model of Four-Stage Auditor Disassociation Process

***Table 1.*** Survey Subject Profile

| *Frequencies:* | *Number of Respondents** |
|---|---|
| Position | |
| Partner | 46 |
| Manager | 35 |
| Gender | |
| Male | 68 |
| Female | 15 |
| Education | |
| Bachelor degree | 67 |
| Masters degree | 16 |
| Married | |
| Yes | 60 |
| No | 15 |
| Children | |
| Yes | 60 |
| No | 23 |
| Attendance at weekly religious services | |
| Zero weeks per year | 24 |
| Every week per year | 18 |
| Other than 0 or 52 weeks per year | 32 |
| Firm size | |
| Big Six/National | 21 |
| Regional/Local | 61 |
| Documented a written disassociation: | |
| Yes | 7 |
| No | 75 |
| Means | *Percentage* |
| Size of audit staff in office | 58.5 |
| Number of audit clients in office | 72.0 |

***Note:*** *Some responses total less than 83 because some subjects omitted certain demographic items.

In addition, we analyze responses by level of responsibility (partner or manager), size of firm (Big Six/national or regional/local), and whether the respondent had documented a disassociation.

A survey instrument designed to obtain responses useful in addressing these research questions was divided into three sections. The first section presented three cases and elicited auditor perceptions regarding the two personal decisions in

***Table 2.*** Frequency Distribution of Responses to Statements Presented in Cases

*Statements Presented*

Statement 1 (S1): I believe that there is a problem here with which the audit partner should deal.

Statement 2 (S2): If the audit partner persists in ignoring this situation I will feel compelled to prepare a written filing under my firm's policy.

| | *(1) Strongly Agree* | *(2) Agree* | *(3) Neutral* | *(4) Disagree* | *(5) Strongly Disagree* | *Mean* | *Standard Deviation* |
|---|---|---|---|---|---|---|---|
| Case A (Inappropriate accounting procedure) | | | | | | | |
| S1 | 47 | 29 | 1 | 2 | 0 | 1.457 | 0.653 |
| S2 | 28 | 36 | 9 | 3 | 3 | 1.949 | 0.986 |
| Case B (Independence problem) | | | | | | | |
| S1 | 21 | 22 | 20 | 14 | 2 | 2.407 | 1.127 |
| S2 | 15 | 14 | 21 | 24 | 5 | 2.873 | 1.223 |
| Case C (Illegal act) | | | | | | | |
| S1 | 49 | 27 | 5 | 0 | 0 | 1.458 | 0.848 |
| S2 | 32 | 31 | 15 | 3 | 0 | 1.864 | 0.611 |

Chi-Square Goodness-of-Fit Test

| | *Chi-Square* | *df* | *P-value* |
|---|---|---|---|
| Case A | 57.371 | 2* | $p < .001$ |
| Case B | 10.068 | 2* | $p < .01$ |
| Case C | 40.291 | 2* | $p < .001$ |

***Note:*** *Because the expected cell frequencies for response categories 4 and 5 were less than five, the responses for categories 3, 4, and 5 were combined.

the whistle-blowing process: (1) existence of a problem and (2) willingness to expose the problem. The second section elicited auditor perceptions regarding the two decisions required of audit firm management in the whistle-blowing process: (3) response to the problem itself and (4) reaction to the auditor who went outside the audit team to document disagreement as to resolution of the problem. The third section requested audit firm data and demographic information, including whether the respondent had ever documented a disagreement.

## DATA DESCRIPTION

The population of interest was all auditors who have had unresolved disagreements with their firms. As a practical matter, however, this population was not observable. Therefore, we used an AICPA data base of members who identified themselves as partners and managers having a primary interest in auditing. Because partners and managers are likely to have more years of experience with their firms, they have more opportunities than staff members to have been involved in an unresolved disagreement. Also, partners and, to a lesser extent, managers have achieved considerable standing in their firms and should feel less threatened by potential retaliation.[4] Thus, if differences are found in this study between detecting a problem and filing a written disassociation at the partner and manager levels, even greater differences are likely to exist for the relatively insecure staff accountants at lower levels.

Anonymous survey instruments were mailed to 869 randomly selected subjects. Eighty-three useable instruments were returned, for a response rate of 10%. Although the response rate appears low, given the sensitive nature of the topic addressed it is not unusual (e.g., see Keenan and Krueger 1992, who achieved an 11% response rate for a similarly sensitive instrument). A summary of demographic information is shown in Table 1.

A test for nonresponse bias was performed using a chi-square test to compare the distribution of responses of early and late respondents on the three dimensions of position, firm size, and filers of written disassociations versus non-filers. The hypothesis of no difference between the two groups could not be rejected at the .05 level.

As in any survey involving sensitive characteristics, demand effects (responses that conform with a preconceived anticipated answer or the respondent's self or professional image) may be present. Although respondents were assured anonymity, they may have responded in a manner consistent with their firm's or the AICPA's policy rather than as they would respond in an actual situation.[5] Also, the pressure and stress that would normally accompany an event requiring consideration of filing a written disassociation cannot be replicated in a survey. However, any bias from these sources would make differences between detecting a problem and filing a written disassociation (Research Question 1) more difficult to detect and, thus, would be a bias favoring the finding of no difference.

## ANALYSIS

Three case situations were written to present problems that an auditor might encounter during an audit. Case A dealt with the valuation of receivables (inappropriate accounting procedure), and Case B dealt with auditor independence (independence may be impaired). These two situations were condensed versions of the facts in two court cases in which an audit firm was sued and lost (Hall and Renner 1988; Causey 1988). Case C dealt with a violation of federal law (illegal act).

For each case, the auditors were asked to indicate on a five-point Likert scale their level of agreement with the following two statements:

***Table 3.*** Frequency Distribution of Auditor Perceptions of Audit Firm Responses to the Filing of a Written Disassociation

*Statements Presented*

Statement 1: If I file a written disassociation my firm would give it a thorough consideration.

Statement 2: If I file a written disassociation I may never know what really happened to it.

Statement 3: By its very existence, my firm's policy makes a contribution to the audit quality of the firm.

Statement 4: If I file a written disassociation my firm would regard it as evidence of my commitment to the quality of the firm's audit services.

Statement 5: If I file a written disassociation I would probably be terminated during the following year.

Statement 6: If I file a written disassociation the probability that I will be given high quality, high profile audit clients in the future will be reduced.

*Statement Response Frequency*

| *Statement* | *(1) Strongly Agree* | *(2) Agree* | *(3) Neutral* | *(4) Disagree* | *(5) Strongly Disagree* | *Binomial Test (normal approximation)* *Mean* | *Z-Score* | *P-Value* |
|---|---|---|---|---|---|---|---|---|
| 1 | 50 | 24 | 4 | 3 | 1 | 1.549 | 5.478 | <.00* |
| 2 | 3 | 2 | 7 | 34 | 36 | 4.195 | 4.576 | <.00* |
| 3 | 22 | 37 | 13 | 3 | 0 | 1.960 | 3.182 | <.00* |
| 4 | 18 | 40 | 19 | 3 | 2 | 2.159 | 1.871 | .031*** |
| 5 | 0 | 4 | 7 | 29 | 42 | 4.329 | 4.801 | <.00* |
| 6 | 2 | 4 | 27 | 31 | 17 | 3.704 | –0.250 | .595 |

***Notes:*** *Significant at the .001 level.
**Significant at the .05 level..

**Statement 1.** I believe that there is a problem here with which the audit partner should deal.

**Statement 2.** If the audit partner persists in ignoring this situation, I will feel compelled to prepare a written filing under my firm's policy.

The auditor responses are presented in Table 2.

**Table 4.** Comparison of Mean Responses by Level of Responsibility Partner versus Manager (1 = Strongly Agree; 5 = Strongly Disagree)

| *Response to* | *Partner* *n* = 46 | *Manager* *n* = 35 | *Kruskal-Wallis P-Value* |
|---|---|---|---|
| Case A | | | |
| Statement 1 | 1.435 | 1.485 | .880 |
| Statement 2 | 1.822 | 2.125 | .347 |
| Case B | | | |
| Statement 1 | 2.467 | 2.235 | .373 |
| Statement 2 | 2.818 | 2.848 | .787 |
| Case C | | | |
| Statement 1 | 1.435 | 1.486 | .642 |
| Statement 2 | 1.667 | 2.147 | .019* |
| Perceptions of audit firm responses | | | |
| Statement 1 | 1.391 | 1.677 | .412 |
| Statement 2 | 4.378 | 4.029 | .200 |
| Statement 3 | 1.956 | 2.400 | .050* |
| Statement 4 | 4.533 | 4.114 | .093** |
| Statement 5 | 3.841 | 3.571 | .343 |
| Statement 6 | 1.902 | 2.000 | .638 |

***Notes:*** *Significant at the .05 level.
**Significant at the .10 level.

The response frequencies from Statement 1 were hypothesized as the expected response frequencies for Statement 2. Because responses to the two statements are not independent, the chi-square goodness-of-fit test is used to examine each case. The distribution of responses is significantly different for each case, and Table 2 indicates that the auditors are less willing to agree that they would document a disassociation than they are willing to agree that a problem exists. This shift on the scale to the right for each case highlights the existence of a "willingness gap" for auditors with respect to the two statements.

The Pearson correlation coefficients between the first and second statements are examined for each case. If the respondents are equally willing to acknowledge a problem *and* to document a disassociation, then the correlation of responses between the two statements would be expected to approach 1.0. The results indicate that while the correlation is positive it is less than 1.0 for each case:

Case A = .4726 Case B = .6916 Case C = .5294

***Table 5.*** Comparison of Mean Responses by Size of Firm Big Six/National versus Regional/Local (1 = Strongly Agree; 5 = Strongly Disagree)

| *Response to* | *Big Six/ National n = 21* | *Regional/ Local n = 61* | *Kruskal-Wallis P-Value* |
|---|---|---|---|
| Case A | | | |
| Statement 1 | 1.474 | 1.459 | .802 |
| Statement 2 | 2.053 | 1.898 | .504 |
| Case B | | | |
| Statement 1 | 2.000 | 2.533 | .070* |
| Statement 2 | 2.600 | 2.965 | .245 |
| Case C | | | |
| Statement 1 | 1.381 | 1.475 | .729 |
| Statement 2 | 2.048 | 1.780 | .190 |
| Perceptions of audit firm responses | | | |
| Statement 1 | 1.600 | 1.541 | .930 |
| Statement 2 | 4.190 | 4.200 | .635 |
| Statement 3 | 2.429 | 2.083 | .125 |
| Statement 4 | 4.190 | 4.383 | .306 |
| Statement 5 | 3.476 | 3.797 | .194 |
| Statement 6 | 2.095 | 1.907 | .506 |

***Note:*** *Significant at the .05 level.

Table 3 presents the distribution of responses to the statements regarding the auditors' perceptions of audit firm decisions in the whistle-blowing process. To empirically assess the degree of consensus present in the auditors' responses to each of the statements, nonparametric binomial tests were performed. "Consensus" was defined to exist when 60% or more of the responses to a statement fell at either "strongly agree/agree" or "strongly disagree/disagree." The normal approximation to the binomial distribution was utilized due to the large sample size. Statements 1, 2, 3, and 5 had $p$-values of <.001; Statement 4 had a $p$-value of .031; and only Statement 6 failed to indicate significant consensus, with a $p$-value of .595. The respondents exhibited consensus in their perceptions that audit firm management would respond positively to the subject of the disassociation (Statements 1, 2, and 3 in Table 3). They appear to believe that the problem causing the disassociation would be given through consideration (Whistle-blowing Decision 3). There was less consensus that audit firm management would respond positively to the auditor who documented the disassociation (Whistle-blowing Decision 4). There was substantial disagreement to the notion in Statement 5 that an

***Table 6.*** Comparison of Mean Responses by Whether Auditor Had or Had Not Documented a Disassociation During an Audit (1 = Strongly Agree; 5 = Strongly)

| *Response to* | *Had Not Documented n = 75* | *Had Documented n = 7* | *Kruskal-Wallis P-Value* |
|---|---|---|---|
| ***Panel A*** | | | |
| Case A | | | |
| Statement 1 | 1.480 | 1.286 | .487 |
| Statement 2 | 1.958 | 1.714 | .777 |
| Case B | | | |
| Statement 1 | 2.425 | 2.429 | .999 |
| Statement 2 | 2.873 | 2.857 | .993 |
| Case C | | | |
| Statement 1 | 1.467 | 1.429 | .999 |
| Statement 2 | 1.863 | 1.714 | .848 |
| Perceptions of Audit Firm Responses | | | |
| Statement 1 | 1.540 | 1.429 | .961 |
| Statement 2 | 4.162 | 4.714 | .108 |
| Statement 3 | 2.189 | 1.714 | .171 |
| Statement 4 | 4.297 | 4.571 | .517 |
| Statement 5 | 3.657 | 4.286 | .084* |
| Statement 6 | 1.940 | 2.000 | .771 |
| ***Panel B*** | | | |
| Position (1 = partner; 0 = manager) | 0.562 | 0.571 | .961 |
| Age (years) | 39.155 | 43.000 | .254 |
| Gender (1 = female; 0 = male) | 0.173 | 0.286 | .465 |
| Education level (1 = bachelors; 0 = masters | 0.840 | 0.429 | .009** |
| Married (1 = married; 0 = single) | 0.827 | 0.714 | .465 |
| Children (1 = yes; 0 = no) | 0.733 | 0.571 | .365 |
| Religion (weeks per year attendance) | 22.636 | 31.571 | .376 |
| Total firm size (1 = regional/local; 0 = Big 6/ national) | 0.743 | 0.714 | .868 |
| Audit staff size (persons) | 60.813 | 41.286 | .567 |
| Audit client total (clients) | 77.333 | 21.167 | .066* |
| Any other staff ever filed? (1 = yes; 0 = no) | 0.107 | 1.000 | .001*** |

***Notes:*** *Significant at .100 level.
**Significant at .010 level.
***Significant at .001 level.

auditor would be terminated in the near future as a result of filing a written disassociation, with 87% of the respondents marking "disagree" or "strongly disagree." Statement 6 presented a more subtle outcome, that of receiving fewer high quality, high profile clients. While over half of the auditors still marked "disagree" or "strongly disagree," one-third marked "neutral," indicating a shift in perceptions that such a result would never occur.

Auditor responses also were analyzed by level of responsibility, firm size, and whether the auditor had filed a written disassociation in an audit. The results by level of responsibility (tested with the Kruskal-Wallis statistic because responses were not normally distributed) are presented in Table 4. Responses were generally not statistically different between partners and managers. One significant difference was found in the response to Statement 2 of Case C. For this illegal act, managers were less likely to indicate a willingness to file a written disassociation. The other notable difference was in Statement 3, about perceptions of audit firm responses. Partners were more likely to believe that the firm would regard the documenting of a disassociation as evidence of a "commitment to the quality of the firm's audit services." To the extent that partners have an idealized view of their firm and lawsuits impact partners more severely than managers, partners are likely to view the procedure more positively.

The analysis by firm size is presented in Table 5. The auditors were grouped as "Big Six/National" and "Regional/Local." Only one significant difference ($p$ = .070) was found out of 12 statements (about what would be expected by chance). Considerable consensus is apparent in this dimension.

The analysis in Table 6, Panel A, compares the responses of auditors who had documented a disassociation with those who had not documented a disassociation in an audit. Only one significant difference ($p$ = .084) was found out of 12 statements.

The analysis of the personal characteristics of the auditors who had and had not documented a disassociation in Table 6, Panel B, indicates a significant difference ($p$ = .009) for education level. Subjects who had documented a disassociation were more likely to have a master's degree than those who had not documented a disassociation. This result is consistent with other whistle-blowing studies. Two potential explanations are: (1) additional education (technical knowledge) may increase an auditor's confidence in the correctness of his or her conclusions and (2) an auditor with an advanced degree may feel better equipped to find another job if the management of the audit firm reacts negatively.

A difference existed between the religious service attendance rates of the two groups with the auditors who had documented a disassociation attending at a rate 40% higher than that of the other subjects. However, because the distribution of attendance was bimodal at the two extremes (attend zero weeks per year and attend 52 weeks per year), the variation was too great for this to be a statistically significant difference.

The most significant difference ($p$ = .001) between the groups was not a personal characteristic but a firm characteristic. After being asked in the instrument if the respondent had ever filed a written disassociation, the respondent was then asked if anyone else at the respondent's firm had ever done so. All who had documented a disassociation and who answered the latter question indicated that another auditor at their firm had also documented a disassociation.

From these results, it could be argued that some firms are better able than others to foster an atmosphere of trust within firm personnel that allows audit team members to come forward with a warning about an unresolved problem.[6] This outcome is significant because of the benefits to audit firms of having audit team members willing to highlight unresolved problems. Firms may wish to emphasize the existence of the disassociation policy in staff training meetings and indicate the benefits to both the firm and the individual auditor of its use. Auditors at levels below that of partner may be encouraged by the fact that only 2% of the partners in this study indicated that they did not believe the audit firm would regard the filing of a written disassociation as evidence of the auditor's commitment to the quality of the firm's audit services. However, as noted earlier, demand effects may be present.

Some respondents who had filed a written disassociation stated that the sources of disagreement involved "recording of a contingent gain on sale of real estate—ultimately resolved in our office with my disassociation," "client issuance of stock

***Table 7.*** Factors Identified in Accounting Whistle-blowing Literature* that Affect the Likelihood of Filing a Written Disassociation on an Audit Engagemen

| *Factor* | *Effect on Likelihood to Disassociate* |
|---|---|
| Existence of reporting policy | Increase |
| Effective response by the firm | Increase |
| Anonymity | Increase |
| Number of levels in the process | Decrease, as number of levels increase |
| Organizational rank of perpetrator | Decrease, as rank increases |
| Participative organizational culture | Increase |
| Seriousness (or materiality) of problem | Increase, as seriousness increases |
| Unambiguous evidence | Increase |
| Expected costs (loss of job, slower promotions, etc.) | Decrease, as potential costs increase |

***Note:*** Factors summarized in Hooks et al. (1994) and Ponemon (1994).

to a foreign trust under circumstances believed to require additional examination and disclosure," and "the issue of revenue recognition." One respondent described the circumstances of a written disassociation involving another member of his or her firm that illustrates the positive effect of the policy.

> Large sales were recorded at an SEC client. When it was determined that the goods had not been shipped, management presented a copy of an agreement with the customer showing the transfer of title. The audit manager was directed to accept the agreement as sufficient evidentiary matter. Disagreeing, he submitted the issue to the Director of A and A (Accounting and Auditing) who ordered additional work to be performed. It was ultimately determined that the agreement was fraudulent. The firm withdrew from the engagement.

Finally, a respondent illustrated the dilemma faced by accounting firms: "We need to encourage people to agree on one solution, but we can't stifle disassociation because it is a good control."

## FACTORS THAT AFFECT THE LIKELIHOOD OF WHISTLE-BLOWING

The accounting whistle-blowing literature (for a summary of this literature, see Hooks et al. 1994; Ponemon 1994) has identified a number of factors that affect the likelihood of whistle-blowing action. The factors are adapted to the auditor's disassociation decision in Table 7.

The information in the table suggests that there are conflicting pressures on auditors who are considering filing a disassociation. The existence of a policy requiring disassociation increases the likelihood that an auditor will disassociate when necessary. Such policies are required by generally accepted auditing standards, as noted earlier; however, different firms may place different emphasis on promoting their policies to firm members. An effective and nonretaliatory response by the firm will increase the likelihood of compliance with the policy in the future, if the response is known by firm employees. Assuming that firms respond to problems effectively, they may wish to publicize, within their organizations, the results of filed disassociations. Similarly, as part of auditor training, audit firms may wish to present examples of legal problems resulting from situations where a written disassociation should have been filed, but was not.

Although desirable, anonymity for the auditor filing the disassociation is not practical in cases of written disassociations. The likelihood of filing a disassociation decreases as the number of levels or links in the process increases. Some firms, particularly larger ones, have numerous levels in the process of responding to disassociations, such as partner in charge of the audit department within the office, office managing partner, regional managing partner, national partner in charge of audit practice, and firm managing partner. The literature suggests that firms could increase the likelihood of filing a disassociation by reducing the number of levels in the process, however, this seems unlikely in public accounting firms.

The likelihood of disassociation decreases as the rank of the perpetrator increases. Therefore, the higher the level within the firm the questioned decision is made (e.g., engagement partner, regional office partner, national office partner), the less likely that someone on the audit team would file a written disassociation. Also, the more participative the organizational culture, the greater the likelihood of disassociation. A more participative organizational culture also appears to be associated with lower incidence of wrongdoing and less serious offenses.

The likelihood of a disassociation increases as the seriousness or materiality of a problem increases and as the ambiguity of the related evidence decreases. Also, as potential costs such as loss of job, loss of prestige, slower promotions, and so forth, increase, the likelihood of a disassociation decreases.

Five of the above factors are controllable by the firm. If management of a firm desires to increase the likelihood of disassociation when appropriate, it should: (1) emphasize the existence of the disassociation policy, (2) publicize within the firm effective responses to the disassociation policy, (3) decrease the number of links in the process, (4) encourage a participative organizational culture, and (5) minimize the expected costs associated with disassociation.

## LIMITATIONS AND CONCLUSION

An important limitation of the study is the potential lack of generalizability because of the low response rate, although the rate is not unreasonable for a survey inquiring about a sensitive topic. The survey in essence asked the subjects to consider the likelihood that they would be "squealers" or "informers" while on an audit team. Negative associations with that role likely resulted in some subjects discarding the survey. Although our test failed to detect evidence of nonresponse bias, realistically the impact of nonresponses can not be conclusively identified.

The results reported in this paper indicate that, for the auditors responding to this survey, there is a greater willingness to agree that accounting and auditing problems exist than there is a willingness to report them. This finding suggests that disagreements by audit team members may not be documented when they occur which increases the chance that a standard audit report will be issued in inappropriate circumstances. Additionally, we find a consensus of perceptions that audit firms respond positively to the subject of a documentation of a written disassociation—that is, the problem is given thorough consideration by the audit firm. While some consensus of perceptions exist that the audit firm will respond positively to the individual auditor who documented the disassociation, there is a lack of consensus as to whether that individual would be assigned to high-profile clients in the future. Little difference in response is found by level of responsibility or size of firm. There is some evidence that certain firms may be more successful in implementing the disassociation procedures than others. Audit firm management should emphasize in training meetings the importance of complying with the disassociation procedure and the benefits to both the firm and the individual auditor of its use.

## ACKNOWLEDGMENT

The authors appreciate comments by Judy Weishar, Don Finn, and participants at the 1994 Mid-Atlantic Regional AAA meeting and the University of Arkansas accounting colloquium on an earlier version of the paper.

## NOTES

1. Note that this disassociation is from the resolution of matters considered by other members of the audit team and not a disassociation from the client because of the treatment of an accounting issue.

2. The degree to which a disassociation represents going public may vary between international multi-office firms and local single-office firms.

3. Hooks et al. (1994) developed a whistle-blowing model that focused on the effect of internal control and external audit functions on fraud detection. Finn and Lampe (1992) presented a whistle-blowing model that was used to test whether different situations would elicit different intended behaviors from auditing students and practitioners. We use the Near and Miceli (1985) model as the basis for our study because of its more general applicability.

4. This standing in the audit firm does not suggest that partners, and particularly managers, are not at risk. Both have made considerable investments in their careers that may be impaired.

5. Marquis et al. (1986) reviewed the literature on response bias and reliability in sensitive topic surveys (e.g., illegal activities, alcoholism, drug use, intimate diseases) and found that, on average, respondents do not withhold or underreport sensitive personal information and that the survey data are unbiased but noisy.

6. Firms may face a difficult task of promoting the benefits of disassociating from the decision of the audit team and at the same time cultivating a sense of teamwork and collegiality.

## REFERENCES

American Institute of Certified Public Accountants. 1995. *Professional Standards*. New York: AICPA.

Causey, D.K., Jr. 1988. The CPA's guide to whistle blowing. *The CPA Journal* (August): 27-37.

Finn, D.W., and J.C. Lampe. 1992. A study of whistleblowing among auditors. *Professional Ethics: A Multidisciplinary Journal* 1(2): 137-168.

Hall, D.H., and A.J. Renner. 1988. Lessons that auditors ignore at their own risk. *Journal of Accountancy* (July): 50-58.

Hooks, K.L., S.E. Kaplan, and J.J. Schultz. 1994. Enhancing communication to assist in fraud prevention and detection. *Auditing: A Journal of Practice & Theory* (Fall): 86-117.

Keenan, J.P., and C.A. Krueger. 1992. Whistleblowing and the professional. *Management Accounting* (August): 21-24.

Marquis, K.H., M.S. Marquis, and J.M. Polich. 1986. Response bias and reliability in sensitive topic surveys. *Journal of the American Statistical Association* 81(June): 381-389.

Miceli, M.P., and J.P. Near. 1988. Individual and situational correlates of whistle-blowing. *Personnel Psychology* 41(2): 267-282.

Near, J.P., and M.P. Miceli. 1985. Organizational dissidence: The case of whistle-blowing. *Journal of Business Ethics* 4: 1-16.

Ponemon, L.A. 1994. Whistle-blowing as an internal control mechanism: Individual and organizational considerations. *Auditing: A Journal of Practice & Theory* (Fall): 118-130.

Securities and Exchange Commission. 1985. *Accounting and Auditing Enforcement Release No. 81*. Securities Exchange Act Release No. 22686. Washington, DC: SEC.

Securities and Exchange Commission. 1993. *Accounting and Auditing Enforcement Release No. 455*. Securities Exchange Act Release No. 32505. Washington, DC: SEC.

Trevino, L.K., and B. Victor. 1992. Peer reporting of unethical behavior: A social context perspective. *Academy of Management Journal* 35(1): 38-64.

# FACTORS THAT AFFECT ETHICAL REASONING ABILITIES OF U.S. AND IRISH SMALL-FIRM ACCOUNTANCY PRACTITIONERS

Nancy Thorley Hill, Kevin Stevens, and Peter Clarke

## ABSTRACT

Several studies have used the Defining Issues Test (DIT) to examine the moral reasoning ability (MRA) of accounting students and accountants in Big Six firms and the factors that may affect MRA. Prior research examining the effect of such variables as nationality and gender on MRA report conflicting results. For example, some research indicates that U.S. accountants in Big Six firms have lower MRA than Big Six accountants in Canada and that women accountants have higher levels of MRA than their male counterparts. The current study builds on prior research by comparing the MRA of sole and small-firm practitioners of accountancy in the United States and Ireland. The study considers the effect of five variables on MRA: nationality, age, gender, social belief (i.e., liberalism, conservatism), and the frequency of ethical decision-making. The results indicate, among other things, that small-firm practitioners in the United States and Ireland have similar levels of MRA and that small-firm practitioners have generally the same levels of MRA as U.S. Big Six practitioners. As suggested in prior

**Research on Accounting Ethics, Volume 4, pages 145-165.**

**ISBN: 0-7623-0339-5**

accounting research, female accountants have relatively higher levels of moral reasoning than males regardless of country. We also discover that the more often a person makes ethical decisions, the more able that person is to make those choices. Furthermore, accountants who have liberal viewpoints have significantly higher levels of MRA than conservatives which confirms prior research. Contrary to prior research using the DIT, the results here indicate that age has a negative relationship with the MRA of U.S. and Irish accountancy practitioners.

## INTRODUCTION

In the last two decades, the public has placed greater attention on ethical issues. Public awareness of business ethics has increased, in part, due to media coverage of improprieties by business leaders in general and accountants in particular. While only a small number of accountants are involved in business improprieties, a large proportion of accountants certainly face ethical dilemmas in daily practice. Given the importance of the attest function of public accountants and the now-global nature of the public accountancy profession, it is critical that accountants in all countries have high levels of ethical reasoning abilities. Previous studies of the moral reasoning abilities (MRA) of accountants in the United States, however, have shown that U.S. accountants have lower moral reasoning skills than other professionals in the United States. Based on prior research, some argue that the accountancy profession may attract individuals with lower MRA than other professions or that the education of accountants (at least in the United States) may not place sufficient emphasis on the importance of ethical issues. On the other hand, it may be that the pressures faced by accountants in public practice erode the capacity for ethical decision making.

Most of the prior research on the MRA of U.S. accountants has focused on certified public accountants (CPAs) in Big Six firms. The work environment of these accountants is highly structured and includes a great deal of institutional support. This differs greatly from the environment of the sole- and small-firm practitioner. In addition to filling a void in prior research, this study examines small-firm CPAs for two reasons.

Small-firm practitioners provide substantial services to individuals and small businesses worldwide. In fact, far more CPAs work in sole and small-firm practices than in Big Six firms. In addition, while accountants in Big Six firms typically specialize in one area (e.g., tax), a small-firm practitioner may perform a variety of accounting functions for clients, ranging from accounting, auditing and tax services to financial planning and consulting. The ethical dilemmas these small-firm practitioners face are as diverse as the functions they perform. Yet, unlike CPAs in Big Six firms, small-firm practitioners do not typically have such resources as peer reviews, discussions with colleagues or in-house seminars to aid them in their ethical decisions. Given the important role CPAs play in business and society and the public's desire for ethical accountants, an examination of the moral reasoning abilities of small-firm practitioners seems warranted.

Also, evidence on the MRA of small-firm practitioners may further delineate the reasons suggested in prior research for the low moral reasoning found for U.S. Big Six accountants. Some have speculated that the accounting education curriculum, the practice of accounting, and/or self-selection into the field of accounting may account for the relatively lower MRA of U.S. accountants. However, if small-firm practitioners have higher MRA than U.S. Big Six accountants, then additional investigation may focus on comparisons of the practice and socialization of accountants in Big Six firms versus small firms as explanations for differing ethical reasoning abilities.

In addition, we include Irish small-firm practitioners in our analysis. Although Canadian public accountants score higher than U.S. accountants in prior studies, it is not clear if U.S. accountants have unusually low MRA or if Canadian accountants have unusually high MRA, when looking internationally at "accountants" as a professional group. An international comparison is important to the field of accounting as the technical training of accountants becomes more similar and the public perceives accountants and accounting advice as generic worldwide.[1] This study has several purposes. First, using a psychometric measure of MRA known as the Defining Issues Test (DIT), we calculate and compare the MRA of U.S. small-firm CPAs to the MRA of Big Six firm CPAs reported in prior research. Next, we include an analysis of the MRA of small-firm practitioners in the United States and in Ireland. To our knowledge, this is the first study to calculate and compare the MRA of small-firm practitioners between countries. We then examine the effect on MRA of two variables (age and gender) where prior accounting research reports conflicting results from other DIT research. For example, most DIT studies report that MRA increases with age, but some accounting DIT studies report the intriguing result that more experienced (hence, older) CPAs in Big Six firms in the United States have lower MRA than their less experienced counterparts. We consider the effect of age on MRA for small-firm practitioners.

Finally, we investigate the impact of two other variables on MRA. Research outside accounting using the DIT reports that individuals with more liberal viewpoints have higher MRA than more conservative individuals. To our knowledge, no published study has examined the effect of this variable on the MRA of accountants; therefore, we seek to extend accounting research by including a measure of social belief. Further, we introduce a new variable to the study of MRA. We seek to determine if a relationship exists between how frequently one makes moral decisions and one's ability to make those decisions. In essence, we wonder if "practice makes perfect" and speculate that the more often a person recognizes and makes ethical decisions, the better that person is at making those choices.

The following section discusses prior research in moral reasoning theory and the development of the Defining Issues Test (DIT). The third section reviews prior studies using the DIT and the impact of independent variables on MRA. The fourth section reports the methodology of the study and research propositions, and the final sections offer results, discussion, and conclusions.

## MEASURES OF MORAL REASONING ABILITY

Research in moral development theory has centered mostly on the work of Lawrence Kohlberg (1969, 1976, 1984). He posits that individuals progress through three levels of moral development, with two stages in each level. At the first level of moral reasoning, the preconventional level, individuals possess an egocentric viewpoint. Ethical behavior stems from the individual's desire to defer to authority in order to avoid punishment. Within Level I, an individual may progress to the point where he/she acts entirely in his/her own best interests while recognizing that others may have a different point of view. The conventional level of moral reasoning, Level II, includes individuals who desire to be considered a good person. People at this level are concerned about others and will act for the welfare of the group with which they identify. Last, individuals at the third level of moral development, the postconventional level, seek to act in accordance with what society believes to be good. In addition, they also believe that fundamental values and rights of the individual exist that supersede society's interests.

Kohlberg and Kramer (1969) assert that progression through these stages is invariant; that is, no individual ever skips a stage or "slips back" to an earlier development level even though progression may cease at any stage. Further, they claim that these stages are the same for all individuals and for all cultures. This strong claim has been at least partially supported by several studies.[2] Following Kohlberg, James Rest (1979) developed the Defining Issues Test (DIT). Researchers in accounting and other disciplines have used the DIT both in the United States and abroad to assess the moral reasoning ability of students, professionals, and adults in general.[3] Essentially, the DIT provides scenarios of moral dilemmas and asks subjects to consider and rank the importance of several questions on the dilemmas. For example, one scenario addresses the issue of whether a relatively poor individual ought to steal medicine for his dying spouse. Twelve possible considerations are listed, and subjects are asked to rank their importance in resolving the dilemma. From these rankings, a Principled score (P score) is calculated.[4] Ranking considerations that show concern for others or reflect a belief in basic fundamental and self-determined values as most important indicates a high level of moral reasoning ability and results in high P scores. In contrast, when one ranks more lower-level considerations (those that focus on narrow self-interest and the avoidance of punishment) as the most important, one necessarily ranks fewer higher-level considerations which leads to low P scores.

## PRIOR RESEARCH

Prior research suggests that the accountancy profession in the United States tends to be rule-oriented, and accountants often see their job as providing definitive answers to questions (Gaa 1992; Ponemon 1992). However, the professional accountant's job also requires the ability to solve diverse and open-ended problems in unstructured settings (Arthur Andersen & Co. et al. 1989). Accountants also have the responsibility, as professionals, to possess and exercise moral as well as technical expertise (Gaa and Ponemon 1993). Arnold and Ponemon (1991) concur and suggest that the level of ethical reasoning ability may be an important factor in the professional judgment to disclose sensitive information.

Although the importance of sound ethical reasoning and expertise is well documented, prior research using the DIT P score as a measure of MRA indicates that U.S. accountants, in general, have lower MRA than would be expected, given age, education, and experience (Armstrong 1987; Ponemon 1988, 1992; Ponemon and Gabhart 1993). However, this result does not hold for accountants in Canada, where both chartered accountants (CAs) and certified management accountants (CMAs) score higher than their U.S. counterparts (Ponemon and Gabhart 1993; Etherington and Schulting 1995; Etherington and Hill 1998).

This study examines the MRA of small-firm accountants in the United States and Ireland and investigates the impact of five variables on MRA: (1) nationality, (2) age, (3) gender, (4) social belief—that is, liberalism/conservatism, and (5) frequency of ethical decision making. Discussions of each of these variables follow.

## Nationality/Culture and MRA

Hofstede (1980) asserts that different societies put different weights on a variety of social behaviors, including ethical decisions. He further argues (1991) that some cultures (e.g., Japan) seek to avoid uncertainty and conflict by closely adhering to the form of rules and regulations (i.e., follow the letter of the law). In contrast, individuals from cultures more tolerant of ambiguity and dissent (e.g., the United States) are more likely to condone behavior, even if outside the letter of the law, if such behavior reflects the spirit of the law (i.e., substance over form). Cohen et al. (1993) discuss the impact of cultural norms on firm-specific ethical values and rules and cite a need for further empirical research examining the impact of ethical issues across cultures.

The internationalization of accounting makes a comparative study of the MRA of accountants across countries important. As businesses and accounting firms expand transnationally, it becomes more likely that in-house accountants and auditors will come from differing nations. Thus, a comparison of MRA of Irish and U.S. accountants must consider both similarities and differences in such factors as professional certification and training. For example, the requirements to be a certified/chartered accountant vary across countries, which may affect technical skill and MRA. Arguably, both providers and users of accounting information have an interest in the qualifications of accountants from different countries. Professional examinations and designations (e.g., the title CPA or CA) provide assurance on the relative technical knowledge and ability of accountants. Surely, providers and users of accounting information also have a strong interest in whether accountants from varying countries have similar notions of what constitutes ethical behavior. However, differences in training and codes of ethics may lead to differing definitions of what is considered to be ethical behavior. For example, U.S. auditors are held to a higher standard of independence than are auditors in the United Kingdom or Germany (Frost and Ramin 1996). The DIT provides a standardized measure of MRA across nations, at least across Western nations.[5]

The DIT has been used in studies conducted in many countries. Rest's (1986) analysis of 20 cross-cultural studies using the DIT suggests that Western cultures report similar results when examining the impact of variables on MRA (e.g., MRA usually increases with age). Few studies focusing on accountants and MRA, however, have been conducted outside the United States. Ponemon and Gabhart (1993) compare Canadian and U.S. accountants from national public accounting firms. Their analysis of U.S. CPAs and Canadian CAs shows, surprisingly, that Canadian CAs have statistically significant higher MRA (P score mean = 44.2) than U.S. CPAs (P score mean = 40.0). Also, while the Canadian public accountants' DIT P scores are similar to those of college students, the U.S. CPAs scored statistically lower than college students.[6] Etherington and Schulting's (1995) examination of Canadian CMAs also finds P scores (mean = 43.5) not significantly different from P scores for college students. In contrast, a recent survey of U.S. CMAs (Etherington and Hill 1998) reports lower-than-expected levels of MRA (P score mean = 39.3).

## Age and MRA

Rest (1986) reports that age is a significant explanatory variable for MRA: as individuals age, their moral reasoning abilities generally increase. Furthermore, Rest reports that in five of six cross-cultural studies, age has a positive effect on MRA. Yet, accounting researchers report that more experienced accountants (e.g., partners) in U.S. Big Six firms (who are presumably older) have lower MRA than the more inexperienced (presumably younger) staff accountants (Ponemon 1988, 1992; Ponemon and Gabhart 1990; Shaub 1994). In studies of Canadian Big Six CAs and Canadian CMAs, upper-level management does not have lower MRA than their subordinates (Ponemon and Gabhart 1993; Etherington and Schulting 1995).

## Gender and MRA

Several researchers have provided empirical support for the validity of Kohlberg's theory of etnical behavior and gender (Nunner-Winkler 1984; Lifton 1985). However, some researchers (Betz et al. 1989; Harris 1990; Cohn 1991; Whipple and Swords 1992) find empirical support for the notion that women are socialized to promote harmonious behavior and, therefore, evaluate ethical assessments of behavior differently than men. Gilligan (1982) argues that gender differences affect ethical perceptions but that Kohlberg's theory, because it places justice as the central tenet of ethics, is biased in favor of males. She asserts that the theory—and, therefore, the DIT derived from the theory—are flawed because women place greater emphasis on other values such as caring and responsibility than do men, which biases the DIT results against women. Interestingly, Thoma (cited in Rest 1986) and Moon (cited in Rest 1986) report that the DIT itself slightly *favors* women but that the gender differences are trivial in effect.

Studies outside of accounting generally find no significant relationship between gender and MRA. For example, Rest (1986) reports that gender differences contributed, on average, less than one percent of the variance in P scores across 56 DIT studies. However, St. Pierre et al. (1990) and Shaub (1994) find that U.S. female accounting students have significantly higher moral reasoning abilities than male accounting students. Furthermore, Shaub (1994) finds that female accountants in the United States have significantly higher moral reasoning abilities than male accountants. Etherington and Schulting (1995) report similar results for Canadian accountants.

## Social Belief and MRA

In general, P scores seem to be correlated with political, social and religious attitudes or beliefs. Prior research indicates that more "liberal" individuals have higher moral reasoning levels than more "conservative" individuals.[7] Given the internationalization of accounting, a question of interest is whether this result holds true both in the United States and abroad. Some have suggested that the public accountancy profession in the United States attracts more conservative individuals than in other countries (Ponemon and Glazer 1990). In addition, Ponemon and Gabhart (1993) and Lampe and Finn (1992) suggest that the accountancy profession may attract individuals with a strong rule-orientation. If rule-orientation and conservatism are related, the relatively low P scores of American accountants and accounting students reported in some studies may be partially explained.

## Frequency of Ethical Decision Making and MRA

Rest (1979) argues that it would be naive to point to any one life experience and conclude that engaging in "x" behavior necessarily leads to "y" change in MRA. Rather, MRA increases with general social development. This means that a host of life experiences (e.g., working in a stimulating environment, becoming involved in one's community, taking an interest in large societal issues, and so on) develop MRA. Nevertheless, some specific interventions do affect MRA. In particular, moral education programs that emphasize discussion of ethical dilemmas do have a definite, albeit modest, positive impact on MRA (Rest 1986).

Gaa and Ponemon (1993) observe that resolution of ethical conflicts is not accomplished by consulting some technical rule, but by acquiring and "exercising" professional, ethical judgment. We posit that if discussing ethical dilemmas helps increase MRA, then so also may the socialization experience of actually making ethical decisions. We argue that one form of general social development is engaging in repeated acts of ethical decision making. Furthermore, we argue that the complex and changing nature of the work and accounting environment is likely to provide accountants involved in small-firm practices with a myriad of ethical dilemmas to be resolved. Thus, perhaps those accountants who recognize that they do make ethical decisions frequently will have higher MRA than those who do not believe they face such dilemmas. Consider two reasons why a small-firm practitioner may believe he or she never or rarely makes ethical decisions: (1) The accountant may be correctly assessing the situation and may indeed rarely make ethical decisions. If so, we suggest that the lack of opportunities for ethical decision making may affect the accountant's ability to make such decisions. Lack of experience may lead to lower MRA. Or (2), as we believe to be more likely the case, the accountant incorrectly perceives the situation and is confronted with ethical dilemmas but simply does not discern the moral implications as they occur in real life situations. This lack of recognition also may lead to lower MRA. We argue that individuals who cannot recognize ethical dilemmas as such are less able to resolve them.

## RESEARCH METHODOLOGY

### Sample and Survey Instrument

We administered the three scenario version of the DIT[8] together with a short questionnaire to a random sample of 1,092 CPAs in sole and small-firm practices throughout the United States and to a random sample of 669 CAs in sole and small-firm practices in Ireland. We received 175 responses from the U.S. accountants (a response rate of approximately 16%) and 151 responses from the Irish accountants (a response rate of approximately 23%). Due to the nature of the ethics test and additional survey questions and to ensure complete anonymity, we did not code or identify the surveys. This precluded sending a second request for participation.[9] Scoring of the DIT identifies those respondents who have unacceptably high inconsistencies in their rankings and who have selected too many "meaningless" responses as important. For analysis purposes, we eliminated the inconsistent and meaningless scores, leaving 131 U.S. and 110 Irish respondents in the analysis.

### Research Propositions

The first purpose of this study is to measure and compare the moral reasoning levels of U.S. small-firm CPAs to Big Six CPAs. As noted above, prior research has focused on CPAs in Big Six firms. However, sole and small-firm practitioners are arguably a much different population from Big Six accountants, regardless of country. Accountants who work in tightly structured, hierarchical Big Six firms have a wealth of decision-aid resources available to them. In particular, an accountant in a large public accounting firm can turn to his or her professional colleagues for guidance in making ethical decisions, participate in in-house training, and rely on advice from departments which specialize in resolving ethical issues. In contrast, small-firm practitioners work alone or with only a few colleagues and often must grapple with ethical quandaries on their own. Without help and direction in resolving ethical issues, small-firm practitioners may have lower MRA than Big Six practitioners.

On the other hand, researchers have suggested that accountants with higher levels of moral reasoning self-select out of public accounting (Ponemon and Gabhart 1993) or fail to be promoted in large public accounting firms (McNair 1991). Additionally, it is suggested that those at the top tend to promote people like themselves (Chatman 1991; Maupin and Lehman 1994; McNair 1991; Ponemon and Gabhart 1993). Big Six firms typically have lock-step retention and promotion policies which may make conformity important. Since small-firm practitioners do not face the same promotion or retention barriers, we may find that these (more independent) CPAs have higher MRA than Big Six practitioners. Proposition 1a examines this relationship.

**Proposition 1a.** The moral reasoning levels of U.S. small-firm CPAs are different from the moral reasoning levels of U.S. Big Six firm CPAs.

We also calculate and compare the MRA of small-firm practitioners in the United States versus Ireland. Cohen et al. (1993) state that Anglo-American societies are especially individualistic, which should lead to similar types of ethical reasoning across Anglo-American cultures. Yet, as noted above, prior research in accounting reports the anomalous result that U.S. accountants in Big Six firms have lower MRA than Canadian accountants in Big Six firms. Most aspects of the accounting environment in the United States and Ireland are similar since Ireland is part of the Anglo-American accountancy tradition (Hanlon 1994). Yet, we do find subtle differences. For example, both the Irish and the U.S. small-firm practitioners in the study are chartered/certified. Thus, both groups have received professional training and undergone rigorous examination. However, to be a CPA, one must have a college degree, and continuing professional education (CPE) is required to retain one's license. In contrast, a college degree is not a prerequisite to be a CA,[10] and continuing professional development is suggested but optional. In order to sit for the final admitting exam (the equivalent of the CPA exam) in Ireland, most candidates must have a minimum of three years experience. In contrast, U.S. students can take the CPA exam as they finish college. Further, the CPA exam has become increasingly more "objective"—that is, it tests candidates largely via multiple-choice and matching questions. In contrast, the final admitting exam (FAE) poses interdisciplinary case studies. Finally, the CPA exam tests candidates on their knowledge of ethics, albeit narrowly defined. In general, it examines ethics (in the Business Law and Professional Responsibilities part of the exam) by posing multiple-choice questions on actions that focus on the Code of Ethics (e.g., would a certain action be a violation of the Code). However, the FAE emphasis on ethics is relatively insignificant.

In both countries, accountants subscribe to Codes of Conduct which proscribe various unethical behaviors and include conflict of interest rules. Further, professional organizations have the power to punish unethical behavior by levying fines and revoking licenses. Also, in Ireland a commission was established to examine the state of financial reporting and auditing within the country in the light of major business scandals and the loss of confidence of both the business community and the public in the accountancy profession (Financial Reporting Commission 1992). Finally, publicly traded firms in either country must have an independent audit committee which oversees matters relating to the financial affairs of the company.

If the workplace and certification practices tend to shape moral reasoning ability, one would expect that the similarities in professional certification and training should lead to similar MRA for Irish and U.S. accountants. Yet, differences in MRA have been shown between U.S. and Canadian Big Six accountants. We wish to determine if this phenomenon repeats itself when the sample includes American and Irish small-firm practitioners. That is, perhaps variables other than training and certification exist that transcend culture and nationality. We test this through Proposition 1b.

**Proposition 1b.** The moral reasoning levels of U.S. small-firm CPAs are different from those of Irish small-firm CAs.

The second purpose of this study is to determine the effect of several variables on DIT P scores. First, we investigate the impact of age on moral reasoning levels. As discussed above, many studies using the DIT report that older individuals have, all else being equal, higher MRA than younger people. However, studies in accounting find that MRA decreases with experience for U.S., but not Canadian, public accountants. Thus, U.S. accountants do not seem to mirror the experiences of other professionals and, perhaps, other accountants. By examining the impact of age on MRA for U.S. and Irish small-firm practitioners, we will be better able to determine if a negative relationship between age and MRA exists for accountants (in general) that does not exist for other professionals or other non-American accountants.

**Proposition 2.** As age increases, moral reasoning levels for small-firm CPAs and CAs also increase.

The next explanatory variable we consider is gender. We examine whether males and females have similar levels of moral reasoning ability. As noted above, gender does not seem to impact MRA for most populations. Yet, some accounting research has found that female accounting practitioners and students have higher MRA than males. Proposition 3 builds on the accounting literature and posits that this is also true in the small-practice workplace and that the gender difference appears in Ireland as well.

**Proposition 3.** The moral reasoning levels of female small-firm practitioners are higher than those of male small-firm practitioners.

As discussed, prior research finds a relationship between social belief and moral reasoning ability and that, in general, individuals with "liberal" viewpoints have higher MRA than "conservatives."[11] Many have argued that the accountancy profession attracts conservative individuals. If so, this may help explain the relatively (compared to other professions) low levels of MRA reported in some studies of U.S. accountants. The proposition to be tested follows:

**Proposition 4.** The moral reasoning levels of CPAs and CAs who rate themselves as liberal are higher than those CPAs and CAs who rate themselves as conservative.

Last, we consider the effect of making ethical decisions on MRA. We hypothesize that the more often one makes ethical decisions, the higher one's MRA becomes. To assess this, we asked the small-firm practitioners how often (on a five-point scale ranging from never through often) they made ethical decisions; we then compared their answers to their P scores. Proposition 5 states our assertion.

**Proposition 5.** CPAs and CAs who report resolving ethical dilemmas frequently have higher moral reasoning levels than CPAs and CAs who do not report resolving ethical dilemmas frequently.

## RESULTS

Table 1 reports the P score means for the respondents. As compared to prior studies of Big Six CPAs' P scores, we posited that the U.S. small-firm CPAs would have different P scores. We find no statistical support for Proposition 1a; there is no difference between the MRA of Big Six and small-firm CPAs. Table 1 shows that the small-firm CPAs have P scores similar to those reported on average in prior research on U.S. Big Six CPAs (36.6 versus 38.8) and are lower than expected for professionals with a college education. Surprisingly, the mean P score for Irish small-firm CAs is not statistically different from P scores for their U.S. counterparts (Proposition 1b). Irish small-firm practitioners appear to have MRA of the same low levels as U.S. small-firm practitioners (34.8 versus 36.6, respectively).

***Table 1.*** DIT P Score Means (std. dev.)

| | *Mean P Scores* | | *(Std. Dev.)* | *N* |
|---|---|---|---|---|
| ***By Country*** | | | | |
| United States | 36.6 | | (15.3) | 131 |
| Ireland | 34.8 | | (15.9) | 110 |
| ***Reported in Prior Research*** | | | | |
| U.S. Big six CPAs | | | | |
| Ponemon and Gabhart (1993) | 40.0 | | (10.1) | 133 |
| Armstrong (1987) | 38.5 | | (15.1) | 119 |
| Ponemon (1992) | 38.1 | | ( 8.1) | 180 |
| Average | 38.8 | * | (10.6) | 432 |
| **Canadian Big Six CAs** [1] | 44.2 | + * | (11.3) | 102 |
| **U.S. CMAs** [2] | 39.3 | * | (16.5) | 468 |
| **Canadian CMAs** [3] | 3.5 | + * | (15.7) | 76 |

***Notes:*** [1]From Ponemon and Gabhart (1993).
[2]From Etherington and Hill (1998).
[3]From Etherington and Schulting (1995).
+Significantly different from U.S. mean (36.6) at $p < 0.01$.
* Significantly different from Irish mean (34.8) at $p < 0.01$.

We next investigate the impact that the other variables have on moral reasoning ability for each country. Table 2 reports P score means for four independent variables by country. While age has been found in prior general population studies to have a positive impact on MRA, accounting research does not confirm the positive effect. Table 2 shows that for both the U.S. and Irish small-firm practitioners, as age increases, levels of moral reasoning *decrease*. For the U.S. respondents, mean P scores for the younger group are 38.4 versus only 31.6 for the older group. Similarly, the Irish respondents report mean P scores of 37.2 for the younger group and 29.2 for the older group. Therefore, we find no support for Proposition 2.

Proposition 3 focuses on the effect of gender on moral reasoning ability. While general research using the DIT finds no gender impact, accounting studies have found that females score higher on the DIT. Our findings concur. As shown in Table 2, women have higher P scores in both countries (44.6 and 47.8 for U.S. and Irish women, respectively, versus 34.2 and 34.5 for U.S. and Irish men, respectively). The only statistically significant difference, however, is for the U.S. respondents. The Irish respondents only included three women; however, we note that all three scored above the sample mean.[12] We also find support for the measure of social belief for the Irish respondents. Although this is a self-evaluated measure, respondents who rated themselves as liberal scored statistically significantly higher than those who rated themselves as conservative (39.7 versus 29.2, respectively). While statistical significance is not found for the U.S. respondents,

***Table 2.*** Mean P Scores (std. dev.)

| | *United States* | *Irish* |
|---|---|---|
| Age | | |
| Ages 23-40 | 38.3 (12.9) *n* = 29 (22%)[1] | 37.2 (16.8) *n* = 49 (45%)[2] |
| Ages 41-50 | 39.2 (16.1) *n* = 60 (46%)[1] | 34.4 (14.5) *n* = 44 (40%) |
| Ages 51-74 | 31.6 (14.6) *n* = 42 (32%) | 29.2 (16.1) *n* = 17 (15%) |
| | *n* = 131 (100%) | *n* = 110 (100%) |
| Gender | | |
| Men | 34.2 (14.1) *n* = 101 (77%)[3] | 34.5 (15.9) *n* = 107 (97%) |
| Women | 44.6 (16.6) *n* = 30 (23%) | 47.8 (6.9) *n* = 3 (3%) |
| | *n* = 131 (100%) | *n* = 110 (100%) |
| Social belief | | |
| Liberal | 41.7 (19.5) *n* = 24 (18%) | 39.7 (14.4) *n* = 35 (33%)[4] |
| Moderate | 36.4 (10.9) *n* = 27 (21%) | 36.7 (16.4) *n* = 29 (27%)[4] |
| Conservative | 35.1 (15.0) *n* = 80 (61%) | 29.2 (15.7) *n* = 43 (41%) |
| | *n* = 131 (100%) | *n* = 107 (100%) |
| Frequency of involvement in ethical dilemmas | | |
| Never or seldom | 34.2 (15.5) *n* = 39 (30%)[5] | 32.6 (14.8) *n* = 46 (42%)[6] |
| Sometimes | 36.0 (15.3) *n* = 60 (46%) | 33.3 (15.7) *n* = 51 (46%)[6] |
| Frequently or often | 41.1 (14.3) *n* = 31 (24%) | 48.8 (14.3) *n* = 13 (12%) |
| | *n* = 130 (100%) | *n* = 110 (100%) |

***Notes:*** [1]Ages 23-40 and 41-50 are significantly different from ages 51-74 at $p < 0.05$.
[2]Ages 23-40 are significantly different from ages 51-74 at $p = 0.09$.
[3]Men are significantly different from women at $p < 0.01$
[4]Significant difference between liberal and conservative and between moderate and conservative at $p < 0.05$.
[5]Significant difference between never and frequently at $p = 0.06$.
[6]Significant difference between never and between sometimes and frequently at $p < 0.01$.

the P score means decrease as the classification goes from liberal to conservative. We also note that 61% of the U.S. respondents classified themselves as conservative and only 18% as liberal. The small number of self-described liberals in the U.S. sample may contribute to the lack of significance for this variable. For the Irish respondents, we find a more equal distribution by classification, with liberals comprising 33% of the sample and conservatives representing 41%.

Finally, we report the results of MRA and the frequency of resolving ethical dilemmas. We find as noted in Proposition 5 that as the number of times one is involved in ethical dilemmas increases, the level of MRA also increases. Table 2 shows that the U.S. respondents' mean P scores increase from 34.2 to 41.1 and Irish respondents' mean P scores increase from 32.6 to 48.8 as the number of ethical dilemmas faced goes from "never or seldom" to "frequently or often." It appears that these respondents learn from or improve their moral reasoning skills as they are faced with more dilemmas in the workplace.

Table 3 reports a multivariate regression where the dependent variable is the P score and the independent variables are age, gender, social belief, and frequency of experience in resolving ethical dilemmas. A multivariate regression analysis is useful for three reasons. First, the regression determines the estimated coefficient for each independent variable while holding the other independent variables constant. Second, the multivariate coefficients show the incremental effect of a change in an independent variable on the dependent variable, the P score.[13] Third, multiple-regression models allow one to determine both how much of the variance is explained (through the $R^2$statistic) and the significance of the overall model (through the $F$-statistic).

The multivariate analysis shows that for the American respondents the P score varies with age, gender, and frequency of resolving ethical dilemmas. The gender coefficient for the U.S. model (10.27) suggests that women's scores are 10 points

***Table 3.*** Regresssion Analysis

| | *U.S. Respondents* | *Irish Respondents* |
|---|---|---|
| *Variables* | *Coefficient (Std. Dev)* | *Coefficient (Std. Dev)* |
| Age | | |
| Age (41-50) | 2.18 (3.25) | −2.21 (3.02) |
| Age (51-74) | −5.92 (3.45)* | −7.78 (4.07)* |
| Gender | | |
| Females | 10.27 (3.04)*** | 13.12 (8.68) |
| Social beliefs | | |
| Liberal | 5.23 (3.35) | 10.31 (3.21)*** |
| Moderate | 0.94 (3.31) | 8.38 (3.42)** |
| Involvement in ethical dilemmas in practice | | |
| Somewhat | 3.29 (3.00) | 1.09 (2.96) |
| Frequently or often | 7.52 (3.52)** | 18.19 (4.54)*** |
| Intercept | 30.68 (3.59)*** | 28.41 (3.17)*** |
| Adjusted $R^2$ | 0.14 | 0.19 |

***Notes:*** * $p < 0.10$.
** $p < 0.05$.
*** $p < 0.01$.

The baseline or omitted variables include:
Age group: 23-40
Gender: Males
Social belief Conservative
Involvement in ethical dilemmas in practice: Never or seldom

higher than men, on average, showing the strong impact of gender on MRA. Further, the oldest group of accountants in the United States have, on average, P scores that are approximately six points lower (–5.92) than the youngest group indicating that MRA declines with age. Finally, the coefficient for frequently or often resolving ethical dilemmas is 7.52, which may indicate that exposure to moral situations in practice does lead to higher MRA.

For the Irish model, the coefficients for liberal (10.31) and moderate (8.38) social beliefs and for being frequently or often involved in ethical dilemmas (18.19) all have positive, strong, and statistically significant impacts on P scores. The coefficient for age for the older group of Irish accountants (–7.78), as the Americans, is negative and statistically significant. The extremely high coefficient for resolving ethical dilemmas shows that experience and practice increase P scores dramatically.

## DISCUSSION

Our proposition that U.S. small-firm CPAs have different P scores than those in Big Six firms is not supported by the data. In addition, these data indicate that as age increases, MRA declines for small-firm practitioners (consistent with DIT studies of U.S. Big Six accountants). Although prior research suggests that accountants with higher MRA either self-select out of Big Six practices or fail to be promoted, it appears that this alone does not explain the relatively lower P scores for Big Six accountants as compared to college students. Small-firm practitioners are not subject to the same promotion barriers but also have lower MRA as compared to college students. In addition, the diverse and more independent nature of the practice of accountancy in these small firms does not, by itself, lead to higher MRA. Perhaps the accounting education curriculum or the rule-oriented practice of accounting limits ethical growth or development of accountants.

The average P score for Irish small-firm accountants is somewhat lower than would be expected given prior DIT results for individuals of similar age and education. We also find, as for the Americans, that as age increases, moral reasoning levels decline. This is in direct contrast to research in moral reasoning levels for other professions and disciplines. Again, the reason may lie in the technically based accounting education curriculum, where, for example, the management of earnings is treated as a technical rather than an ethical issue. It could also be due to self-selection into the field of accounting and/or the rule-oriented nature of the practice of accountancy. Curiously, as noted earlier, these results are not found to be true for Canadian accountants, which suggests that the practice of accounting and/or the business climate in Canada does not limit or constrain ethical development. More specifically, Davidson and Dalby's (1993) study of Canadian public accountants suggests that Canadian CAs are more apt to disregard rules than the average population.

While prior research outside of accounting does not find any difference in moral reasoning ability for gender, many accounting studies (this one included) reveal that a gender difference exists for accounting professionals. The strong result we find for the gender variable is interesting for the following reason. The statistically significant difference between male and female mean P scores is not due to higher than average P scores for women,[14] but, rather, to lower than average P scores for men. This finding has important implications. Since, in general, the accountancy profession has a greater percentage of men than women (as shown in descriptive statistics for prior research samples and in this sample of accountants), the average P scores for accountants in general would necessarily be lower than the average for college students.[15] Perhaps as more women enter the field of accounting, the average MRA of accountants will increase. This does not, however, change the fact that male accountants reason at a lower level than women, at least as measured by the DIT. Further research may investigate whether male accountants, in general, are more rule-oriented and, therefore, have lower MRA.

As the globalization of business and the practice of accounting expands, further research investigating the link between conservatism, rule-orientation, and MRA is necessary for all cultures. While liberal and conservative social beliefs have been investigated in research outside of accounting, few have specifically examined the measure empirically as it relates to the DIT. We find strong support for this variable for the Irish respondents. Liberal individuals score statistically higher than conservatives on the DIT. We also find directional support for the U.S. respondents. Those that rated themselves as conservative scored lower on the DIT. Given that conservatives, in general, have lower MRA than liberals and that the majority of U.S. practitioners rated themselves as conservative, the overall MRA for the sample is negatively affected. Unlike prior research, which examined college students and adults, this sample of U.S. small-firm practitioners is more homogenous in social belief which may explain, in part, the lower MRA we find for U.S. accountants.

Finally, we find that as the number of ethical dilemmas faced by the respondents increases, the level of moral reasoning ability also greatly increases, particularly for the Irish. Since these small-firm CPAs and CAs typically do not have in-house training and senior partners on which to rely, they must often solve ethical dilemmas on their own. As the number of ethical dilemmas increases, it appears that practitioners gain from past experience in wrestling with various issues, as suggested by the higher levels of MRA.

We also speculated earlier about the information revealed by those small practitioner accountants who indicate they have "never or seldom" been involved in ethical dilemmas. First, if it is true that these practitioners do not often encounter ethical dilemmas, they cannot benefit from "practice" in resolving ethical dilemmas. When an ethical conflict does occur, they may be unable to deal with it effectively. Second, while it is possible that some small-firm accountants are seldom involved in ethical conflicts, it seems unlikely that approximately one-third of the U.S. respondents and 40% of the Irish respondents have rarely been involved in ethical dilemmas. Rather, we suggest that those who do not recognize dilemmas as ethical questions also erroneously view complex issues as either black or white and easily resolvable by rules. In either case, continuing education programs and professional societies have an important role to play in providing support and training in ethical decision making for small-firm CPAs and CAs.

## CONCLUSIONS

This study relies on the DIT. Therefore, the results reported are only as valid as the DIT itself. Further, the study may suffer because the ethical "decisions" are contrived and are not necessarily what a person would do if confronted with a real-world ethical dilemma. Finally, we emphasize that the DIT attempts to measure the moral reasoning ability of an individual. It cannot, and does not, measure how ethical that person is. However, it should be noted that the benefit of using the DIT is that it provides a measure of MRA that is widely acknowledged to be valid and reliable. It also allows for comparisons across samples.

The results reported here indicate that it is not only Big Six accountants in the United States who have lower-than-expected levels of MRA but also small-firm practitioners. Furthermore, small-firm accountants in Ireland also operate at low levels of moral reasoning, which may reflect similarities in training and certification between Ireland and the United States.

We also note the counter-intuitive result that MRA decreases with age. Gender is strongly associated with MRA, and female accountants have much higher levels of MRA than their male counterparts. Views on social policies also matter, as liberals have higher MRA than conservatives. Finally, experience in resolving ethical dilemmas seems to increase moral reasoning abilities.

These results may indicate that individuals within the accountancy profession in both the United States and Ireland are not as able to resolve ethical dilemmas as they ought to be. However, there may be ways to alleviate the problem. First, accounting should attract and retain females in the profession.[16] Second, accounting programs and employers need to attract individuals with more liberal viewpoints. For example, schools of accounting could place greater emphasis on programs that provide career change opportunities to individuals with liberal arts degrees, and accounting firms could increase their recruitment of individuals from a wider range of academic backgrounds. Finally, practitioner-focused journals have played a positive role in disseminating information about and increasing the awareness of ethics. Given that, in general, the more often a person resolves ethical dilemmas, the better that person becomes at the process, it seems appropriate for professional societies and institutes to continue to emphasize ethics and provide training in ethical decision making

## NOTES

1. For example, the international accounting firm Arthur Andersen periodically brings accountants from around the world to a central location in the United States to provide standardized training.

2. See Snarey (1985) for a review of this research.

3. See Rest (1990) for a manual describing how the DIT works, how to administer it, and how the scores are developed.

4. The P score can range from zero to 95. A score of zero indicates that all answers were considerations at the lower four stages (Levels 1 and 2); a score of 95 indicates all responses were stage five and six considerations (Level 3).

5. One barrier to using the DIT cross-nationally is correctly translating the DIT from English to other languages. For example, Moon reports (cited in Rest 1986) that differences in Korean versions of DIT instruments affect P scores. Rest concludes (1986) that it is uncertain whether differences in P scores between Western and non-Western countries are due to cultural differences or lack of fluency in English. However, various versions of the English language DIT (e.g., adapted for Australian populations) show high internal consistency, indicating the reliability of the DIT.

6. Rest (1979) reports average P scores for 2,479 college students of 42.3. Since this average includes all college years, college graduates would likely score higher, on average, than 42.3. Rest also reports a P score average of 53.3 for 183 graduate students.

7. See Rest (1986) for a complete review of DIT studies on liberal/conservative attitudes (pp. 148-161).

8. Rest (1990) provides six moral reasoning vignettes. We chose the three-story version of the DIT to make the DIT results comparable across countries. For example, one excluded vignette focused on the Vietnam war, which is arguably less relevant to an Irish audience than a U.S one. We also slightly adapted the wording of some vignettes in order to employ commonly used terms. For example, we substituted the word "pharmacist" for "druggist." Rest reports consistent and reliable DIT results when using a DIT instrument that is only slightly different than the original DIT instrument. Rest (1990) reports that scores obtained from the three story version of the DIT closely correlate (96%) with scores obtained from the longer version.

9. Given that completion of the DIT takes between 20 to 40 minutes and that only one request for participation was made, we were satisfied with the response rate. This rate of response is considered to be relatively high by those who survey this type of audience. For example, Korn/Ferry (a well-known institutional investor survey firm) obtained a 16.1% response rate to a survey of outside directors, a rate described by Korn/Ferry as excellent and as evidence that the survey was meaningful to the audience (1993).

10. In recent years, it has become more common for charted accountants to hold undergraduate degrees, and now over 90% of those sitting for the final admitting exam have a degree.

11. This does not imply that liberals are necessarily more ethical than conservatives.

12. While only 3% of the sample small-firm practitioners in Ireland are women, women in all accounting positions (industry, small-firm practices, and Big Six firms) comprise about 15% of the population. However, an increase in the hiring of women has been noted. For example, in 1988, 31% of the newly hired graduates were women (Barrett and Granleese, 1990). Still, the career paths of women accountants in Ireland are not as favorable as those of males (Barker and Monks, 1995).

13. The formal assumptions of multivariate regression do not include causality. Justification for a "causal" interpretation of regression coefficients must be based on theory external to the regression (Berry 1993).

14. The mean P scores for women in this sample (44.6 for the U.S. and 47.8 for the Irish) are not statistically different from the mean P score of college students (42.3) (Rest, 1979).

15. We assume the average college student sample is made up equally of men and women.

16. See Pillsbury, Capozzoli, and Ciampa (1989) for a discussion of alternatives for attracting and retaining women in the accountancy profession in the United States. For an Irish perspective, see Barker and Monks (1995) on the progression of women in accounting in Ireland.

## REFERENCES

Armstrong, M. 1987. Moral development and accounting education. *Journal of Accounting Education* (Spring): 27-43.

Arnold, D.F., Sr., and L.A. Ponemon. 1991. Internal auditors' perceptions of whistle-blowing and the influence of moral reasoning: An experiment. *Auditing: A Journal of Practice & Theory* 10(2): 1-15.

Barker, P., and K. Monks. 1995. Women in accounting: Career progression. *The Irish Accounting Review* (Spring): 1-27.

Barrett, T.F., and J. Granleese. 1990. A comparative analysis of the social characteristics, Job satisfaction and personality of ACA's in three organisational settings. *Irish Accounting Association Proceedings of the Annual Conference*, 91-107. Dublin: The Irish Accounting and Finance Association.

Berry, W.D. 1993. *Understanding Regression Assumptions*. Newbury Park, CA: Sage.

Betz, M., L. O'Connell, and J.M. Shepard. 1989. Gender difference in proclivity for unethical behavior. *Journal of Business Ethics* 8: 321-324.

Chatman, J.A. 1991. Matching people and organizations: Selection and socialization in public accounting firms. *Administrative Science Quarterly* 36: 459-494.

Cohen, J.R., L.W. Pant, and D.J. Sharp. 1993. Culture-based ethical conflicts confronting multinational accounting firms. *Accounting Horizons* 73(September): 1-13.

Cohn, C. 1991. Chiefs or Indians—Women in accountancy. *Australian Accountant* (December): 20-30.

Davidson, R.A., and J.T. Dalby. 1993. Personality profiles of Canadian public accountants. *International Journal of Selection and Assessment* (April): 107-116.

Etherington, L.D., and N.T. Hill. 1998. Ethical development of CMAs: National and international comparison. In *Research on Accounting Ethics,* Vol. 4, ed. L.A. Ponemon. Greenwich, CT: JAI Press.

Etherington, L.D., and L. Schulting. 1995. Ethical development of management accountants: The case of Canadian CMAs. In *Research on Accounting Ethics,* Vol. 1, ed. L.A. Ponemon, 237-253. Greenwich, CT: JAI Press.

Financial Reporting Commission. 1992. *Report of the Commission of Inquiry into the Expectations of Users of Published Financial Statements*. Dublin: The Institute of Chartered Accountants in Ireland.

Frost, C.A., and K.P. Ramin. 1996. International auditing differences. *Journal of Accountancy* 181(4): 62-68.

Gaa, J. 1992. Discussion of a model of auditors' ethical decision processes. *Auditing: A Journal of Practice and Theory* 11(Supplement): 60-66.

Gaa, J., and L. Ponemon. 1993. Toward a theory of moral expertise: A verbal protocol study of public accounting professionals. Working paper, McMaster University and State University of New York at Binghamton.

Gilligan, C. 1982. *In a Different Voice*. Cambridge, MA: Harvard University Press.

Hanlon. G. 1994. *The Commercialisation of Accountancy: Flexible Accumulation and the Transformation of the Service Class*. Basingstoke, UK: Macmillan Press.

Harris, J.R. 1990. Ethical values of individuals at different levels in the organizational hierarchy of a single firm. *Journal of Business Ethics* 9: 741-750.

Hofstede, G. 1980. *Culture's Consequences*. Beverly Hills, CA: Sage.

Hofstede, G. 1991. *Cultures and Organizations: Software of the Mind.* New York: McGraw Hill.

Kohlberg, L. 1969. Stage and sequence: The cognitive developmental approach to socialization. In *Handbook of Socialization Theory and Research,* ed. D. Goslin, 347-480. Chicago, IL: Rand McNally.

Kohlberg, L. 1976. Moral stages and moralization: The cognitive-developmental approach. In *Moral Development and Behavior,* ed. T. Likona, 31-53. New York: Holt, Rinehart and Winston.

Kohlberg, L. 1984. *The Psychology of Moral Development.* San Francisco, CA: Harper and Rowe.

Kohlberg, L., and R. Kramer. 1969. Cognitive-developmental theory and the practice of collective moral education. In *Group Care: An Israeli Approach,* eds. M. Wolins and M.Gottesman, 93-120. New York: Gordon & Breach.

Korn/Ferry Organizational Consulting. 1993. *Reinventing Corporate Governance: Directors Prepare for the 21st Century.* Boston, MA: Korn-Ferry.

Lampe, J.C., and D.W. Finn. 1992. A model of auditors' ethical processes. *Auditing: A Journal of Practice and Theory* 11(Supplement): 33-59.

Lifton, P.D. 1985. Individual differences in moral development: The relation of sex, gender, and personality to morality. *Journal of Personality* 53(2): 306-334.

Maupin, R.J., and C.R. Lehman. 1994. Talking heads: Stereotypes, status, sex-roles and satisfaction of female and male auditors. *Accounting, Organizations and Society* 19: 427-437.

McNair, C.J. 1991. Proper compromises: The management control dilemma in public accounting and its impact on auditor behavior. *Accounting, Organizations and Society* 16: 635-653.

Nunner-Winkler, G. 1984. Two moralities? A critical discussion of an ethic of care and responsibility vs. the ethic of rights and Justice. In *Morality, Moral Behavior and Moral Development,* eds. W.M.Kurtines and T.A. Gewirtz, 348-361. New York: Wiley.

Arthur Andersen & Co., Arthur Young, Coopers Lybrand, Deloitte Haskins & Sells, Ernst & Whinney, Peat Marwick Main & Co., Price Waterhouse, and Touche Ross. 1989. *Perspectives on Accounting Education: Capabilities for Success in the Accounting Profession.* New York: Arthur Andersen & Co. et al.

Pillsbury, C., L. Capozzoli, and A. Ciampa. 1989. A synthesis of research studies regarding the upward mobility of women in public accounting. *Accounting Horizons* (March): 63-70.

Ponemon, L.A. 1988. A cognitive-developmental approach to the analysis of certified public accountants' ethical judgments. Unpublished Ph.D. dissertation, Union College, Schenectady, NY.

Ponemon, L.A. 1992. Ethical reasoning and selection-socialization in accounting. *Accounting, Organizations and Society* 17(3/4): 239-258.

Ponemon, L.A., and D. Gabhart. 1990. Auditor independence judgments: A cognitive developmental model and experimental evidence. *Contemporary Accounting Research* (Fall): 227-251.

Ponemon, L.A., and D. Gabhart. 1993. *Ethical Reasoning in Accounting and Auditing.* Vancouver, Canada: CGA Canada Research Foundation.

Ponemon, L.A., and A. Glazer. 1990. Accounting education and ethical development: The influence of liberal learning on students and alumni in accounting practice. *Issues in Accounting Education* (Fall): 195-208.

Rest, J. 1979. *Development in Judging Moral Issues.* Minneapolis, MN: The University of Minnesota Press.

Rest, J. 1986. *Moral Development: Advances in Theory and Research.* New York: Praeger.

Rest, J. 1990. *Manual for the Defining Issues Test.* Minneapolis, MN: The University of Minnesota Press.

Shaub, M.K. 1994. An analysis of the association of traditional demographic variables with the moral reasoning of auditing students and auditors. *Journal of Accounting Education* 12(1): 1-26.

Snarey, J.R. 1985. Cross-cultural universality of social-moral development: A critical review of Kohlbergian research. *Psychological Bulletin* 97(2): 202-232.

St. Pierre, K., E. Nelson, and A. Gabbin. 1990. A study of the ethical development of accounting majors in relation to other business and nonbusiness disciplines. *The Accounting Educators' Journal* (Summer): 23-35.

Whipple, T.W., and D.F. Swords. 1992. Business ethics judgments: A cross-cultural comparison. *Journal of Business Ethics* 11: 671-678.

# THE LOSS OF AUDITOR INDEPENDENCE:
## PERCEPTIONS OF STAFF AUDITORS, AUDIT SENIORS, AND AUDIT MANAGERS

Terry J. Engle and Terry L. Sincich


## ABSTRACT

This paper presents the results of a research project that focused upon the degree to which auditors were perceived to be violating Rule 101 of the AICPA's *Code of Professional Conduct*. Data were gathered by questionnaire survey that was mailed to 2,000 randomly selected AICPA members who indicated to the AICPA that they were currently in public accounting, practicing in the capacity of an auditor, with the title of staff auditor, senior, or manager. The questionnaire recipients were presented with 15 activities that are violations of Rule 101 and were instructed to provide their best estimate of the frequency with which each of the ethical violations had occurred at their present firm during the previous year. The paper presents overall descriptive statistics and the results of a variety of statistical contrasts based upon the size of the public accounting firm employing the respondents, and the rank of the respondents


**Research on Accounting Ethics, Volume 4, pages 167-184.**

**ISBN: 0-7623-0339-5**

(i.e., staff auditor, senior, or manager). Overall, the perceived occurrence rate for a majority of the 15 independence-related ethical violations was small but problem areas were revealed. The auditors most commonly believed that the amount of professional fees obtained from audit clients, and management advisory services being provided to audit clients, were inappropriately influencing audit judgments. Other salient findings included a positive relationship between the size of public accounting firms and the percentage of auditors perceiving ethical violations and a pattern where audit seniors were most likely to believe that independence-related ethical violations were occurring at their public accounting firm. The paper also includes recommendations as to how the findings of this study can be effectively utilized by future researchers to help improve compliance with Rule 101 of the AICPA's *Code of Professional Conduct.*

## INTRODUCTION

Auditor independence has been justly recognized as an essential prerequisite to an effective external audit. This notion was best articulated by Mautz and Sharaf (1961, 204) when they noted, "The significance of independence in the work of the independent auditor is so well established that little justification is needed to establish this concept as one of the cornerstones in any structure of auditing theory."

In an effort to maintain an acceptable level of auditor independence, the American Institute of Certified Public Accountants (AICPA) has promulgated a *Code of Professional Conduct* (1992) that contains detailed independence-related guidance for its membership. This guidance is also looked to by the judicial system and other societal institutions as they judge the professional behavior of public accountants acting in the capacity of an external auditor.

While leaders of the public accounting profession in general, and the AICPA in particular, seek to promote auditor independence, it is quite likely that several factors are inhibiting their efforts. One of these factors is the complexity of the independence regulations in the AICPA's *Code of Professional Conduct.* The subject of independence is addressed in the *Code of Professional Conduct* through Rule 101 and its related interpretations and rulings. Rule 101 states: "A member in public practice shall be independent in the performance of professional services as required by standards promulgated by bodies designated by Council." While Rule 101 is relatively straightforward, and deceptively simple, the operational complexity of the independence-related guidance is in the interpretations and rulings that were promulgated to help implement Rule 101. For example, Interpretations 101-1 and 101-9 contain requirements relating to the employment status and financial interests of auditors, their immediate families, unrelated dependents, and nondependent close relatives. These requirements vary depending upon the public accountants job specialty (e.g., auditor, consultant, tax specialist), hierarchical position within the public accounting firm (e.g., staff auditor, senior, manager), timeframe when the activity occurred (e.g., "During the period of a professional engagement or at the time of expressing an opinion . . ."), and type of financial transaction (e.g., direct or indirect, material or immaterial). While all of this detailed guidance is designed to promote independence, the dynamic nature and complexity of the independence regulations introduces the real possibility that individual public accountants are finding it very difficult to remain cognizant of all of the detailed rules. This lack of awareness introduces the possibility of unintentional violations.

In addition to the complexity of the independence-related rules contained within the AICPA's *Code of Professional Conduct*, evidence exists that other factors may be adversely affecting the independence of auditors. For example, researchers have produced evidence indicating that auditor independence can be adversely affected by a highly competitive public accounting environment (e.g., Shockley 1981). In addition, the phenomenon of public accounting firms providing both auditing and consulting services has also been suspected as detrimental to auditor independence (e.g., Commission on Auditors' Responsibilities 1978; Shockley 1981; Schulte 1965). The risk of litigation and the threat of losing an audit client may be negative influences as well. For example, Farmer, Rittenberg, and Trompeter (1987) conducted an experiment involving auditors from seven of the (then) Big Eight public accounting firms and accounting students. The results indicated that litigation and the potential loss of a client may be influencing auditor independence. The results indicated that auditors "agreed with the client most often when the risk of client loss was high and the risk of litigation was low" (Farmer et al. 1987, 8). Further evidence that independence problems exist were revealed with a study by Pearson (1987). The researcher gathered evidence by surveying 250 CPAs working for the (then) Big Eight and 250 CPAs working for non-Big Eight firms. The questionnaire recipients were individual practitioners, partners or shareholders/officers of public accounting firms. An analysis of the recipients responses led the researcher to conclude "that there are a number of CPAs who perceive independence deficiencies in the U.S. audit practice, and there are a number of CPAs who admit to personal independence deficiencies" (Pearson 1987, 286).

The public accounting profession has recognized that individual public accounting firms must have an effective quality control system in place to mitigate the many threats to auditor independence. Over the years, the AICPA has provided guidance in this area primarily through the efforts of the AICPA Quality Control Standards Committee. This committee was a senior technical committee of the AICPA that was charged with the responsibility of issuing Statements on Quality Control Standards. Statement on Quality Control Standards No. 1, issued in 1979, contained a description of nine elements that a firm should consider in establishing its quality control system. One of those elements was titled "Independence." Note that the AICPA's Auditing Standards Board (ASB) is currently responsible for promulgating the Statements on Quality Control Standards. In 1996, the ASB reduced the nine elements that were contained in Statement on Quality Control Standards No. 1 to five elements. One of the five elements addresses independence.

The AICPA Quality Control Standards Committee also issued a guide titled: *Quality Control Policies and Procedures for CPA Firms: Establishing Quality Control Policies and Procedures.* This guide contains some specific independence-related policies and control procedures that were intended to be helpful to all public accounting firms in their efforts to insure their adherence to the independence requirements of the AICPA. The recommended policies and procedures are not mandatory or comprehensive, but "members of the AICPA and member firms of the division for CPA firms should be aware that they may be called upon to justify departures from the guide" (AICPA Quality Control Standards Committee 1980).

The existence of all of this quality control guidance is incongruent with the evidence that tends to indicate that significant problems exist with auditor independence. One explanation for this phenomenon is that the independence-related quality control procedures that public accounting firms are implementing, based upon AICPA guidance, are not working at an acceptable level.

The importance of auditor independence to the success of the external auditing function mandates further inquiry to separate conjecture from reality on this issue. Evidence is needed to understand the nature and magnitude of the problem and to formulate effective solutions if necessary.

The remainder of this paper presents the findings of a research project that should provide important insights into this important issue. The project was specifically designed to assess the magnitude and characteristics of existing independence problems as perceived by a large national sample of public accountants acting in the capacity of an auditor.

## RESEARCH METHOD

### Questionnaire Development

Data were gathered by questionnaire survey. The questionnaire recipients were presented with 15 activities that are violations of Rule 101 of the AICPA's *Code of Professional Conduct.* The list of unethical activities was intentionally designed to include some of the relatively simple, as well as the some of the more complex, activities that have been deemed unethical under Rule 101. For each of the activities, the recipients were requested to provide their best estimate of the frequency with which that activity had occurred at their present public accounting firm during the previous year. The recipients were told that their estimates were to be based upon their firsthand personal observations as well as their understanding of their firm's policies and the behavior of the firm's professional employees.

The questionnaire also contained 12 procedures that are specifically designed to promote adherence to the independence requirements of the AICPA. These procedures are adaptations of the recommended independence-related quality control procedures that were promulgated by the AICPA Quality Control Standards Committee in *Quality Control Policies and Procedures for CPA Firms—Establishing Quality Control Policies and Procedures* (1980). For each of the 12 procedures, the individuals were asked to indicate whether they personally were required to comply with the procedure while working in the capacity of an auditor for their current employer during the previous year.[1] Survey recipients were also asked to provide demographic information about themselves and their firm.[2,3]

## Sample Selection

The sampling procedure utilized the membership listing of the AICPA as of June, 1991. The surveys were mailed to 2,000 randomly selected individuals who were members of the AICPA who had indicated to the AICPA that they were currently working in public accounting with the title of staff, senior or manager, and that the their main interest was auditing. A total of 897 completed instruments were returned, resulting in an overall response rate of 45%. To validate the accuracy of the AICPA's demographic information about the questionnaire recipients, the recipients were asked to provide information about their employment status and employment history in the demographic section of the questionnaire. The researchers utilized the 746 questionnaires in which the respondents indicated that they were currently classified as an auditor, had a job description of staff accountant, senior, or manager (or equivalent), and had been employed by their current employer for a minimum of one year. The resulting usable response rate was 37%.[4]

The researchers restricted the analyses to staff accountants, seniors and, managers (or the equivalent) in the desire to exclusively focus on the experiences and perceptions of auditors in the field who are "in the trenches" on a day-to-day basis. The actual experiences and perceptions of these individuals may be quite different from the partners' (or their equivalent) view of reality. The analyses only incorporated responses from individuals who stated that they have been employed by their present employer for at least one year to increase the likelihood of a reasonably informed response.

## Data Analysis

The data analyses included both descriptive and statistical analytical procedures. The descriptive analyses focused on the overall rate at which the perceived violations of the AICPA's ethics Rule 101 occurred in the sample. In addition, the analyses concentrated on the breakdowns in the perceived rates of violations by firm size and rank variables.

In addition to the descriptive statistical analyses, statistical tests of independence were utilized to determine whether the perceived violation rate is significantly related to either the size of the public accounting firm employing the auditor or the rank of the auditor. For each of the 15 unethical activities (violations), a Pearson chi-square analysis for a two-way contingency table was employed to test the following null hypotheses:

**Hypothesis 1.** The rate at which the violation is perceived to have occurred is independent of the number of professionals employed by the public accounting firm.

**Hypothesis 2.** The rate at which the violation is perceived to have occurred is independent of the rank of the auditor.

For violations where statistical significance was detected (i.e., where the null hypothesis was rejected), an *a posteriori* analysis was conducted by constructing simultaneous confidence intervals on the difference between two violation rates for all possible pairs of firm size or rank categories. These simultaneous confidence intervals were constructed to provide insight into the magnitude of the percentage of auditors who perceived violations across firm size and rank categories.

## RESULTS

Overall, the respondents indicated that the proportion of auditing activity where violations of the AICPA's ethics Rule 101 occurred is relatively low, but significant violations were believed to exist. This phenomenon is evident from the data presented in Table 1.

For each of the 15 violations of ethics Rule 101, Table 1 provides a trichotomous categorization of the percentage of respondents who indicated that a particular violation did not occur within their firm during the previous year, the percentage who indicated that the violation occurred between 1% and 10% of the time, and the percentage of respondents who indicated that the violation occurred more than 10% of the time. Table 1 (as well as the other tables in this paper) contains brief summaries of each of the 15 specific independence-related ethical violations. The complete descriptions that were provided to the respondents are listed in Appendix A.

The data clearly shows that the occurrence rate for most of the violations is relatively small. For 13 of the 15 violations, 90% or more of the respondents indicated that the prohibited activity was not occurring at all within their firm. When violations were reported, the respondents typically indicated that it was occurring 1%-10% of the time. It must be noted that an accurate interpretation of this finding must be made in the appropriate environmental context. Any violation of the

**Table 1.** Frequency of Ethical Violations

| Violation Number | Description | *Percentage of Respondents* n | 0% | 1-10% | > 10% |
|---|---|---|---|---|---|
| V1 | Fees Inappropriately Influencing Judgments | 744 | 76.3 | 19.0 | 4.7 |
| V2 | Financial Interests Inappropriately Influencing Audit Judgments | 743 | 95.6 | 4.3 | 0.1 |
| V3 | Relatives or Dependents Associated with Clients Inappropriately Influencing Audit Judgments | 743 | 96.0 | 3.7 | 0.3 |
| V4 | MAS Inappropriately Influencing Audit Judgments | 737 | 87.7 | 9.5 | 2.8 |
| V5 | Auditor with Financial Interest | 742 | 92.6 | 6.6 | 0.8 |
| V6 | Spouse or Dependent with Financial Interest | 739 | 90.6 | 8.0 | 1.4 |
| V7 | Nondependent with Material, Known, Financial Interest | 735 | 91.0 | 7.6 | 1.4 |
| V8 | Prohibited Loans | 741 | 98.0 | 2.0 | 0.0 |
| V9 | Estate or Trust with Financial Interest in Audit Client | 736 | 95.3 | 4.3 | 0.4 |
| V10 | Auditor with Prohibited Loan from Audit Client | 741 | 95.7 | 3.8 | 0.5 |
| V11 | Spouse or Dependent with "Significant Influence" or in an "Audit Sensitive" Position | 742 | 94.6 | 5.1 | 0.3 |
| V12 | Auditor is a Promoter, Underwriter, Employee, etc. | 743 | 97.6 | 2.2 | 0.3 |
| V13 | Auditor is a Trustee for Pension or Profit-Sharing Trust | 742 | 97.6 | 2.4 | 0.0 |
| V14 | Nondependent with "Significant Influence" or in an "Audit Sensitive Position" | 742 | 93.8 | 5.5 | 0.7 |
| V15 | Partner or Manager's Nondependent Relative with "Significant Influence" | 741 | 90.9 | 8.6 | 0.5 |

independence-related ethical standards is cause for concern due to the fact that the maintenance of auditor independence is essential to an effective auditing function.

The respondents indicated that the most commonly occurring prohibited activities were Violations 1, 4, 5, 6, 7, and 15. Focusing on the phenomenon of professional fees that are obtained from a client inappropriately influencing audit judgments (Violation 1), the respondents indicated that this was the most common of the 15 prohibited activities that were investigated. Nineteen percent of the respondents indicated that this occurred 1%-10% of the time at their firm during the previous year, while and another 5% of the responding auditors thought that phenomenon had occurred over 10% of the time.

Turning to Violation 4, 12% of the auditors reported some instances where management advisory services (MAS) that were being provided to audit clients had

**Table 2.** Analyses by Firm Size

| Violation Number | | Percent Indicating that Some Violations Occurred: 2-9 Firms (n = 110) | 10-99 Firms (n = 261) | 100> Firms (n = 95) | Big 6 Firms (n = 280) |
|---|---|---|---|---|---|
| V1 | Fees Inappropriately Influencing Judgments | 18.2 | 21.1 | 22.1 | 28.8* |
| V2 | Financial Interests Inappropriately Influencing Audit Judgments | 1.8 | 2.3 | 5.3 | 7.2** |
| V3 | Relatives or Dependents Associated with Clients Inappropriately Influencing Audit Judgments | 0.9 | 3.1 | 4.2 | 6.1* |
| V4 | MAS Inappropriately Influencing Audit Judgments | 11.9 | 8.1 | 12.8 | 16.4** |
| V5 | Auditor with Financial Interest | 1.8 | 4.6 | 6.3 | 12.6*** |
| V6 | Spouse or Dependent with Financial Interest | 0 | 5.0 | 8.4 | 17.4*** |
| V7 | Nondependent with Material, Known, Financial Interest | 0 | 6.0 | 9.5 | 16.1*** |
| V8 | Prohibited Loans | 0.9 | 0.4 | 2.1 | 4.0*** |
| V9 | Estate or Trust with Financial Interest in Audit Client | 1.8 | 4.3 | 7.5 | 5.4 |
| V10 | Auditor with Prohibited Loan from Audit Client | 0 | 0.8 | 4.2 | 9.4*** |
| V11 | Spouse or Dependent with "Significant Influence" or in an "Audit Sensitive" Position | 0 | 1.5 | 9.5 | 9.8*** |
| V12 | Auditor is a Promoter, Underwriter, Employee, etc. | 0 | 1.5 | 4.2 | 3.6** |
| V13 | Auditor is a Trustee for Pension or Profit-Sharing Trust | 0.9 | 2.3 | 4.2 | 2.5 |
| V14 | Nondependent with "Significant Influence" or in an "Audit Sensitive Position" | 0.9 | 3.8 | 10.5 | 9.1*** |
| V15 | Partner or Manager's Nondependent Relative with "Significant Influence" | 0 | 5.0 | 9.5 | 16.7*** |

***Notes:*** *P- value for chi square less than .10.
**P-value for chi square less than .05.
***P-value for chi square less than .01.

inappropriately influenced audit judgments regarding those same clients. The data relating to Violations 5, 6, and 7 revealed that 7% of the respondents thought there were instances where there were prohibited financial holdings by auditors themselves, and 9% thought that there were prohibited financial holdings by the auditors' spouse, dependents, or nondependent close relatives. Focusing on Violation 15, 9% of the auditors reported that there were instances where proprietors, partners, shareholders, or managers violated the AICPA's independence rule because of the combination of their organizational position and the fact that they had a nondependent close relative who could exert "significant influence" over an audit client's operating, financial, or accounting policies.

To glean a more comprehensive understanding of the extent that the independence-related ethical violations are occurring, contrasts were made based upon the size of the respondent's public accounting firm and the rank (e.g., staff auditor, senior, manager) of the respondent.

## Analyses Based Upon Firm Size

Table 2 contains the firm size comparisons. The firms were grouped into the following four categories based upon the number of professional employees within the organization: Those with 2-9 professional employees (2-9 Firms); Those employing 10-99 professional employees (10-99 Firms); Non-Big Six firms employing 100 or more professional employees (100> Firms); and the Big Six firms. For each of the 15 ethical violations, Table 2 presents the percentage of respondents who indicated that the particular ethical violation had occurred, to some extent, at their public accounting firm during the previous year. An analysis of the descriptive statistics revealed a positive relationship between the size of the public accounting firm and the amount of perceived independence-related ethical violations. This relationship is displayed with nine of the 15 ethical violations that were investigated. It should be noted that while this positive relationship exists, perceived ethical violations by Big Six auditors were much more common than the number of perceived violations by auditors from the other three categories of firms. Focusing on the responses from Big Six auditors, 10% or more of the respondents thought that six of the 15 ethical violations had occurred at their firm during the previous year. The most common phenomena was the amount of professional fees inappropriately influencing audit judgments (Violation 1), MAS inappropriately influencing audit judgments (Violation 4), professional employees (and/or their spouses or dependents) improperly having direct or material indirect financial interests in audit clients (Violations 5 and 6), nondependent close relatives of professional employees assigned to particular audits having material financial interests in the audit clients and the professional employees being aware of the relationships (Violation 7), and a partner, proprietor, shareholder, or manager who was not working on a particular audit but was in the office performing a significant portion of the audit, having a nondependent close relative who could exert "significant influence" over the client's operating, financial, or accounting policies (Violation 15).

The chi-square test was used to test null Hypothesis 1 for each of the 15 ethical violations. Statistically significant differences, at the .10 level, were present for 13 of the 15 violations. Thus, the differences isolated with the descriptive statistical analyses were supported.

Table 3 presents the results of the *a posteriori* analysis that was conducted to isolate the specific sources of the overall statistical significance. The table displays

**Table 3.** Bonferroni Pairwise Comparisons Based on Firm Size

| Violation Number | Description | Contrast | Intervals of Difference in Percentages of Violations* LCL | UCL |
|---|---|---|---|---|
| V2 | Financial Interests Inappropriately Influencing Audit Judgments | 2-9—Big 6 | −10.6 | −0.1 |
| | | 10-99—Big 6 | −9.6 | −0.1 |
| V4 | MAS Inappropriately Influencing Audit Judgments | 10-99—Big 6 | −15.6 | −0.9 |
| V5 | Auditor with Financial Interest | 2-9—Big 6 | −17.0 | −4.5 |
| | | 10-99—Big 6 | −14.7 | −1.7 |
| V6 | Spouse or Dependent with Financial Interest | 2-9—10-99 | −8.6 | −1.4 |
| | | 2-9—100> | −15.9 | −0.9 |
| | | 2-9—Big 6 | −23.3 | −11.4 |
| | | 10-99—Big 6 | −19.5 | −5.4 |
| V7 | Nondependent with Material, Known, Financial Interest | 2-9—10-99 | −8.6 | −1.4 |
| | | 2-9—100> | −17.4 | −1.5 |
| | | 2-9—Big 6 | −22.0 | −10.2 |
| | | 10-99—Big 6 | −18.0 | −4.2 |
| V8 | Prohibited Loans | 10-99—Big 6 | −6.9 | −0.3 |
| V10 | Auditor with Prohibited Loan from Audit Client | 2-9—Big 6 | −14.0 | −4.8 |
| | | 10-99—Big 6 | −13.5 | −3.8 |
| V11 | Spouse or Dependent with "Significant Influence" or in an "Audit Sensitive" Position | 2-9—100> | −17.4 | −1.5 |
| | | 2-9—Big 6 | −14.5 | −5.1 |
| | | 10-99—Big 6 | −13.3 | −3.1 |
| V12 | Auditor is a Promoter, Underwriter, Employee, etc. | 2-9—Big 6 | −6.5 | −0.6 |
| V14 | Nondependent with "Significant Influence" or in an "Audit Sensitive Position" | 2-9—100> | −18.3 | −1.0 |
| | | 2-9—Big 6 | −13.3 | −3.0 |
| V15 | Partner or Manager's Nondependent Relative with "Significant Influence" | 2-9—10-99 | −8.5 | −1.4 |
| | | 2-9—100> | −17.4 | −1.5 |
| | | 2-9—Big 6 | −22.6 | −10.7 |
| | | 10-99—Big 6 | −18.6 | −4.8 |

***Note:*** *Only the intervals with significant difference between percentages (at the overall .05 level) are shown.

the comparisons that yielded statistically significant differences.[5] For each of the 15 ethical violations, simultaneous confidence intervals on the difference between two percentages were constructed for all possible pairs of firm size categories. To protect against an inflated Type I error rate, the overall alpha risk was set at the .05 level for each violation using the familiar Bonferroni adjustment.

**Table 4.** Analyses by Rank

| Violation Number | Description | *Percent Indicating Some Violations Occurred* Staff (n = 33) | Senior (n = 187) | Manager (n = 525) |
|---|---|---|---|---|
| V1 | Fees Inappropriately Influencing Judgments | 12.1 | 28.9 | 22.6* |
| V2 | Financial Interests Inappropriately Influencing Audit Judgments | 6.1 | 6.5 | 3.6 |
| V3 | Relatives or Dependents Associated with Clients Inappropriately Influencing Audit Judgments | 3.0 | 5.4 | 3.6 |
| V4 | MAS Inappropriately Influencing Audit Judgments | 9.1 | 12.9 | 12.4 |
| V5 | Auditor with Financial Interest | 6.2 | 10.3 | 6.5 |
| V6 | Spouse or Dependent with Financial Interest | 3.1 | 9.2 | 9.8 |
| V7 | Nondependent with Material, Known, Financial Interest | 3.1 | 8.7 | 9.4 |
| V8 | Prohibited Loans | 3.1 | 3.2 | 1.5 |
| V9 | Estate or Trust with Financial Interest in Audit Client | 3.2 | 5.5 | 4.6 |
| V10 | Auditor with Prohibited Loan from Audit Client | 0 | 6.5 | 3.8 |
| V11 | Spouse or Dependent with "Significant Influence" or in an "Audit Sensitive" Position | 0 | 4.3 | 6.1 |
| V12 | Auditor is a Promoter, Underwriter, Employee, etc. | 0 | 4.8 | 1.7** |
| V13 | Auditor is a Trustee for Pension or Profit-Sharing Trust | 3.0 | 3.2 | 2.1 |
| V14 | Nondependent with "Significant Influence" or in an "Audit Sensitive Position" | 3.0 | 8.0 | 5.6 |
| V15 | Partner or Manager's Nondependent Relative with "Significant Influence" | 3.0 | 8.6 | 9.8 |

***Notes:*** *P-value for chi- square less than .10.
**P-value for chi- square less than .05.

The *a posteriori* analysis confirmed the results of the descriptive data analysis in revealing that the vast majority of the statistically significant differences were caused by a larger percentage of Big Six auditors reporting the existence of unethical independence-related activity at their firms. In addition, the observation of an overall positive relationship between firm size and the percentage of auditors with the perception that unethical activity was occurring was also confirmed with the *a posteriori* analysis. All of the contrasts that proved statistically significant involved a greater percentage of auditors from the larger of the two firm categories (that were compared) believing that ethical violations had occurred at their firm.

### Analyses Based Upon Rank

The responses from individual auditors possessing the title of staff auditor, senior, or manager (or equivalent) were contrasted and the results are presented in Table 4. The analysis of the descriptive statistics revealed that the perceptions of staff auditors, seniors, and managers were relatively consistent, but two patterns should be noted. Seniors tended to perceive more ethical violations when compared with the perceptions of auditors from the other rank categories and staff auditors tended to exhibit the lowest level of perceived ethical violations. The phenomenon for staff auditors is evident with 12 of the 15 violations, while the phenomenon for seniors is present with 11 of the 15 violations.

An overall analytical review across ranks revealed that the most common unethical activities were the amount of professional fees inappropriately influencing audit judgments (Violation 1), MAS inappropriately influencing audit judgments (Violation 4), and professional employees (or their dependents or spouses) improperly having direct, or material indirect, financial interests in audit clients (Violations 5 and 6).

Two additional observations from the analysis of the descriptive data are noteworthy. First, while the percentage of staff auditors with the perception of independence-related wrongdoing is generally lower than the percentage of seniors and managers with the same belief, significant numbers of staff auditors are of the opinion that certain ethical problems exist. Secondly, a significant number of audit seniors and managers reported the existence of Violations 7, 14, and 15 that deal with the auditors' nondependent close relatives. These nondependent relatives either had material financial interests in clients or were connected to audit clients in such a way that they could exert "significant influence" over the clients' operating, financial or accounting policies or possess the status of an employee in a position that could be considered "audit sensitive."

The relative consistency in the perceptions of the staff auditors, seniors, and managers was supported by the chi-square tests of hypothesis H2. Statistically significant differences, at the .10 level, were present with only two of the 15 ethical violations.

Table 5 presents the results of the *a posteriori* analysis for auditor rank that utilizes the Bonferroni simultaneous confidence intervals. The table includes contrasts that revealed statistically significant differences. It should be noted that in two of the three rank comparisons where statistically significant differences are present, a larger percentage of seniors perceived ethical wrongdoing.

## SUMMARY, CONCLUSIONS, AND FUTURE RESEARCH

The results of this research project contain both good and bad news for members of the public accounting profession. The good news is that the analyses revealed

**Table 5.** Bonferroni Pairwise Comparisons Based on Auditor-Rank

| | | | *Intervals of Differences in Percentages of Violations** | |
|---|---|---|---|---|
| *Violation Number* | *Description* | *Contrast* | *LCL* | *UCL* |
| V1 | Fees Inappropriately Influencing Audit Judgments | Staff—Senior | −32.5 | −1.0 |
| V12 | Auditor is a Promoter, Underwriter, Employee, etc. | Staff—Senior | −8.6 | −1.1 |
| | | Staff—Manager | −3.1 | −0.4 |

***Note:*** *Only the intervals with a significant difference between percentages (at the overall .05 level) are shown.

that a large majority of practicing auditors do not perceive that the professional employees at their public accounting firms participated in a majority of the prohibited independence-related activities that were investigated in this study. The bad news is that the data analysis also revealed the public accounting profession's record is perceived to be far from perfect.

While all of the perceived ethical violations that were discovered should be of concern, the most problematic area relates to the perception that auditors are jeopardizing their independence by engaging in activities that hold the potential for financial gain. This broad problematic area can be subdivided, first, into perceptions about what may be viewed as the "commercial topics" of audit fees and the providing of MAS to audit clients, and second, into perceptions about the jeopardizing of auditor independence through involvement in prohibited financial transactions.

While both subdivisions of behavior are significant, the data analyses indicated that problems with the "commercial topics" are far more common. For example, out of the 15 ethical violations that were focused upon in this study, the most common perception (possessed by 24% of the respondents) is that the amount of professional fees obtained from a client is inappropriately influencing audit judgments pertaining to that client. The second most common perception (held by 12% of the respondents) is that the providing of MAS to audit clients is inappropriately influencing audit-related judgments. The contrasts by firm size and by rank of respondent supported the overall number one and two rankings of the audit fee and MAS issues. These findings lend support to the well established hypotheses that the size of audit fees, and the providing of MAS to audit clients, can lead to the loss of auditor independence. Further insight into the seriousness of the problem, and the auditor mind-set that may be contributing to it, can be obtained by reflecting on a comment that one respondent placed on his/her questionnaire. The auditor, who identified him/herself as an audit senior working for a 100> firm, noted: "All the rules and regulations in the world will never achieve independence, due to the obvious—we get paid directly by our audit clients. That's the bottom line! The rest is pure (B_ _ _ S_ _ _)! Don't let anybody ever tell you anything different. Remember, business is business!"

The second salient subdivision of unethical activity involves prohibited financial transactions by auditors, or their spouses, dependents, or nondependent close relatives. For example, it was discovered that 7% of the auditors believed that there were instances within the previous year where professional employees within their firm had prohibited direct, or material indirect, financial interests in audit clients, and 9% believed that there were instances where the spouses or dependents of these professional employees held these prohibited investments. The firm size and rank comparisons generally confirmed these findings.

Added insights can be obtained by analyzing the firm size and rank comparisons further. Focusing on the firm size contrasts, the descriptive data revealed two overall patterns. First, a positive relationship existed between the size of the firm and the percentage of auditors who perceived independence-related violations. The phenomenon existed for nine of the 15 violations that were investigated in this study. Second, the perception of independence violations is much more prevalent among auditors at Big Six firms. The level of perceived independence violations at Big Six firms was a definite outlier in comparison to the perceived violations in the other three categories of firms.

The statistical analyses supported the observation that the magnitude of perceived independence problems varied significantly based upon the size of the public accounting firm to which the auditor belonged. The initial testing involving the chi-square procedure produced statistically significant differences, at the .10 level or below, with 13 of the 15 violations. The a posteriori analyses, which attempted to isolate specific sources of the overall statistical significance, produced results that supported the initial finding of a positive relationship between independence violations and firm size. In addition, the a posteriori analyses confirmed the fact that significantly more problems were being reported by auditors from the Big Six firms. These findings are surprising because larger firms (particularly, the Big Six) have significantly more resources to implement formal quality control programs and they face more severe scrutiny due to extensive involvement with SEC clients.

The contrasts based upon the rank of the auditor revealed relative consistency, but the descriptive analysis indicated that auditors with the rank of senior tended to more commonly believe that independence-related violations were occurring at their firm. In addition, staff auditors tended to perceive the lowest level of ethical difficulty.

Collectively, the findings from this research clearly reveal that there are a significant number of auditors who are currently in the practice of public accounting who believe that violations of Rule 101 of the AICPA's *Code of Professional Conduct* have occurred in the very recent past. This disturbing reality exists in spite of the fact that the AICPA, many individual public accounting firms, and numerous other prestigious members of the accounting profession, have promulgated detailed independence-related quality control guidance to promote auditor independence. While the majority of practicing auditors believe that independence-related ethical violations are not occurring at their firms, the amount of perceived violations that was indicated by this study should be the cause of significant concern. The importance of auditor independence to the external auditing function, and the severe societal penalties for the loss of auditor independence, justify a very low materiality threshold for noncompliance with Rule 101 of the AICPA's *Code of Professional Conduct*.

Future researchers should build upon this study in an attempt to mitigate the problems that exist with auditor independence. While this research project focused on the extent to which specific types of prohibited activities are occurring, future research should focus on the underlying reasons for the ethical lapses and controls that should be implemented to effectively deal with the problems. This important point can be clarified with a specific example. This research project has revealed that an unacceptably large number of auditors believed that the amount of fees obtained from audit clients, as well as the providing of MAS to audit clients, were inappropriately influencing audit judgments in certain instances (thereby compromising auditor independence). To productively use this finding, it must be realized that the obtaining of audit fees from audit clients and the providing of lucrative MAS services are a fact of life that is unlikely to change without drastic modifications to the commercial environment in which public accounting firms currently operate. Since most members of the accounting profession believe that public accounting firms should remain free to operate within a competitive free enterprise system, public accounting firms must effectively address the specific threats that this commercial environment presents to auditor independence.

Future researchers should attempt to understand the demographic profiles and underlying motivations of individuals who compromise their independence because of the size of audit fees or the amount of MAS being provided to audit clients. Researchers should also test the effectiveness of some of the independence-related quality control procedures that are currently recommended by the AICPA. Armed with that knowledge, researchers could then develop and test additional, more innovative, quality control procedures that were designed to mitigate these specific "commercial" threats to auditor independence.

In conclusion, it must be emphasized that the current study focused upon the perceptions of auditors and those perceptions can differ from reality. While this point is valid, it does not negate the importance of the project's findings. The perceptions of a large number of practicing auditors regarding the subject of auditor independence are of extreme importance due to the fact that the public accounting profession has the obligation to maintain auditor independence both in fact, and in appearance.

# APPENDIX A: ETHICAL VIOLATIONS

1. The amount of professional fees obtained from a client inappropriately influenced audit judgements pertaining to that client.

2. Financial interests (e.g., investments, loans) involving auditors and a client (or officers of the client) inappropriately influenced audit judgements pertaining to the client.

3. An auditor's audit-related judgements pertaining to a client were inappropriately influenced by the fact that relatives or dependents of the auditor were associated with the audit client (e.g., as an employee).

4. Management advisory services being provided to a client by your firm inappropriately influenced audit-related judgements made by auditors of your firm while auditing that same client.

5. A professional employee had a direct financial interest of any amount, or a material indirect financial interest, in an audit client. An example of a direct financial interest is ownership of an audit client's stock while an example of a material indirect interest would be a material investment in a mutual fund that has materially invested in an audit client.

6. A spouse (or cohabitant), or dependents, of a professional employee had a direct financial interest of any amount, or a material indirect financial interest, in an audit client.

7. A professional employee who participated in the audit of a client, had a nondependent close relative that had a financial interest in the audit client that was material to the close relative, and the professional employee had knowledge of the material financial interest. Examples of nondependent close relatives include nondependent children, stepchildren, brothers or sisters, grandparents, parents, parents-in-law, and the respective spouses of these individuals.

8. A professional employee lent an audit client, or its officers, directors or principle stockholders, any amount of money. The professional employee may or may not have received a written note evidencing the loan.

9. A professional employee was a trustee of any trust or the executor or administrator of any estate that was committed to acquiring any direct, or material indirect, financial interest in the client.

10. A professional employee received an unsecured loan, that was not a home mortgage, from an audit client that was material to the net worth of the professional employee.

11. A spouse (or cohabitant), or dependent, of a professional employee who participated in the audit was employed by the audit client in a "audit sensitive position" or a position that allowed "significant influence" over the client's operating, financial, or accounting policies. A person's activities are generally considered to be "audit-sensitive" if his or her activities are normally an element of or subject to significant internal accounting controls. Examples of an "audit sensitive" position

include a cashier, internal auditor, accounting supervisor, purchasing agent, or inventory warehouse supervisor. Examples of positions that can exert "significant influence" include promoters, underwriters, voting trustees, general partners, directors, chief financial officers, chief executive officers, chief operating officers, or chief accounting officers.

12. A professional employee was connected with an audit client as a promoter, underwriter or voting trustee, as a director or officer, or in any capacity equivalent to that of a member of management or of an employee.

13. A professional employee was a trustee for any pension or profit-making trust for an audit client.

14. A professional employee who participated in an audit had a nondependent close relative who could exercise "significant influence" over the client's operating, financial, or accounting policies or a nondependent close relative that was employed by the client in a position that could be considered "audit sensitive."

15. A professional employee who was classified as a proprietor, partner, shareholder, or manager (or its equivalent), was located in the office of your firm that was participating in a significant portion of an audit for a client. The professional employee was not assigned to any of the professional services provided to this client, but he or she had a nondependent close relative who could exercise "significant influence' over the operating, financial, or accounting policies of the client.

***Note:*** The questionnaire recipients were asked about the percentage of the time that each of the 15 events occurred within a time period that would make involvement with the event a violation of Rule 101 of the AICPA's *Code of Professional Conduct.* The questionnaires included specific descriptions of the time periods.

## NOTES

1. This paper does not utilize the data that was obtained about the respondent's exposure to the independence-related quality control procedures.

2. The questionnaire was pretested by three collegiate auditing professors of the associate or full professor rank and 12 auditors currently in the practice of public accounting. Several minor changes were made to the survey based upon the feedback from the auditing professors and practicing public accountants.

3. The possibility of nonresponse biasing was addressed by utilizing the research results referred to by Oppenheim (1966). Responses were categorized as being "early" or "late" and the responses of the two groupings were compared. The results of the analysis indicated that the demographic characteristics of the two categories of respondents were virtually identical. In addition, an item-by-item contrast (early versus late) was performed for each of the 12 independence-related quality control procedures and for each of the 15 ethical violations, utilizing chi-square independence tests. Only two quality control procedures, and only one ethical violation, at the .05 level, yielded statistically significant differences between the early response percentages and the late response percentages. The similarities of the demographic characteristics and the other responses indicated that the inferences made in this study were not materially affected by nonresponse biasing.

4. The sampling plan involved mailing second requests to individuals who did not initially respond. The second requests did not contain survey control numbers. Responses were received from 136 respondents where the respondents name could not be determined from information in the demographic section of the survey. To guard against the remote possibility that a significant number of these individuals would have responded twice, all analyses reported in this paper were run both with and without the 136 surveys without names. The omission of these second requests had an immaterial affect on all of the reported conclusions in this paper.

5. Two violations (i.e., V1 and V3) that had significant chi-square values in Table 2 did not yield significant contrasts. This is due to the necessary adjustment in the alpha level for the multiple Bonferroni confidence intervals.

## REFERENCES

American Institute of Certified Public Accountants (AICPA), Quality Control Standards Committee. 1979. *Statement on Quality Control Standards 1.* New York: AICPA.

American Institute of Certified Public Accountants, Quality Control Standards Committee. 1980. *Quality Control Policies and Procedures for CPA Firms: Establishing Quality Control Policies and Procedures.* New York: AICPA.

American Institute of Certified Public Accountants (AICPA). 1992. *Code of Professional Conduct.* New York: AICPA.

Commission on Auditors' Responsibilities (Cohen Commission). 1978. *Report, Conclusions and Recommendations.* New York: AICPA.

Farmer, T.A., L.E. Rittenberg, and G.M. Trompeter. 1987. An investigation of the impact of economic and organizational factors on auditor independence. *Auditing: A Journal of Practice and Theory* 7(Fall): 1-14.

Mautz, R.K., and H.A. Sharaf. 1961. *The Philosophy of Auditing.* Sarasota, FL: AAA.

Oppenheim, A.N. 1966. *Questionnaire Design and Attitude Measurement.* New York: Basic Books.

Pearson, M.A. 1987. auditor independence deficiencies & alleged audit failures. *Journal of Business Ethics* 6: 281-287.

Schulte, A.A. 1965. Compatibility of management consulting and auditing. *The Accounting Review* XL(July): 587-593.

Shockley, R.A. 1981. Perceptions of auditors' independence: an empirical analysis. *The Accounting Review.* LVI (October): 785-800.

# CORPORATE SOCIAL RESPONSIBILITY IN ACCOUNTING:
## A REVIEW AND EXTENSION

Bernadette M. Ruf, Krishnamurty Muralidhar, and Karen Paul

**ABSTRACT**

The demand for more social reporting is being influenced by the increase in social investors, the expanded role of accountants in society, and the change in corporate managements' focus on multiple stakeholders groups. Prior research has indicated that users of social information have been dissatisfied with the information provided in financial reports. This study identifies eight corporate social performance (CSP) issues that are prevalent in the literature and investigates the differences between how the providers of social information (management accountants) and the users of such information (social investors) view these issues. The results indicate that accountants and social investors differ in the importance they place on CSP issues. Ethical implications of the findings are discussed.

**Research on Accounting Ethics, Volume 4, pages 185-200.**

**ISBN: 0-7623-0339-5**

## INTRODUCTION

Increases in investments in social funds and environmental reporting regulations have led researchers to revisit the long-debated topic of corporate social responsibility. Since the early 1970s, opponents of social reporting have argued against mandatory disclosure of social information due to the lack of consensus on how to define corporate social performance (CSP), how to measure it, and the cost of producing social information (Rockness and Williams 1988). Today, social investment funds and special interest groups are assessing CSP and using their assessments to screen stocks and to call for changes in corporate policy and governance (Domini et al. 1992).

Research on the usefulness of social information has examined the relationship between CSP, social disclosure, and economic or stock market performance (see Ullmann 1985; Mathews 1987; Pava and Krauz forthcoming). A major theme of this research is that corporate social reporting is merely an extension of financial reporting for profit seeking investors (Gray et al., 1988). Hence, investors' concerns with CSP are limited to the extent that social performance is related to or indicative of financial performance. Owen (1990) argued:

> such a focus is profoundly disturbing for those who consider that aspects of corporate performance beyond the purely financial are worthy of attention, and that corporate social reporting has a potentially major role to play in promoting accountability towards a corporate constituency wider than that of the speculative investor (p. 249).

Investors concerned with corporate behavior beyond purely financial performance are often referred to as social investors or ethical investors. Survey results have shown that social investors tend to use company reports for investment decisions in spite of the perceived social information deficiencies in these reports (Rockness and Williams 1988; Harte et al. 1991). Such a situation is especially disturbing considering that there has been a significant increase in social investing over the past decade (Rockness & Williams 1988; Owen 1990; Harte et al. 1991). In the United States, social investments were estimated at $825 billion in 1991, compared to about $40 billion in 1985 (Shapiro 1992). Increases in social investing also occurred in Canada and Great Britain (Sousa-Shields 1992; Miller 1992). Hence, it is reasonable to assume that the demand for relevant social information will increase in the future.

The need for increased social reporting is also being influenced by other factors. In response to the increase in publicized business scandals at the national and international levels, the accounting profession has expanded the responsibility of the accountant to not only the organization they serve but also their profession and the public (Institute of Management Accountants [IMA] 1984). Increased emphasis on accountant's responsibility to the community or society raises the issue that accountants are ethically obligated to provide social information to internal and external users. In addition, corporate managers are recognizing that "corporate success will depends on management's ability to satisfy not just its investors and employees, but the entire range of interests that make up society" (Aoi 1994, 26). When multiple stakeholders, interests, and values are in conflict and law is unclear, as with corporate social responsibility, ethical issues are ever-present (Trevino 1986). As companies recognize the importance of meeting multiple stakeholders needs, accountant will increasingly be faced with the dilemma of what and how to report on social information.

This paper attempts to further our progress towards corporate social reporting by identifying the importance placed on certain social performance issues and by identifying differences in how preparers (management accountants) and users (social investors) view various social issues. For management accountants to provide pertinent information on social performance, identifying the social information needs of corporate constituents is important. Given the costs associated with producing social information, there is also a need to know the importance placed on these social issues. How management accountants view various corporate social issues is relevant since they will be responsible for collecting and reporting the future social information. Furthermore, management accountants have an obligation to provide clear and complete reports to not only the organization they work for, but also to their profession and the public (IMA 1982).

Finally, identifying how accountants and social investors differ in their opinion on the importance of these social issues has several implications. First, agreement between the groups would suggest that there is congruence in information requested and provided, correspondence not found in prior research (Rockness and Williams 1988). Disagreement between these two groups may suggest where accountants should become more involved in providing social information to corporate constituents.

This paper is organized as follows. The next section provides a historical review of corporate social performance criteria identified in the literature. The methodology employed is presented in the following section. The next section presents the results, followed by the conclusions in the final section.

## CORPORATE SOCIAL PERFORMANCE ISSUES

Although practitioners and academics agree that entities should be held socially responsible, defining social responsibility has been an ongoing debate. Attempts at measuring corporate social performance (CSP) began in the 1960s. Abt Associates, Inc., was one of the first management consulting firms to market a social audit. The purpose of the social audit was to measure social benefits and costs to employees, communities, clients, and the general public. The audit provided both objective and subjective measures on health, security, equality, environment, and so forth (Abt 1972). David Linowes proposed a similar social disclosure format as Abt, but restricted the disclosure of information to expenditures that were voluntary by the company to improve welfare of employees and the public, product safety, or environmental conditions (Belkaoui 1984).

In the 1970s, several professional organizations sponsored research projects to identify CSP issues. Ernst and Ernst (1978) (currently Ernst and Young) identified six general CSP issues based on a survey of social involvement data disclosed in annual reports by 100 of Fortune 500 companies. These issues include environment, energy, fair business practice, human resources, community involvement, and product. The Research and Policy Committee of the Committee for Economic Development (1971) proposed 10 CSP issues, including economic growth and efficiency, education, employment and training, civil rights and equal opportunity, urban renewal and development, pollution abatement, conservation and recreation, cultural arts, medical care, and government. In 1974, another taxonomy of social responsibility was presented in a report by the Committee on Accounting for Corporate Social Performance by the National Association of Accountants (currently the Institute of Management Accountants). It identified four major areas of social responsibility: community involvement, human resources, physical resources and environmental contributions, and product or service contributions. These issues were based on providing benefits to four groups: the general public, the employee, the consumer, and the environment.

The most comprehensive list of CSP issues was proposed by the American Institute of Certified Public Accountants (AICPA) in 1977. Areas of significant social concern were the environment, nonrenewable resources, suppliers of purchased goods and services, products, services and customers, and the community. For each of these social areas, a breakdown was provided on what specific information to provide and the source of that information.

More recent CSP assessments have been motivated predominantly by demands of universities, church groups, and investment firms. The most extensive development and use of CSP assessments is by social investment firms. A descriptive survey of seven managers from the largest socially responsible mutual funds in the United States identified 11 social criteria used in rating corporations on CSP (Rockness and Williams 1988).[1] Rockness and Williams found that no one social criterion was listed by all respondents, although seven of the social criteria were listed by a majority of the respondents. The seven social criteria identified, in order of frequency, were environmental protection, equal employment opportunity, particularly women and minorities, treatment of employees, business with repressive regimes (notably South Africa), innovativeness/quality of products or services, social contribution to community, and defense contracts.

In an extension of Rockness and Williams's study, Harte et al. (1991) surveyed 11 social fund managers in Great Britain to identify the importance that fund managers placed on CSP issues. When making social investment decisions, eight social issues were identified as qualifiers and eight issues as disqualifiers. Respondents rated each CSP issue as "of little or no importance," "important," or "vital or very important." The fund managers specified as "very important" social issues:

**Table 1.** Summary of Social Issues

| *Current Study* | *Harte et al. (1991)* | *Rockness et al. (1988)* | *AICPA (1977)* | *NAA (1974)* | *COEP (1971)* | *Ernst & Ernst (1970)* |
|---|---|---|---|---|---|---|
| Community relations | Contribution to community | Contribution to community, charity | Corporate citizenship | Community involvement | Urban renewal and development<br>Cultural arts education, conservation and recreation | Community involvement |
| Employee relations | Promote employee welfare | Treatment of employees | Human resources | Human resources | Employment training, medical care | Human resources |
| Environmental | Environmental record and awareness | Environmental protection | Environment<br>Nonrenewable resources | Physical resources and environmental contribution | Pollution abatement | Environment<br>Energy |
| Product/Liability | Socially beneficial products and services<br>Good customer relations | Innovativeness/quality of products or services consumer protection and product purity | Product<br>Suppliers of goods and services<br>Customer service | Product or service contribution | Economic growth and efficiency | Product |
| Women/Minorities | Equal employment opportunity | Equal employment opportunity | Employee opportunity and equity | | Civil rights and EOP | Fair business practice |
| Nuclear | | Involvement with nuclear power | | | | |
| Military | Manufacture armaments | Defense contracts | | | | |
| South Africa | Repressive regimes<br>Manufacture or sell alcohol, tobacco and gambling<br>Involvement with fur trade, animals | Repressive regimes | | | | |

environmental awareness, employee relations, contributions to the community, the production of socially beneficial products and services, and equal employment opportunity. This study provides insight in to the importance of social performance issues, but not the relative importance of these issues.

CSP assessments can also be found in social database ratings services. In the United States, social databases are available from a variety of sources including the Council on Economic Priorities, Investor Responsibility Research Center, DataCenter, Interfaith Center on Corporate Responsibility, and Nuclear Free America. Social databases available in Great Britain include Ethical Investment Research Information Service (EIRIS) and Merlin Research Unit. In Canada, Social Investment Organization (SIO) has recently been established as a national social investment information clearing house.

While numerous CSP issues have been identified in the literature (see Table 1), five issues that have been prevalent since the 1970s are: corporate responsibility to the consumer, the employee, the community, the environment, and to civil rights or equal opportunity. Three other social performance issues—military, nuclear power, and South Africa—became popular in the 1980s and are often used in stock screening with social investment funds. Based on this review, eight issues of CSP were selected for the current study: community relations, employee relations, environmental performance, product development and liability, women/minority policies, generation of revenues from military, nuclear power, and South Africa.[2] Definition of the eight social issues were based on the criteria used by Kinder, Lydenberg, Domini & Company (KLD) social rating service. KLD, formed in 1984, provides ratings on eight CSP issues for 800 publicly traded companies.[3] Its rating system is based on earlier attempts to measure CSP such as the Council of Economic Priorities' *Rating America's Corporate Conscience* (Lydenberg, Marlin, and Stub 1986). Based on KLD's definitions, the following definitions were used in the current study:

- *Community Relations* refers to corporate response to community by donations, contribution to the economically disadvantaged and support of job training.
- *Employee Relations* refers to corporate policies of no-layoff plan, hiring and promoting disabled, cash profit sharing, and good union relations.
- Environmental refers to corporate development, processing, and use of products or services that minimize environmental damage or are environmentally safe.
- *Product/Liability* refers to corporate efforts in research and development, reputation of high quality products, and avoidance of producing harmful products.
- *Women/Minority* refers to corporate hiring and promoting of women and minority employees, including addressing family concerns such as child care and elder care.

- *Nuclear Power* refers to the percentage of utilities generating power from nuclear power plants.
- *Military* refers to a corporate generation of revenue from the production of weapons.
- *South Africa* refers to the equity interest or ownership in South Africa.

## METHODOLOGY

To assess the relative importance of these eight social issues, managerial accountants and social investors were surveyed. Management accountants were selected because they represent current and potential future providers of social information. Management accountants are responsible for coordinating the collection of data and preparing proposed budgets. Management accountants were randomly selected from a southeast membership listing of the Institute of Management Accountants.

Two groups of social investors were selected: members of the Interfaith Center on Corporate Responsibility (ICCR) and investors in the Domini Social Equity Fund. While both these groups reflect investors who are trying to influence corporate behavior through their investments, they differ in their approaches. The ICCR group takes an active approach in influencing corporate behavior by issuing shareholder resolutions.[4] The ICCR was the first group to sponsor social shareholder resolutions and has been active for more than 25 years.[5]

The Domini Social Equity Fund selects stocks based on the Domini Social Index.[6] This group plays a more passive role in terms of influencing corporate behavior. Investments are based on financial and social performance. Investors influence corporate behavior by avoiding corporate stock that has poor social performance or by investing in stock with good social performance. Hence, for the purpose of this paper, ICCR members are considered social activists and Domini Social Equity Fund investors are considered social investors.

### Instrument

The survey instrument was developed using the Analytic Hierarchy Process (AHP). This approach for determining the relative importance of different issues of corporate social performance has been suggested by Arrington et al. (1982), Wokutch and Fahey (1986), and Ruf et al. (1993). The AHP, developed by Saaty (1980), uses a pairwise comparison process that provides a systematic means to quantify decision-maker's perceptions when the criteria are primarily qualitative (Saaty 1980). The AHP has been successfully applied in a variety of multi-decision-making settings, such as government, accounting, management information systems, and so forth (for a review, see Shim 1989; Apostolou and Hassell 1993).

Participants were asked to complete a two-part questionnaire. The first part of the questionnaire was designed to obtain weighted estimations from the respondents on the eight CSP issues: community relations, employee relations, environmental, product development and liabilities, women/minority, nuclear power, military, and South Africa. To ensure that the respondents understood what these social issues represented, the definition of each issue was provided in the beginning of the questionnaire. To establish the weights of each attribute, the questions were designed based on the AHP's weight generation scheme. For each pairwise comparison, respondents were asked to first identify whether the two social issues were of equal importance or to indicate which social issue was of more importance. If the respondents identified one social issue as more important than the other, they were then asked to assess the relative importance of the social issue identified as more important based on Saaty's (1980) nine-point scale. The second part of the questionnaire obtained demographic and descriptive information on the respondents. The questionnaire was validated in a prior study (see Ruf et al. 1993).

## Participants

Two hundred questionnaires were sent to management accountants and 73 questionnaires were sent to social activists. Two weeks later, a follow-up letter was sent. From the 200 mailings to management accountants, five questionnaires were

**Table 2.** Descriptive Statistics

| | *Total Sample (n = 277)* | *Management Accountant (n = 42)* | *Social Investor (n = 194)* | *Social Activist (n = 41)* |
|---|---|---|---|---|
| Average Age (years) | 46 | 45 | 45 | 52 |
| Gender | | | | |
| Female | 52% | 24% | 56% | 63% |
| Male | 48% | 76% | 44% | 37% |
| Education | | | | |
| Some college | 6% | 5% | 8% | 2% |
| Bachelor | 30% | 31% | 34% | 10% |
| Masters | 42% | 48% | 36% | 61% |
| Advanced | 22% | 16% | 22% | 27% |
| Income | | | | |
| < than 25,000 | 14% | 2% | 11% | 39% |
| 25,001-50,000 | 37% | 30% | 41% | 29% |
| 50,001-75,000 | 25% | 21% | 26% | 25% |
| > than 75,000 | 24% | 47% | 22% | 7% |

returned due to wrong addresses, four were returned as unusable replies, and 42 (26%) were returned as usable responses. Of the 400 social investors, eight surveys were returned incomplete, six were returned with incorrect addresses, and 194 (49%) were returned as usable. From the 73 social activist mailings, two were returned unusable and 41 (59%) usable.

Table 2 provides descriptive statistics of the respondents. One hundred and forty-five of the respondents were female and 132 were males. The ages of the respondents ranged from 26 to 77 with an average of 46 years. About 22% of the respondents had completed an advanced college degree, 42% had begun or completed a masters degree, 30% had a bachelors degree, and 6% had some college courses. Twenty-four percent of the respondents reported annual family income greater than $75,000, 25% had incomes between $50,000 and $75,000, 37% had incomes between $25,000 and $50,000, and 14% had income levels below $25,000. Note that the social activist group tends to be older, have more education, and hence less income than the management accountants and the social investors. Furthermore, the social activist group is composed of more females than the other groups. The composition of the social investor group differs from the management accountants in having more females and a lower income level.

## RESULTS

To calculate the social dimension weights, each participant's responses were entered into a matrix. The eigen values of the matrix were then computed using IMSL (1989). The eigen-vector associated with the largest eigen-value was computed. This eigen-vector was normalized so that the sum of the eigen-vectors was 1.0. To evaluate the consistency of the participants' responses, the input matrices of the respondent were combined into a single input matrix, using geometric means as suggested by Saaty (1980). The inconsistency ratio was 0.01, which is well within the acceptable level. The overall relative importance of each social issue was calculated by averaging the individuals' weights for each social issue.[7] These average weights represent the relative importance of each social issue as perceived by the participants in this survey.

The weights derived for each social dimension are presented in Table 3. For the management accountants, product/liability was considered the most important issue with a weight of 25%, followed by employee relations at 19%. The relative importance placed on the next three issues were environment (14%), women/minorities (13%), and community relations (11%). The three social issues considered least important when assessing CSP were nuclear power (7%), military (6%), and South Africa (5%). The combined weights of the three least important social issues is less important than either of the two most important social issues

***Table 3.*** Statistical Result

| | *Management Accountants* | | *Social Investors* | | *Social Activists* | |
|---|---|---|---|---|---|---|
| *Social Issues* | *Weights* | *Ranks* | *Weights* | *Ranks* | *Weights* | *Ranks* |
| Community Relations | 0.11 | 5 | 0.08** | 7 | 0.08** | 7.5 |
| Employee Relations | 0.19 | 2 | 0.12** | 5 | 0.14** | 4 |
| Environment | 0.14 | 4 | 0.20** | 1 | 0.15 | 2 |
| Product/Liability | 0.25 | 1 | 0.15** | 2 | 0.12** | 6 |
| Women/Minorities | 0.13 | 3 | 0.14 | 4 | 0.16 | 1 |
| Nuclear Power | 0.07 | 6 | 0.10* | 6 | 0.08 | 7.5 |
| Military | 0.06 | 7 | 0.14** | 3 | 0.14** | 4 |
| South Africa | 0.05 | 8 | 0.07 | 8 | 0.13** | 4 |

***Notes:*** *Weights that are significantly different from the management accountants at the 0.05 level.
**Weights that are significantly different from the management accountants at the 0.01 level.

(product/liability and employee relations). This indicates that management accountants clearly differentiate between the importance of these social issues.

The most important issue for the social investors was the environment (20%). The relative importance of the next three issues were product/liability (15%), women/minorities (14%), and military (14%). This was followed by employee relations (12%)and nuclear power (10%). The least important issues were community relations (8%) and South Africa (7%). Like the management accountants, the social investors differentiated between the importance of the social issues.

For the social activists, the most important social issues were women/minorities and environmental, with weights of 16% and 15%, respectively. The next two social issues of importance were employee relations and military. These issues were given similar weights of 14%. Close in weight to these issues were product/liability (12%) and South Africa (13%). The least important issues were community relations (8%) and nuclear power (8%). Compared to the management accountants, the social activists did not differentiate as much between the social issues.

To determine whether the management accountants differed in the importance placed on these social issues from the social investors and the social activists, statistical analysis was performed. The results presented in Table 3 show that the management accountants' weights significantly differ from the social investors on all the social issues except for women/minorities and South Africa. Social investors placed more emphasis on environment ($t = 3.60$, $p < .01$), military ($t = 4.07$, $p < .01$), and nuclear power ($t = 2.23$, $p < .05$), while management accountants placed more emphasis on product/liability ($t = -4.75$, $p < .01$), employee relations ($t = -4.73$, $p < .01$), and community relations ($t = -3.45$, $p < .01$). The negative values in the test statistic value show that the mean weights of the management accountants was higher than those of social investors.

The social activists significantly differed from management accountants on the following social issues: community relations, employee relations, product/ liability, military, and South Africa. Similar to the findings of the social investors, management accountants placed more emphasis on product/liability ($t$ = -5.20, $p$ < .01), employee relations ($t$ = -2.87, $p$ < .01), and community relations ($t$ = -2.86, $p$ < .01) than did the social activists. The social activists placed more emphasis on military ($t$ = 4.74, $p$ < .01) and South Africa ($t$ = 4.22, $p$ < .01) than did the management accountants. Weights assigned to environment, women/minorities, and nuclear power were not significantly different among the groups.

Another way to examine the importance of the social issues is by considering the rank order of the different issues. The weights of each issue can be used to determine the rank order. Included in Table 3 is the social issues ranks. The degree to which the ranks of the management accountants and the social investors and the social activists agree is determined by Spearman's Rank Order Correlation ($r_S$), a measure of the degree of correspondence between the ranks of the sample observation (Daniel, 1985). The $r_S$ between the ranks of the accountants and the social activists corrected for ties is a relatively low value of .15. A hypothesis test of the form:

**Hypothesis 0.** The two sets of ranks are independent

**Hypothesis 1.** The two sets of ranks have a positive relationship

was conducted. The null hypothesis is not rejected (p > 0.1), leading to the conclusion that there is insufficient evidence to suggest any relationship between the ranks of the management accountants and the social investors and between the ranks of the management accountants and the social activists.[8]

The lack of agreement between the users and providers on the specific social issues provides some insight into the incongruence between the groups. Management accountants place more emphasis on issues that are internal to the organization or under managements direct influence such as product/liability and employee relations. Social investors place more emphasis on environmental issues. This may be indicative of the increase in environmental regulations and lawsuits that threaten a company's future liabilities and hence future profits. Another possible explanation is that these results reflect the social orientation of this specific group. While the social activists did not differentiate much among the various social issues, social issues that were weighted higher than by the management accountants tended to be related to human rights issues, such as military and South Africa.

The findings also reveal that management accountants placed more emphasis on the community relations dimension than did both the social investors and social activists. One explanation for this finding, provided by Steven Lydenberg from

Kinder, Domini, and Lydenberg Corp., is that social investors may perceive philanthropy as a means of advertisement rather than as altruistic behavior. Another finding of interest is that all three groups assigned similar weights to women/minority issues. This finding may be indicative of the time period, that is the awareness of equal opportunity.[9]

## DISCUSSION

Accountants' responsibility have changed from a focus that use to be mainly on shareholders to a focus on multiple stakeholder groups. This shift in focus has come about for several reasons. First, fueled by the increase in business scandals, the accounting profession has taken certain initiatives to increase public trust in the profession. For example, the IMA charged management accountants with obligations not only to the organization they serve but also to their profession and the public to maintain the highest standards of ethical conduct (1984). The American Institute of Certified Public Accountants (AICPA) expanded the CPA's role to safeguarding the public interest (1988). Similar charges have been carried out in Great Britain by the Council of Chartered Institute of Management Accountants (CIMA) (the organization includes CPAs and management accountants)[10] and in Canada by the Canadian Institute of Chartered Accountants (CICA).[11] What effects these changes have had on accountants' perceptions of their role or ethical decisions are unknown. The current study does not directly address the perception of accountants role in society. However, the finding that accountants place heavy emphasis on social issues internal to the organization may suggests that accountant do not acknowledge their greater role in society.

Second, corporate managers are realizing that to remain competitive, companies must balance the needs of the shareholders' and other stakeholder groups, such as employees, customers, public interest groups, and the community. Increased emphasis on accountants' responsibility to the community or society raises the issue of what ethical obligation do accountants have to provide social information to internal and external users. Ruf (1997) argues that in order for accountants to be responsive to society, they must first be aware of current social issues and the importance of the various social issues to the relevant stakeholder groups. This paper provides preliminary evidence on how management accountants and two stakeholder groups (social activists and social investors) perceive the importance of eight social issues. Accountants' emphasais on social issues internal to the organization may be an indication of their lask of awareness of the concerns of other stakeholder groups. Increasing accountants' awareness of various social issues will enable them to develop a clearer definition of corporate social responsibility and to provide more relevant information to both internal and external users. Ruf (1997) argues that in order for accountants to be responsive to society, they must first be aware of current social issues and the importance of the various social issues to the relevant stakeholder groups.

Given that interested constituencies tend to use company annual reports for social investment decisions, there is a need for the accounting profession to reevaluate what social information is important to various stakeholder groups (Rockness and Williams 1988; Harte et al. 1991). The survey results provide strong evidence that accountants and two social investor groups perceive CSP differently, both in terms of the degree to which a dimension is important and in terms of the order of importance. These findings suggest that management accountants must become more aware of which issues are more salient to the various stakeholder groups.

The findings also reveal that management accountants placed significantly less weight on environmental issues than did the social investors. Further, management accountants weight environmental issues similar to community relations and women/minorities issues. These findings raise concern given that the cost of environmental compliance is estimated to be more than $200 billion (Garvin 1993), clean-up costs for toxic waste sites and clear air provisions are estimated at $100 billion and $19 billion, respectively (Rittenberg et al. 1992), and mass tort claims are beginning to threaten corporate survival (Dupree and Jude 1995). While some organizations, such as the IMA, are in the process of implementing voluntary guidelines for measuring and reporting environmental performance, accountants have been slow to respond to the changing environmental needs of businesses. As argued by Skalak et al. (1993/1994), companies need to be proactive in informing the public of its environmental achievements beyond compliance and beyond industry standards.

## CONCLUSION

This paper provides more of a forum for raising issues than resolving issues. The resistance to the development of social reporting has largely been due to the difficulty in arriving at a consensus on what social issues are of importance to warrant the additional costs companies take on as a result of producing such information. To meet the changing business environment, accountants need to expand their role and "develop measures for social, environmental, and ethical impacts and for corporate governance and accountability" (Epstein 1993, 26). Furthermore, accountants must be aware of social information needs of various stakeholders to meet their ethical obligations to society.

While this study provides insight on the importance of the social issues to the two social investor groups, the differences in the results with respect to social investors and social activists illustrate the dilemma management must deal with when addressing multiple social issues. Understanding information needs of various stakeholder groups will aid management in efficiently allocating its resources, as well as providing users with more relevant information. Further research is needed to determine whether other stakeholder groups, such as creditors and consumers, differ in the relative importance assigned to the social issues.

Finally, reporting on a company's social performance may have an unforseen positive impact on the company. If a company is performing well on one or more of the social issues and the public is not aware, the company maybe forgoing some potential benefits, such as potential employees, consumers, and/or investors. Although it is difficult to estimate lost benefits due to not disclosing, there is support for potential benefits from disclosing information. For example, Blacconiere and Patten (1993) showed that chemical companies with extensive environmental disclosure had less of a negative market reaction to the 1984 Union Carbide chemical leak in Bhopal, India, than did chemical companies with less disclosure. Further research is needed to estimate potential benefits of disclosing social information.

## ACKNOWLEDGMENT

A previous version of this paper was presented at the 1994 ABO conference: Frontiers in Behavioral Research, San Antonio, Texas.

## NOTES

1. Rockness and Williams listed 12 social criteria in their article. Because two of their listed criteria, charity and special economic or social contributions to the community, appear to be the same construct, we combined them.

2. Investment in South Africa was still considered a major issue when this survey was conducted in 1992. Political changes in South Africa have occurred since.

3. This data base is currently being used by a number of large investment firms that maintain portfolios of socially screened funds, such as TIAA-CREF and Merrill Lynch.

4. To file a shareholder resolution, proponents must hold a minimum of one thousand dollars worth of a stock for one year. The resolution is sent to the CEO or corporate secretary with proof of ownership.

5. ICCR represents 25 Protestant denominations and 230 Roman Catholic orders and dioceses.

6. Domini Social Index is produced by KLD social rating service. This is the first broad market stock index in the United States to monitor corporate social performance based on multiple social constraints (Kurtz et al. 1992).

7. The weights were also derived from the single input matrix created by using geometric means. There was no statistical difference between the two sets of weights. Since the use of a single input matrix does not allow for statistical analysis, the average of weights derived from individuals was used.

8. When correction for ties is not applied, the value of $r_s$ is 0.16. This value is still not significant.

9. Note the policy changes currently taking place with affirmative action may have an impact on weights assigned to women/minority issues.

10. The CIMA state that the accountant's obligation is to the "collective well-being of the community of people and the institutions the profession serves" (CIMA 1992).

11. The CICA have taken a stronger stand on defining the changing role of accountants. In the chartered accountants' mission statement, they state that accountants are "to provide services that

enable decisions to be made on the effective allocation and efficient use of resources. . . . CA exercise their functions with expert knowledge and judgement, practice with integrity objectivity and accountability, and are committed to their responsibility to the public" (CICA 1986, 19).

## REFERENCES


Abt, C. 1972. Managing to save money while doing good. *Innovation* 27(January): 38-47

American Institute of Certified Public Accountants (AICPA). 1977. *The Measurement of Corporate Social Performance.* New York: AICPA.

Apostolou, B., and J. Hassell. 1993. An overview of the analytic hierarchy process and its use in accounting research. *Journal of Accounting Literature* 12: 1-27.

Aoi, J. 1994. To whom does the company belong?: A new managment mission for the information age. *Journal of Applied Corporate Finance* (Winter): 25-31.

Arrington, C.E., R.E. Jensen, and M. Tokutani. 1982. Scaling of corporate multivariate criteria: Subjective composition versus the analytic hierarchy process. *Journal of Accounting and Public Policy* 1(1): 95-123.

Blacconiere, W.G., and D.M. Patten. 1994. Environmental disclosures, regulatory costs, and changes in firm value. *Journal of Accounting and Economics* 18(3): 357-377.

Belkaoui, A. 1984. *Socio-Economic Accounting.* Westport, CT: Greenwood Press.

Canadian Institute of Chartered Accountants (CICA). 1986. Long-range strategic palnning committe. Toronto, Ontario: CICA.

Committee for Economic Development (CED). 1971. *Social Responsibilities of Business Corporations.* New York: CED.

Council of the Chartered Institute of Management Accountants (CIMA). 1992. *Ethical Guidelines.*, London, UK: CIMA.

Cossette, G. 1994. Ethics under the gun. *CMA Magazine* (March): 23-26.

Daniel, W.W. 1985. *Applied Nonparametric Statistics.* Boston, MA: Houghton-Mifflin.

Domini, A., P. Kinder, and S. Lydenberg. 1992. What is social investing? Who are social investors? In *The Social Investment Almanac,* eds. P. Kinder, S. Lydenberg, and A. Domini, 5-7. New York: Henry Holt.

Dupree, C.M., and R.K. Jude. 1995. Coping with environmental and tort claims. *Management Accounting* (March): 27-31.

Epstein, M.J. 1993. The expanding role of accountants in society. *Management Accounting* (March): 22-26.

Ernst and Ernst. 1978. *Social Responsibility Disclosure.* New York: Ernst & Ernst.

Garvin, A.O. 1993. The 12 commandments of environmental compliance. *Industrial Engineer* (September): 18-22.

Gray, R.H., D.L. Owen, and K.T. Maunders. 1988. Corporate social reporting: Emerging trends in accountability and the social contract. *Accounting, Auditing, and Accountability* 1(1): 6-20.

Harte, G., L. Lewis, and D. Owen. 1991. Ethical investment and the corporate reporting function. *Critical Perspectives on Accounting* 2: 227-253.

IMSL Stat/library. 1989. *Problem Solving Software Systems*, Vol. 3: *Fortran Subroutines for Statistical Analyses.* Houston, Tx: IMSL.

Institute of Management Accountants (IMA). 1982. *Statement of Management Accounting No. 1B: Objectives of Management Accounting.* Montvale, NJ: IMA.

Kurtz, L., S. Lydenberg, and P.D. Kinder. 1992. The Domini social index: A new benchmark for social investors. In *Social Investment Almanac*, eds. P. Kinder, S. Lydenberg, and A. Domini, 287-322. New York: Henry Holt & Company.

Lydenberg, S.D., A.T. Marlen, and S. Stub. 1986. *Rating American Corporate Conscience.* Reading, MA: Addison-Wesley.

Mathews, R. 1987. Socially responsibility accounting disclosure and information content for shareholders: A comment. *British Accounting Review* 19(2): 161-167.

Miller, A. 1992. Social investment in the United Kingdom. In *Social Investment Almanac*, eds. P. Kinder, S. Lydenberg, and A. Domini, 660-668. New York: Henry Holt & Company.

Owen, D.L. 1990. Towards a theory of social investment: A review essay. *Accounting, Organizations and Society* 15(3): 249-265.

Pava, M.L., and J. Krausz. Forthcoming. The association between corporate social-responsibility and financial performance: The paradox of social cost. *Journal of Business Ethics.*

Rittenberg, L.E., S.F. Haine, and J.J. Wegrandt. 1992. Environmental protection: The liability of the 1990s. *Internal Auditing* (Fall): 13-25.

Rockness, J., and P.F. Williams. 1988. A descriptive study of social responsibility mutual funds. *Accounting, Organization and Society* 13(4): 397-411.

Ruf, B.M. 1997. Corporate social responsibility: The role of the accountant. In *The Blackwell Encyclopedia on Management.* Oxford, UK: Blackwell.

Ruf, B.M., K. Muralidhar, and K. Paul. 1993. Eight dimensions of corporate social performance: Determination of relative importance using the analytic hierarchy process. In *Academy of Management Best Paper Proceedings*, 326-330, Atlanta, GA.

Saaty, T.L. 1980. *The Analytic Hierarchy Process.* New York: McGraw Hill.

Saaty, T.L. 1986. Axiomatic foundation of the analytic hierarchy process. *Management Science* 32: 841-855.

Shapiro, J. 1992. The movement since 1970. In *Social Investment Almanac*, eds. P. Kinder, S. Lydenberg, and A. Domini, 8-23. New York: Henry Holt & Company.

Shim, J.P. 1989. Bibliographical research on the analytic hierarchy process (AHP). *Socio-Economic Planning Sciences* 23(4): 161-167.

Skalak, S.L., W.G. Russell, M. Robinson, G. Miller, and D. Casey. 1993/1994. Proactive environmental accounting and world-class annual reports. *Journal of Corporate Accounting and Finance* (Winter): 177-196.

Smith, T. 1992. Shareholder activism. In *Social Investment Almanac*, 108-114. New York: Henry Holt & Company.

Sousa-Shields, M. 1992. Social investment in Canada. In *Social Investment Almanac,* eds. P. Kinder, S. Lydenberg, and A. Domini, 638-643. New York: Henry Holt & Company.

Trevino, L.K. 1986. Ethical decision making in organizations: A preson-situation interactionist model. *Academy of Managment Review* 11(3): 601-617.

Ullmann, A. 1985. Data in search of a theory: A critical examination of the relationships among social performance, social disclosure, and economic performance of U.S. firms. *Academy of Management Review* 10(3): 540-557.

Wokutch, R.E., and L. Fahey. 1986. A value explicit approach for evaluating corporate social performance. *Journal of Accounting and Public Policy* 5(3): 191-214.

# CPA VALUES ANALYSIS:
## TOWARD A BETTER UNDERSTANDING OF THE MOTIVATIONS AND ETHICAL ATTITUDES OF THE PROFESSION

Thomas E. Wilson, Jr., Dan R. Ward, and
Suzanne Pinac Ward

**ABSTRACT**

Perhaps at no other time have accountants been less confident of their legal responsibilities and their ethical accountability to themselves, to individual clients, and to society in general. This study investigated the personal value systems of practicing CPAs as a fundamental step in the understanding of their individual motivations and their compatibility with societal mores and, ultimately, assessing and improving the ethical climate and perception of the profession. The Rokeach Value Survey was used to assess the respondents' values. The instrumental values profile indicates that CPAs perceive "honest," "responsible," and "loving" to be types of conduct which are both individually and socially desirable. The terminal values profile infers that CPAs possess personal-oriented values such as "self-respect" and "family security"

**Research on Accounting Ethics, Volume 4, pages 201-210.**

**ISBN: 0-7623-0339-5**

rather than social-oriented ones such as "social recognition" and "a world of beauty." The findings in the current study provide additional evidence of an achievement-oriented value system for practicing CPAs. Several demographic characteristics significantly affected the rankings of terminal and instrumental values. Respondent gender, position, experience, age, and income were all correlated with the rankings of several values. Future research is needed to address the relation between personal values and demographic factors, including those incorporated in this study.

## INTRODUCTION

A distinctive combination of experiences and learning, human behavior requires a balancing of considerations both internal and external to the individual. The resulting behavior patterns are as unique and divergent as the individuals themselves. As part of one's internal considerations, values are a static and enduring influence and guide to an individual's behavior, attitudes, and self-image. A knowledge and understanding of the personal value systems of the individual members of a profession should provide insight not only into their personal behavior but also into the commonalities and differences forming the overall behavioral direction and ethical attitudes of the profession.

The intricate combinations of personal, technical, organizational, and societal value systems often create conditions involving ethical dilemmas and value judgments for accounting professionals. The widely publicized white collar crimes and scandals of the 1980s and 1990s have caused an unprecedented increase in litigation involving accountants and have caused many to question the ability of the profession to regulate itself. Ethical confidence in the profession's past actions has eroded as society has increased its demands for higher standards of moral behavior from those in positions of trust. Perhaps at no other time have accountants been less confident of their legal responsibilities and their ethical accountability to themselves, to individual clients, and to society in general. The primary purpose of this study is to investigate the personal value systems of practicing CPAs as a fundamental step in the understanding and evaluating of their individual motivations, compatibility with societal mores, and, ultimately, assessing and improving the ethical climate and perception of the accounting profession.

## VALUES

Values have been defined as "deeply held and enduring beliefs about preferred end-states of existence" (Rokeach 1973). Viewed not only as a controlling drive in life, Brockhaus and Horwitz (1986, 32) assert that "values are an organized conception of nature." While comparable to attitudes, values are, by nature, more general and less likely to be associated with any specific item (England 1967). Rokeach (1968), who categorized values and attitudes/beliefs from a hierarchical perspective, divided values into two sets: terminal and instrumental. This structuring resembles an inverted pyramid with the number of terminal values held being the smallest and attitudes/beliefs the largest. Rokeach (1968, 17) stated that:

> (a)n instrumental value is therefore defined as a single belief which always takes the following form: *I believe that such-and-such a mode of conduct (e.g., honesty, courage) is personally and socially preferable in all situations with respect to all objects.* A terminal value takes comparable form: *I believe that such-and-such an end-state of existence (e.g., salvation, a world at peace) is personally and socially worth striving for.* Only those words or phrases that can be meaningfully inserted into the first sentence are instrumental values, and only those words or phrases that can be meaningfully inserted into the second sentence are terminal values.

Analogous to a philosophy and stable in nature, values are often deeply entrenched and, thus, a relatively permanent part of one's inner self (England 1967). The intensity of fidelity to a value depends on the context in which the value is scrutinized; as a result, dysfunctional or deviant values may possibly be modified (Baker 1976). Values embraced by each individual influence his or her behavior and, accordingly, the behavior of the organization(s) to which they belong (England 1967; Coye 1986). Professions correspond to business organizations in this regard, with individual members' values influencing the behavior of the profession. This, in turn, defines the professional image presented to the public (Swindle and Phelps 1984). The public evaluates this image and assesses the trust and credibility of the profession.

Critical as values are to individuals, organizations, and professions, individuals often do not know the specific values they hold. Moreover, these same individuals frequently fail to act in accordance with known or stated values (Cochran 1983; Wrench 1969). The resulting conflict may produce value anxiety (e.g., dissension or guilt) which curtails the decision-making effectiveness of an individual (Hall 1973). Ultimately, this abatement of individual effectiveness will be reflected in diminished organizational capability and a tainted professional image.

The remainder of this paper is presented as follows. The next section reviews selected prior research into values. Then, the methodology of the current study is presented, followed by the study's results. The paper closes with a summary and discussion of some implications of the findings.

## SELECTED PRIOR RESEARCH

Values have been the subject of much research. However, relatively few studies have focused on the values of professionals and, in particular, those of accountants. Liedtka (1989) found that managers most often encountered conflict when faced with clear organizational values but uncertainty regarding the congruence of those values with their own personal ones. Posner and Schmidt (1993) found that managers with clear concepts of both their personal and organizational values had positive attitudes toward their work as well as the ethical behavior of their colleagues and employers.

Fagenson (1993) used the Rokeach Value Survey in a study of the personal value systems of male and female entrepreneurs and managers. The study found that gender had little impact on personal value systems, although women ranked the value of "equality" higher than men, while men ranked the value of "family security" higher than women. The study did find, however, that managers and entrepreneurs had extensively different personal value systems.

In a study of the value profiles of junior/senior accounting students and business executives, Swindle and Phelps (1984), utilizing the Rokeach Value Survey, found that the two profiles were significantly related as a whole. However, they found that nearly half of the individual values showed significantly different rankings between the two groups. They concluded that accounting students and business executives have different value systems, resulting in potential problems for newly hired accountants.

The results of studies utilizing attitude and opinion questionnaire approaches strengthen the assertion that accountants and non-accountants have different personalities and traits (Baker 1976). Nevertheless, studies following a methodology of personality testing have concluded that these differences are not statistically significant (Baker 1976). An early study of accounting majors' traits conducted by Thielens (1966) for the AICPA indicated that most accounting majors came from low-income Catholic families with parents who had, at most, a high school education. When asked what traits distinguished them from students in general, a significant number of accounting majors characterized themselves as cooperative and quiet. However, DeCoster and Rhode's (1971) research did not support the stereotype of the accountant as antisocial, submissive, unemotional, and detached. Baker (1976), in a comparison of the values of accounting majors and non-accounting majors using the Rokeach Value Survey, found statistical differences for eight of 36 values included in the study. Although refusing to conclude that this provided evidence of different value systems between the two groups, he noted that such a possibility did exist.

Swindle, Phelps, and Broussard (1987) conducted one of the few studies investigating the personal value systems of accountants at the professional level. They examined the personal value systems of practicing certified public accountants (CPAs) via the Rokeach Value Survey. The study concluded that CPAs have personal-oriented values rather than social-oriented ones, thus not accentuating "the values that, according to some authorities, are characteristic of today's society" (p. 6). The authors provided three possible explanations for these findings: (1) CPA's personal values and society's values do not overlap, (2) the portrayal of society's values is incorrect, or (3) the respondents included in the study are not representative of the population of CPAs.

## METHODOLOGY

The primary data for this study were collected via a survey of 733 certified public accountants engaged in public practice in the United States. Responses were received from 195 subjects, resulting in a 26.6% response rate. Of the responding CPAs, 59 were female and 136 were male.

The Rokeach Value Survey, which measures values in a variety of settings, was used to assess the respondents' values. The survey lists two separate types of values: 18 instrumental values and 18 terminal values. The instrumental values comprise means to goals or modes of conduct that the individual believes are personally and socially desirable (Baker 1976). Terminal values are concerned with general goals or end states of existence which one believes are personally and socially worth striving for (Baker 1976). The Rokeach Value Survey has been found to be both a valid and a reliable instrument in the measurement of personal value systems. The Value Survey has been widely used in sociology, psychology, management, and other disciplines (e.g., Fagenson and Coleman 1987; Feather 1984).

Participants were asked to rank each set of values from 1 to 18, with 1 being the most important and 18 the least important. As the rankings constitute an ordinal level of measurement, the appropriate measure of central tendency is the median (Siegel 1956). The median rank for each value was used to develop the value profile for each group.

***Table 1.*** Characteristics of Respondents

| *Characteristic* | *Female*[a] | *Male*[b] |
|---|---|---|
| Firm Type | | |
| International | 28.8% | 17.9% |
| National | 0.0% | 1.5% |
| Regional | 10.2% | 14.2% |
| Local | 61.0% | 66.4% |
| Position | | |
| Partner/Sole Proprietor | 23.7% | 69.4% |
| Manager | 52.5% | 20.1% |
| Senior | 16.9% | 7.5% |
| Staff | 5.1% | 3.1% |
| Other | 1.7% | 0.7% |
| Years as a CPA | 6.3 | 11.8 |
| Age | 30-39 | 30-39 |
| Income | $40,000-$49,999 | $50,000-$59,999 |

***Notes:*** [a]N = 59.
[b]N = 136.

The median test was used to determine whether the value rankings of female CPAs differed from the value rankings of male CPAs. In addition, the effect of selected other demographic factors (position in firm, experience, age, and income) on the value rankings was tested using the median test. A significance level of .05 was used throughout the study.

## RESULTS

Demographic data for the respondents, CPAs in public practice, are presented in Table 1. The composite female respondent is a manager in a local CPA firm. In her thirties, she has been a CPA for 6.3 years with an annual income between $40,000 and $49,999. A partner (or sole proprietor) of a local CPA firm, the thirty-something composite male respondent has been a CPA for 11.8 years and has an annual income of $50,000-59,999.

### Instrumental Values Profile

Table 2 presents the instrumental values profile for the subjects in this study. CPAs perceived the values of "honest," "responsible," and "loving" as the most

***Table 2.*** Instrumental Values Profile (Relative Rankings by Median)

| *Ranking* | *Instrumental Value* | *Median* |
|---|---|---|
| 1 | Honest | 2.0 |
| 2 | Responsible | 4.0 |
| 3 | Loving | 6.0 |
| 4 | Capable | 7.0 |
| 5 | Helpful | 9.0 |
| 5 | Independent | 9.0 |
| 5 | Forgiving | 9.0 |
| 5 | Self-Controlled | 9.0 |
| 9 | Broadminded | 10.0 |
| 9 | Cheerful | 10.0 |
| 9 | Intellectual | 10.0 |
| 9 | Ambitious | 10.0 |
| 9 | Logical | 10.0 |
| 14 | Polite | 11.0 |
| 15 | Imaginative | 11.5 |
| 16 | Clean | 13.0 |
| 17 | Courageous | 14.0 |
| 18 | Obedient | 17.0 |

important instrumental values. The least important values were "clean," "courageous," and "obedient."

Analysis of the instrumental values profile in light of selected demographic factors revealed several interesting results. Between male and female respondents, the only significant difference was in the ranking of the value "independence." Female CPAs felt that this value was of much less importance than males.

The responses were then analyzed to determine if the instrumental value rankings varied by respondent position. Significant differences were noted for the values "broadminded" and "obedient," with CPAs who were partners or sole proprietors feeling these two values to be less important than their colleagues in other positions. The instrumental value "self-controlled" also varied significantly by respondent position; CPAs below the level of manager ranked it as less important than did their more highly placed colleagues.

The instrumental values profile was then analyzed to determine if the respondents' number of years experience as a CPA affected the rankings. Significant relations were noted for two values, "broadminded" and "self-controlled." CPAs with more than 10 years of experience felt "broadminded" to be less important than did less experienced CPAs, while CPAs with fewer than 10 years experience did not rank "self-controlled" as highly as their more experienced colleagues.

***Table 3.*** Terminal Values Profile
(Relative Rankings by Median)

| *Ranking* | *Terminal Value* | *Median* |
|---|---|---|
| 1 | Self-Respect | 4.0 |
| 1 | Family Security | 4.0 |
| 3 | Happiness | 5.0 |
| 4 | Freedom | 7.0 |
| 4 | A Sense of Accomplishment | 7.0 |
| 4 | Inner Harmony | 7.0 |
| 7 | Mature Love | 8.0 |
| 7 | True Friendship | 8.0 |
| 7 | Wisdom | 8.0 |
| 7 | A Comfortable Life | 8.0 |
| 11 | Salvation | 10.0 |
| 12 | A World at Peace | 12.0 |
| 13 | Equality | 13.0 |
| 13 | Pleasure | 13.0 |
| 15 | National Security | 14.0 |
| 16 | An Exciting Life | 15.0 |
| 16 | A World of Beauty | 15.0 |
| 16 | Social Recognition | 15.0 |

The instrumental values profile was unaffected by respondent income. When the rankings were analyzed with respect to respondent age, the relation was significant only for "broadminded," which respondents over the age of 40 viewed as less important than did younger CPAs.

## Terminal Values Profile

The terminal values profile for the respondents is presented in Table 3. CPAs ranked "self-respect," "family security," and "happiness" as their top three terminal values. The lowest-ranked values included "a world of beauty," "social recognition," and "an exciting life." The rankings, revealing an emphasis on personal-oriented values rather than society-oriented values, are consistent with the results reported by Swindle, Phelps, and Broussard (1987).

When the results were analyzed by respondent gender, three terminal values indicated significantly different rankings. Female CPAs ranked the values "social recognition" and "wisdom" as more important than did male CPAs. Surprisingly, "equality" was more important to male CPAs than to female CPAs. Analysis to determine if the terminal values rankings varied by respondent position found a significant relation for "pleasure" and "social recognition." Both of these terminal values were viewed as less important by CPAs at or below the rank of senior accountant.

The rankings of two terminal values were affected by the respondents' number of years experience as a CPA. CPAs with 11-20 years of experience felt "mature love" to be less important than did both their more and less experienced colleagues. The ranking of the value "pleasure" was somewhat puzzling, as the value was felt to be less important by CPAs with a great deal of experience (> 20 years) or relatively little experience (< 5 years) than by CPAs with a moderate amount of experience (5-20 years).

A similar pattern was observed when the terminal values were analyzed with respect to respondent age. The value "family security" was ranked as less important by the oldest (>50 years) and youngest (<30 years) CPAs than by CPAs in their 30s and 40s. The oldest and youngest age groups also placed a higher value on "self-respect" than did other CPAs. CPAs under 40 years of age ranked the value "mature love" more highly than did older CPAs. Respondents in their 50s felt "social recognition" to be more important than other CPAs in the sample.

The terminal values profile was then analyzed to determine the effect of respondent income on the value rankings with two significant relations found. CPAs earning $50,000 or more annually felt "equality" to be less important than did CPAs with lower incomes. Also, the ranking of "social recognition" was positively correlated with income, with higher earning CPAs viewing the value as more important than lower earning CPAs.

## SUMMARY AND DISCUSSION

This study provides insight into the value systems of certified public accountants. The instrumental values profile indicates that CPAs perceive "honest," "responsible," and "loving" to be types of conduct which are both individually and socially desirable. The terminal values profile infers that CPAs possess personal-oriented values such as "self-respect" and "family security" rather than social-oriented ones such as "social recognition" and "a world of beauty." This parallels the values profile of CPAs developed by Swindle, Phelps, and Broussard (1987). The findings in the current study provide additional evidence of an achievement-oriented value system for practicing CPAs. Several demographic characteristics significantly affected the rankings of terminal and instrumental values. Respondent gender, position, experience, age, and income were all correlated with the rankings of several values. Future research is needed to address the relation between personal values and demographic factors, including those incorporated in this study.

Identifying the relative importance of the values to CPAs is critical to an understanding of the basic motivations of practicing CPAs. Knowing that CPAs rank "honesty" and "responsibility" highly indicates that CPAs should be more inclined to act in accordance with these values. Furthermore, the personal orientation of CPAs supplies knowledge concerning the "self" viewpoint and focus of practitioners. Thus, an awareness of the underlying value systems of CPAs provides a vital framework for the integration of often conflicting individual and professional values with those of society in general. An understanding of the differenccs and similarities in existing value systems can help improve ethical training, alter accountant behavior, and, subsequently, increase the stature of the accounting profession. The future of the accounting profession will ultimately depend on the compatibility and synthesis of individual, professional, and societal values.

## REFERENCES

Baker, C. 1976. An investigation of differences in values: Accounting majors vs. nonaccounting majors. *The Accounting Review* (October): 886-893.

Brockhaus, R., and P. Horwitz: 1986. The psychology of the entrepreneur. In *The Art and Science of Entrepreneurship*, 25-48. Cambridge, MA: Ballinger.

Cochran, L. 1983. Implicit versus explicit importance of career values in making a career decision. *Journal of Counseling Psychology* 30(2): 188-193.

Coye, R. 1986. Individual values and business ethics. *Journal of Business Ethics* 5: 45-49.

DeCoster, D., and J. Rhode. 1971. The accountant's stereotype. *The Accounting Review* (October): 651-664.

England, G. 1967. Personal value systems of american managers. *Academy of Management Journal* (March): 53-68.

Fagenson, E. 1993. Personal value systems of men and women entrepreneurs versus managers. *Journal of Business Venturing* 8: 409-430.

Fagenson, E., and L. Coleman. 1987. What makes entrepreneurs tick: An investigation of entrepreneurs' values. In *Frontiers of Entrepreneurship Research*, eds. N. Churchill, J. Hornaday, B. Kirchhoff, O. Krasner, and K. Vesper, 202-203. Wellesley, MA: Babson College.

Feather, N. 1984. Masculinity, femininity, psychological androgyny and the structure of values. *Journal of Personality and Social Psychology* 47: 604-620.

Hall, B. 1973. *Value Clarification as a Learning Process*. New York: Paulist Press.

Liedtka, J. 1989. Value congruence: The interplay of individual and organizational value systems." *Journal of Business Ethics* 8: 805-815.

Posner, B., and W. Schmidt. 1993. Values congruence and differences between the interplay of personal and organizational value systems. *Journal of Business Ethics* 12: 341-347.

Rokeach, M. 1973. *The Nature of Human Values*. New York: Free Press.

Siegel, S. 1956. *Nonparametric Statistics for the Behavioral Sciences*. New York: McGraw-Hill.

Swindle, B., and L. Phelps. 1984. Corporate culture: What accounting students are not taught. *Northeast Louisiana Business Review* (Fall/Winter): 37-43.

Swindle, B., L. Phelps, and R. Broussard. 1987. Professional ethics and values of certified public accountants. *The Woman CPA* (April): 3-6.

Thielens, W. 1966. *Recruits for Accounting: How the Class of 1961 Entered the Profession*. New York: AICPA.

Wrench, D. 1969. *Psychology: A Social Approach*. New York: McGraw-Hill.

# POLITICAL PRESSURE AND ENVIRONMENTAL DISCLOSURE: THE CASE OF EPA AND THE SUPERFUND

Martin Freedman and A.J. Stagliano


## ABSTRACT

This paper is an analysis of SEC Form 10-K disclosures of liabilities under the Comprehensive Environmental Response, Compensation and Liability Act. The study assesses disclosure levels of SEC-registered companies that are known to be Superfund potentially responsible parties. In a comparison of two static states, disclosures are examined before and after an attempt by the Environmental Protection Agency to prompt SEC enforcement of toxic waste site cleanup cost reporting. Since future profitability and cash flows can be impacted materially by liabilities associated with cleaning up toxic waste sites, disclosure of the nature and extent of involvement with Superfund sites is of interest to a firm's stakeholders. The method of analysis used in this study was to categorize the disclosures of 140 Superfund-involved firms based on the content of their disclosures. These disclosures were weighted based on their perceived usefulness to statement users. Disclosures in 1987, the year prior to the EPA attempt to assure disclosure, are compared with disclosures in 1989 and 1990.


**Research on Accounting Ethics, Volume 4, pages 211-224.**

**ISBN: 0-7623-0339-5**

The results of the analysis show that the null hypothesis of no difference in disclosure over the time period studied is not accepted. Although there are a number of possible reasons that could account for the observed increase in disclosure, the fact that the EPA threatened to encourage SEC enforcement action might have been the principal motivation. One implication of this research is that statement preparers, reviewers, and readers must be more diligent in demanding full disclosure of the financial impacts of Superfund and other environmental legislation.

## INTRODUCTION

In 1989, the U.S. Environmental Protection agency (EPA) provided a list to the U.S. Securities and Exchange Commission (SEC) of those firms that were potentially liable for the cleanup of toxic waste sites. Apparently, there was an expectation on the part of the EPA that the SEC would more diligently enforce the appropriate disclosure laws to ensure that these firms reported their liabilities under Superfund (Marcus 1989). Why the EPA took the lead and wanted to assure disclosure of this information is not known specifically, but the significant social and political implications of this action are clear. From the standpoint of furnishing financial statement readers with useful information, the SEC certainly should have enforced the rules mandating such disclosures, since Superfund will require American businesses to spend more than $500 billion to clean up hazardous waste sites (Eisner & Company 1991). Not only may individual companies be materially affected by Superfund, but the cleanup has significant social impact for the communities in close proximity to toxic waste sites.

In this study, an analysis is provided of SEC Form 10-K disclosures for companies affected by Superfund. This analysis is conducted in a comparative context before and after the EPA sent the list of firms to the SEC. The paper is structured as follows. First, the background, including detail on the applicable laws, is presented along with a review of some relevant accounting literature. The research methodology is then described. Next, the empirical results are presented and analyzed. Finally, the conclusion, limitations, and implications are provided.

## ENVIRONMENTAL REGULATIONS AND FINANCIAL DISCLOSURE

The Comprehensive Environmental Response, Compensation and Liability Act (CERCLA) was enacted in 1980. In 1986, CERCLA was strengthened by the Superfund Amendments and Reauthorization Act (SARA). Brought together under the Code of Federal Regulations in Title 40 (captioned "Protection of the Environment"), these laws are known more commonly as Superfund. What they require is the cleanup of hazardous dump sites by parties responsible for the wastes. Responsible parties may include generators or transporters of waste, as well as past or present owners or operators of the sites. All of these potentially responsible parties are made strictly liable for the cleanup cost of sites that are included on a list promulgated annually by the EPA called the National Priorities List. To ensure recovery, liability under Superfund is joint and several among the parties.

Responsible parties can be ordered by the EPA to perform a cleanup, sued to recover the costs of any remediation financed by the EPA, or they may voluntarily settle with the federal government for their share of the cleanup costs. The individual liability that a responsible party can incur for cleanup costs at a given site is limited to $50 million by the Superfund legislation. There is no limit on the costs that might be incurred as a result of civil suits arising from damage claims filed by individuals harmed by hazardous waste sites.

Disclosure of a material potential liability for any environmental expenditure is required in the annual filing (Form 10-K) that a publicly reporting firm must make with the SEC. According to SEC Regulation S-K (Reg. Sec. 229.103), disclosures dealing with the discharge of materials into the environment should be provided in the legal proceedings section of Form 10-K whenever:

> A governmental authority is a party to each proceeding and such proceeding involves potential monetary sanctions, unless the registrant reasonably believes that such proceedings will result in no monetary sanctions, or in monetary sanctions, exclusive of interest and costs of less than $100,000, provided however that such proceedings which are similar in nature may be grouped and described generically.

Based on Statement of Financial Accounting Standards No. 5 (FASB 1975), a material liability must be accrued (i.e., given accounting recognition) if its existence is "probable" and its amount can be estimated reasonably. This action places the liability on the balance sheet and the concomitant expense in the income statement. If a material liability has been incurred, but a reasonable estimate cannot be made, it must be described in the footnote disclosures to the financial statements.

For many U.S. hazardous waste sites (e.g., Maxey Flats, KY; Operating Industries, CA; Mountain View, CA), hundreds of potentially responsible parties have been identified. It has been difficult for these companies to determine the precise extent of their liability. However, since the total cleanup cost for the toxic waste sites, in many cases, is between $50 and $100 million, recognizing and reporting some nominal portion of this cost, or describing a potential liability with an estimate of the aggregate cost to all parties, would not seem to be an onerous disclosure requirement. The fact that firms were not reporting their potential liabilities, or were reporting in ways that were not deemed satisfactory to the EPA, has implications to users of financial statements. Investors demand reliable disclosure of financial data. Other stakeholders, too, have a right to be informed about the financial health and production externalities of businesses.

Disclosure of the nature of the involvement of a corporation with toxic waste sites is of interest to users of financial statements. Investors, creditors, employees, customers, and suppliers are concerned when a company is, or may be, materially affected by the cleanup of toxic waste sites. Even if a company is not currently affected at the materiality threshold level, knowing that it bears some responsibility for waste site cleanup may alter the relationship between a firm and its stakeholders. Furthermore, stakeholders who have a particular interest in corporate social responsibility would like to know the firm's role with toxic waste sites and how effective the firm has been in cleaning up its share of the environmental damage. The community in which a Superfund site is located also would be interested in learning the nature of the problem and the status of the cleanup-especially since the health and safety of its members may be at risk.

The literature is replete with research papers that discuss social disclosure in general (see, e.g., Ingram 1978; Anderson and Frankle 1980; Trotman and Bradley 1981). There also are a number of studies that deal more specifically with environmental disclosures (Belkaoui 1976; Freedman and Jaggi 1986; Ingram and Frazier 1980; Shane and Spicer 1983; Wiseman 1982). In terms of the empirical literature in accounting, however, there appears to be only one article that deals with hazardous waste disposal (Rockness, Schlachter, and Rockness 1986).

Disclosures made by 21 chemical firms in their 1980 through 1983 annual reports were examined in the Rockness, Schlachter, and Rockness study. These authors concluded that few of the firms made any disclosures concerning hazardous waste disposal, either in general or specifically about CERCLA cleanups. Because this analysis was completed just after the passage of CERCLA and before the 1986 SARA amendments, it is possible that firms (or their auditors) were not aware of the potential implications of hazardous waste cleanup requirements on future financial performance. Another not unlikely possibility is that they chose to ignore reporting the financial implications until they received unequivocal instructions from the SEC.

Since SARA was enacted in 1986, the first year that Superfund became completely operative was 1987. With the full extent of the law known in 1987, there was no longer a possibility that changes in legislation could affect the liability. Therefore, disclosure should have begun for Superfund's impact in fiscal periods beginning after 1986. Yet, EPA's 1989 action suggests that it was not satisfied with the amount and/or quality of disclosures made about Superfund. In the view of the EPA at least, firms did not seem to be complying with the spirit of the disclosure requirements.

## METHOD OF STUDY

In this study, an analysis is made of the 1987 Form 10-K disclosures about firms' involvement with Superfund cleanups. These are compared to disclosures made by the same firms in 1989 and 1990.

The following null hypothesis is posited:

**Hypothesis 1.** There is no difference in Superfund disclosures made on SEC Form 10-K in 1987 compared to these disclosures in 1989 and 1990.

It is expected that the EPA's action, described above, will be one of the factors that provides impetus to increase the extensiveness of Superfund disclosures. Accordingly, there should be greater disclosure exhibited in 1989 and 1990 than in 1987. This is significant since rejection of the null hypothesis will show that political pressure can impact the judgmental position that a business firm takes in dealing with controversial financial disclosure matters.

## Sample

Although the EPA provided the SEC with a list of firms that should have been reporting on Superfund, public access to that listing has been denied. Generating a sample of firms that had some involvement with Superfund (i.e., potentially responsible parties) for 1987 required a number of steps. First, the National Priorities List (NPL) for 1987 was obtained; it contains a cataloging of 770 nongovernment Superfund sites. Included as part of the information provided in the NPL is identification of the current site owner. Based on that data, it was possible to determine that 48 publicly reporting companies were owners of one or more NPL-listed sites.

The owner of the site is only one of many possible responsible parties. Any firm that generated waste at the site or transported waste to the site is a potentially responsible party. To identify those other than site owners who might be involved, the complete 1987 SEC Form 10-K filings were searched. This electronic search was conduced on the NEXIS database to obtain a list of firms that reported anything about Superfund or CERCLA. The result was identification of 169 companies that mentioned something about Superfund in their annual SEC filing. The Form 10-K of these companies, along with some of the 48 site owner companies not included among the 169, were reviewed to determine what was disclosed. Upon reading the Form 10-Ks, some firms were eliminated since they had no involvement in Superfund liabilities (e.g., environmental service companies) and some were added (e.g., Exxon) since they were mentioned as potentially responsible parties by other companies.

Certain industries had a particularly strong representation in the sample. It seemed useful to include in the analysis the Form 10-K of firms from these industries: chemicals, oil refining, electric utilities and electronics. The 1987 sample consisted of 140 firms that were potentially liable for Superfund costs. For 1989 and 1990, the reports of these same 140 firms were analyzed.

## Content Analysis

The method of analysis utilized in this study was to categorize disclosures based on the content of what was disclosed and then to weight the disclosure based on its expected importance to users. This method is different from the standard content analysis technique of quantifying the disclosure by enumerating words, lines, or sentences. How much is said is not as important as what is said. Classification into established categories of social disclosure creates a better understanding of the message conveyed by the report. This variation of content analysis has been applied successfully in prior research of this nature (see Wiseman 1982; Freedman and Wasley 1990).

Based on an analysis of the disclosures, they were classified on the following criteria:

1. Disclosure that the company is a Superfund potentially responsible party (PRP).
2. Mention of specific Superfund sites and/or the number of sites for which the company is a PRP.
3. Mention of whether the company is, or is expected to be, materially impacted by the cost of site cleanup.
4. Monetary disclosure as to the expected cost of site cleanup.

The categories chosen were based on the panorama of disclosure variations that actually were found in the Form 10-Ks examined. Since the extensiveness of disclosure before and after the EPA intervention is the important variable considered in this study, it is necessary to weight these disclosures. One possibility is to give each of the disclosure items an equal weight and defend that on the basis that since it is impossible to arrive at an unbiased weighting scheme each category should be treated the same. However, it is clear that some disclosures provide more information than others to users and equal weighting does not do justice to that fact. The benefit of disclosure is not only that it exists but that it provides useful decision-relevant information to stakeholders.

The weighting scheme chosen takes into account the potential differential user benefit of the disclosure categories given above. The weights are:

| | *Type of Disclosure* | *Points* |
|---|---|---|
| Category 1: | Disclosure of something about Superfund | 2 |
| Category 2: | Disclosure of the number of sites | 3 |
| | Disclosure of specific site locations | 3 |
| Category 3: | Disclosure of materiality | 1 |
| Category 4: | Disclosure of monetary consequences | 4 |

Stating that the company is a potentially responsible party under Superfund conveys information, but it does not inform a reader whether this is a minor problem with a few sites or a major problem involving large potential expenses. Disclosing the number of sites provides information as to how much litigation may be involved, but it does not describe the extensiveness (or cost) of the potential cleanup. Usually, when the company provides information about specific sites, it makes a fairly involved disclosure as to the nature of the problem and whether the company is a major or minor player. Finally, disclosing the potential liability in dollars, and whether the company judges this to be material or not, helps the reader decide the relative importance of PRP status to the future of the company.

This weighting scheme is one of many possible weighting schemes that could be developed for these disclosures. It would be possible to construct a different weighting scheme and justify it. For this particular study, one in which the extensiveness of disclosure before and after an event is examined, the method selected provides a useful metric to quantify the extent of the disclosure. It has the benefit of simplicity of construction and interpretation and is virtually free of scorer bias.

***Table 1.*** Index Scores for Firms Not Disclosing PRP Status in 1987

| *Company* | *1989* | *1990* |
|---|---|---|
| American Petrofina | 0 | 0 |
| Amoco | 3 | 0 |
| Burlington Northern | 9 | 0 |
| Chrysler | 6 | 10 |
| Circle K | 0 | 6 |
| Diamond Shamrock | 0 | 0 |
| DuPont | 7 | 7 |
| Exxon | 6 | 0 |
| General Mills | 0 | 0 |
| Holly Corp. | 0 | 0 |
| Litton Industries | 0 | 3 |
| Lockheed | 2 | 2 |
| Morton Industries | 12 | 12 |
| Murphy Oil | 0 | 3 |
| Phillips Petroleum | 0 | 0 |
| Rockwell International | 0 | 9 |
| Southern Company | 0 | 2 |
| Teledyne | 0 | 5 |
| Tesoro Petroleum Corp. | 6 | 0 |
| Thiokol | 0 | 0 |
| Union Electric | 6 | 0 |
| Whitehall Corp. | 0 | 0 |
| Mean Score ($n$ = 22) | 2.6 | 2.7 |

**Table 2.** Index Scores for Firms Not Disclosing PRP Site in 1987

| *Company* | *1987* | *1989* | *1990* |
|---|---|---|---|
| Allied Signal | 6 | 3 | 9 |
| Amerada Hess | 3 | 3 | 3 |
| American Electric Power | 2 | 6 | 6 |
| Ashland Oil | 3 | 3 | 6 |
| Bethlehem Steel | 3 | 10 | 6 |
| Browning-Ferris Industries | 3 | 6 | 6 |
| Cooper Co. | 3 | 6 | 0 |
| Digital Equipment | 3 | 3 | 3 |
| Dominion Resources | 2 | 2 | 3 |
| Dover Corp. | 3 | 3 | 3 |
| Duke Power | 3 | 6 | 9 |
| Eastern Utilities Associates | 3 | 2 | 12 |
| Echo Bay Mines | 3 | 3 | 3 |
| Ecolab | 3 | 6 | 6 |
| FMC Corporation | 3 | 6 | 0 |
| Freeport-McMoRan | 3 | 3 | 7 |
| Fruit of the Loom | 3 | 3 | 2 |
| Iowa-Illinois Gas & Electric | 3 | 3 | 6 |
| Kaneb Services Inc. | 2 | 2 | 2 |
| Kraft General Foods | 3 | 3 | 3 |
| Maxus Energy | 3 | 8 | 10 |
| Merck & Co. | 3 | 3 | 0 |
| Mobil Corp. | 3 | 3 | 3 |
| Millipore Corporation | 3 | 3 | 10 |
| Quantum Chemical | 2 | 2 | 3 |
| SPS Technologies Inc. | 6 | 0 | 0 |
| United Technologies | 3 | 3 | 6 |
| Valhi Inc. | 3 | 6 | 5 |
| Waste Management | 3 | 6 | 6 |
| Westinghouse Electric | 3 | 3 | 6 |
| Whirlpool Corporation | 3 | 3 | 3 |
| Mean Score ($n$ = 31) | 3.1 | 3.9 | 4.7 |

## RESULTS

The disclosure scores of each company for 1987, 1989, and 1990 are shown in Tables 1 through 4. To provide a readily discernible comparison of the change in disclosure over the period, the tables are classified based on the disclosure

***Table 3.*** Index Scores for 1987 PRP Site Disclosers Providing No Quantitative Data

| *Company* | *1987* | *1989* | *1990* |
|---|---|---|---|
| Advanced Micro Devices | 5 | 0 | 3 |
| Amax Inc. | 9 | 13 | 13 |
| American Cyanamid | 8 | 9 | 9 |
| AMR | 9 | 9 | 6 |
| Baker Hughes | 9 | 9 | 9 |
| Borden, Inc. | 5 | 6 | 6 |
| Boston Edison | 5 | 6 | 6 |
| Centerior Energy | 6 | 6 | 9 |
| Central & Southwest | 8 | 8 | 9 |
| Chemical Waste Management | 6 | 6 | 6 |
| Cincinnati Gas & Electric | 6 | 6 | 6 |
| Consolidated Edison | 6 | 6 | 13 |
| Crane Co. | 9 | 9 | 13 |
| Dexter Corp. | 6 | 6 | 9 |
| Duquesne Light | 6 | 7 | 8 |
| EMHart | 6 | 6 | 9 |
| Ford Motor Co. | 6 | 2 | 2 |
| General Public Utility | 5 | 13 | 13 |
| Gulf States Utilities | 8 | 9 | 12 |
| Hecla Mining Co. | 6 | 6 | 5 |
| Hinderliter Industries Inc. | 6 | 10 | 10 |
| Homestake Mining | 8 | 9 | 9 |
| IMO Industries Inc. | 6 | 3 | 6 |
| Inland Steel | 8 | 6 | 9 |
| Kerr-McGee Corporation | 5 | 5 | 10 |
| Keystone Consolidated Industries | 8 | 9 | 12 |
| KN Energy | 8 | 13 | 12 |
| Knight-Ridder, Inc. | 5 | 6 | 9 |
| LTV | 5 | 12 | 13 |
| Long Island Lighting Co. | 5 | 5 | 13 |
| Martin Marietta | 9 | 9 | 6 |
| Metropolitan Edison Co. | 6 | 8 | 12 |
| Monsanto | 6 | 2 | 10 |
| Newmont Mining | 8 | 12 | 13 |
| Occidental Petroleum | 5 | 9 | 10 |
| Oklahoma Gas & Electric | 6 | 6 | 13 |
| PacifiCorp | 6 | 6 | 6 |
| Pacific Power & Light | 5 | 9 | 5 |

*(continued)*

**Table 3** (Continued)

| *Company* | *1987* | *1989* | *1990* |
|---|---|---|---|
| Public Service Enterprise Group | 9 | 9 | 9 |
| Public Service of Colorado | 8 | 8 | 8 |
| Questar Corp. | 6 | 6 | 13 |
| Raytech | 9 | 0 | 0 |
| Raytheon | 9 | 0 | 8 |
| Rohr Industries | 6 | 6 | 9 |
| San Diego Gas & Electric | 5 | 5 | 12 |
| Stepan Co. | 6 | 5 | 8 |
| Tacoma Boatbuilding Co. | 9 | 9 | 9 |
| Tosco Corp. | 6 | 6 | 8 |
| TRW Inc. | 6 | 3 | 3 |
| UNC Inc. | 5 | .5 | 13 |
| Union Pacific | 8 | 5 | 5 |
| Univar Corp. | 5 | 10 | 10 |
| Witco | 9 | 7 | 3 |
| Mean Score (n = 53) | 6.7 | 6.9 | 8.7 |

category in which the firm was placed for 1987. Except for those firms that made quantitative and site-specific disclosures in 1987 (Table 4), it should be clear that from 1987 to 1990, firms increased their disclosures.

Both parametric (*t*-tests) and nonparametric (Wilcoxin matched-pairs signed-rank test and Friedman two-way ANOVA) procedures were conducted to detect any statistical difference in the mean disclosure (as determined by the index described above) over the time periods. Since there was no detectable dissimilarity in the outcomes, the more robust *t*-test results are shown in Table 5.

For the sample as a whole, there is no statistically significant difference between the mean disclosures for 1989 compared with 1987. However, the difference between 1987 and 1990 is statistically significant at the .05 level. The difference in the means for the 1989 to 1990 comparison has a *t*-score at the .14 probability level.

Based on the results for the complete sample, the general null hypothesis that there is no difference in disclosure over the time period is not accepted. Although the disclosure level was not much greater in 1989 as compared to 1987, disclosures made in 1990 were significantly greater than those made in 1987. The strength of this difference leads to tentative rejection of the null hypothesis.

The sample was segmented to identify those firms that made the least disclosure in 1987. After a mean disclosure level for this subgroup was computed, a comparison with the two later years' disclosure index was made. Panel B of Table 5 contains the results of this test. It is fairly evident that those firms that made the

**Table 4.** Index Scores for 1987 PRP Site Disclosers Providing Quantitative Data

| *Company* | *1987* | *1989* | *1990* |
|---|---|---|---|
| AO Smith | 9 | 9 | 0 |
| ARCO Chemical | 13 | 9 | 10 |
| Armco | 9 | 10 | 13 |
| Atlantic Richfield | 13 | 13 | 13 |
| AVX Corp. | 9 | 12 | 10 |
| Baltimore Gas & Electric | 9 | 9 | 10 |
| Beard Oil | 13 | 0 | 0 |
| Berry Petroleum | 12 | 6 | 0 |
| Buckeye Partners | 9 | 10 | 2 |
| Cabot Corp. | 12 | 6 | 0 |
| Carolina Power & Light | 9 | 9 | 10 |
| Central Louisiana Electric | 9 | 10 | 6 |
| Charter Co. | 12 | 12 | 10 |
| Compudyne Corp. | 8 | 13 | 13 |
| Consolidated Rail Corp. | 10 | 6 | 13 |
| Ferro Corp. | 12 | 2 | 0 |
| General Electric | 9 | 10 | 6 |
| General Motors | 10 | 10 | 10 |
| Green Mt. Power | 10 | 10 | 13 |
| Grumman Corp. | 12 | 6 | 6 |
| Hercules Inc. | 9 | 3 | 7 |
| Kaiser | 9 | 6 | 13 |
| Manville Corp. | 10 | 5 | 6 |
| National Presto Industries | 12 | 5 | 2 |
| Northrop Corp. | 10 | 10 | 10 |
| Olin Corp. | 10 | 10 | 10 |
| Plymouth Rubber Inc. | 12 | 12 | 10 |
| Philadelphia Electric Co. | 12 | 12 | 13 |
| Publicker Industries Inc. | 10 | 10 | 12 |
| Smith International Inc. | 12 | 13 | 13 |
| SmithKline Beecham | 13 | 6 | 6 |
| UGI Corp. | 10 | 10 | 3 |
| USX | 10 | 12 | 13 |
| Winn-Dixie Stores, Inc. | 13 | 13 | 13 |
| Mean Score ($n$ = 34) | 10.6 | 8.8 | 8.1 |

**Table 5.** *T*-Tests for Differences in Index Score Means

| | *1989 versus 1987* | *1990 versus 1987* | *1990 versus 1989* |
|---|---|---|---|
| | ***Panel A: All Companies*** | | |
| *T*-statistic | .47 | 1.96 | 1.48 |
| Probability | (.56) | (.05) | (.14) |
| Sample size | 140 | 140 | 140 |
| | ***Panel B: Least Disclosing 1987 Companies*** | | |
| *T*-statistic | 3.45 | 3.90 | .81 |
| Probability | (.001) | (.001) | (.42) |
| Sample size | 53 | 53 | 53 |

least disclosure in 1987 increased their disclosure significantly in comparisons of both the 1987 to 1989 and 1987 to 1990 periods. From 1989 to 1990, there was not a significant increase in disclosure; all of the major change may well have occurred by 1989.

Although there are a number of possible reasons for the increase in disclosure for both the sample as a whole and for those firms that disclosed the least in 1987, the fact that the EPA threatened to encourage SEC enforcement of the disclosure rules might have been the principal one. This seems to be the case since those firms that disclosed the least made the most dramatic changes in the extensiveness of their disclosure.

It is also possible that the least-disclosing firms were aware that firms that disclosed more did not seem to be penalized in any way. Therefore, since disclosing did not have negative consequences, and not disclosing might have some, many of these firms may have decided to change their stance on disclosure. Still, of the 22 firms that made no disclosure in 1987, 13 made no disclosure in 1989, and 12 continued with this posture of no disclosure in 1990. Clearly these firms had no fear of either the EPA or the SEC claiming that their reporting was inadequate. On the positive side, of the 31 firms that disclosed that they were a potentially responsible party in 1987 but did not describe the site(s) for which they were responsible, 12 provided information about the sites or at least stated the number of sites in 1989. With the help of some political influence, progress is being made.

## CONCLUSION

Providing information about a firm's responsibility for toxic wastes is of interest to the company's stakeholders. Despite the need for this information, and the fact that it is legally required for publicly held companies to disclose it, some firms appear to make little or no disclosure effort. Presumably, as a result of this lack of disclosure, the EPA provided the SEC with a list of firms that were not complying with the disclosure laws concerning toxic wastes and asked the SEC to enforce the rules.

Based on the results of this study, it appears that firms that provided little or no disclosure prior to the EPA's action have significantly increased their disclosure about toxic wastes. Whether or not this was as a result of the EPA's action is open to conjecture. Still, the essential point is that firms have increased their disclosure and, in a sense, are owning up to their responsibilities regarding the environment.

It also is important to note that despite this increase in disclosure a number of firms still are not reporting about toxic wastes in their annual financial statements. Furthermore, many of the firms that provide information do not provide enough for even a sophisticated reader of financial statements to make a meaningful assessment of the company's risk concerning toxic wastes and environmental damage.

Finally, as can be noted from the data gathered in this research, companies that are similarly situated regarding environmental cleanup liabilities, and impacted in the same way by the Superfund law, report this event in a myriad of ways. Differential reporting would not be expected given the rather explicit legislative and professional reporting requirements. The results of this examination suggest that statement preparers (management), reviewers (independent auditors), recipients (SEC), and readers (equity owners and other interested or affected stakeholders) must be more diligent in the pursuit of full disclosure in connection with the financial impacts of Superfund and other environmental legislation.

## REFERENCES


Anderson, J., and A. Frankle. 1980. Voluntary social reporting: An iso-beta portfolio analysis. *The Accounting Review* (July): 167-179.

Belkaoui, A. 1976. The impact of the disclosure of the environmental effects of organizational behavior on the market. *Financial Management* (Winter): 26-31.

Eisner & Company. 1991. Trends and developments. *Newsletter* (January): 1-3.

Financial Accounting Standards Board (FASB). 1975. *Statement of Financial Accounting Standards No. 5: Accounting for Contingencies*. Stamford, CT: FASB.

Freedman, M., and B. Jaggi. 1986. An analysis of the impact of corporate pollution disclosures included in annual financial statements on investors' decisions. In *Advances in Public Interest Accounting*, Vol. 1, eds. M. Neimark, T. Tinker, and B. Merino. 193-212. Greenwich, CT: JAI Press.

Freedman, M., and C. Wasley. 1990. The association between environmental performance and environmental disclosure in annual reports and 10-Ks. In *Advances in Public Interest Accounting*, Vol. 3, eds. M. Neimark, T. Tinker, and B. Merino. 183-194. Greenwich, CT: JAI Press.

Ingram, R. 1978. An investigation of the information content of (certain) social responsibility disclosure. *Journal of Accounting Research* (Autumn): 270-285.

Ingram, R., and K. Frazier. 1980. Environmental performance and corporate disclosure. *Journal of Accounting Research* (Autumn): 614-622.

Marcus, A. 1989. Firms ordered to come clean about pollution. *Wall Street Journal* (November): B1, B8.

Rockness, J., H. Schlachter, and H. Rockness. 1986. Hazardous waste disposal, corporate disclosure, and financial performance in the chemical industry. In *Advances in Public Interest Accounting*, Vol. 1, eds. M. Neimark, T. Tinker, and B. Merino. 167-192. Greenwich, CT: JAI Press.

Shane, J., and B. Spicer. Market response to environmental information produced outside the firm. *The Accounting Review* (July): 521-538.

Trotman, K., and G. Bradley. 1981. Association between social responsibility disclosure and characteristics of companies. *Accounting, Organizations and Society* 6(4): 355-362.

Wiseman, J. 1982. An evaluation of environmental disclosures made in corporate annual reports. *Accounting, Organizations and Society* 7(1): 53-63.

# ETHICAL DEVELOPMENT OF CMAs: A FOCUS ON NON-PUBLIC ACCOUNTANTS IN THE UNITED STATES

Lois Deane Etherington and Nancy Thorley Hill

## ABSTRACT

This study remedies a gap in accounting ethics research by examining ethical circumstances and the level of ethical development for a large sample of Certified Management Accountants (CMAs). Prior research focusing on Certified Public Accountants (CPAs) suggests that those with high moral reasoning may not choose public accounting initially, may leave public accounting voluntarily, or may be forced out due to failure to be promoted. This study hypothesizes that professional certified accountants, CMAs, who are not employed in public accounting will have higher ethical reasoning abilities than CPAs in public accounting. Results of the study show, however, that the ethical reasoning ability of CMAs employed in non-public accounting is as low as that of CPAs. There is also no significant difference in moral reasoning found between CMAs who had formerly worked in public accounting and those who had never been employed by a public accounting firm. The study also examines the relationship between ethical reasoning ability and gender, social belief, and position held within the firm. Results show that males and those who

**Research on Accounting Ethics, Volume 4, pages 225-245.**

**ISBN: 0-7623-0339-5**

classify themselves as conservative have significantly lower moral reasoning than women and those self-classifying as liberals or moderates. When the moral reasoning of U.S. CMAs is compared to that of Canadian CMAs, the Canadian CMAs' moral reasoning is found to be higher. The study also examines the frequency of ethical dilemmas faced by management accountants and reports to whom they turn when confronted with ethical dilemmas.

## INTRODUCTION

Recent years have brought increased public concern to ethical issues in many aspects of society and business. Research on ethics in accounting has taken several streams, among them whistle-blowing (Arnold and Ponemon 1991; Finn and Lampe 1992), auditing (Lampe and Finn 1992; Loeb 1971, 1984; Ponemon 1990; Shaub et al. 1994), and accounting education (Armstrong 1987; Cohen and Pant 1989; Loeb 1988, 1990). Some ethics research has examined ways to measure ethical compliance, focusing on factors that affect accountants' ethical behavior (Coreless and Parker 1987; Farmer et al. 1987).

Another body of work has utilized cognitive-developmental psychology to examine the moral reasoning and judgments of accountants in public practice (e.g., Armstrong 1987; Ponemon 1988, 1990, 1992; Ponemon and Gabhart 1993). This latter work is based on Kohlberg's theory of moral reasoning that posits a stage structure of moral development that consists of three levels and six stages (Kohlberg 1976). Based on Kohlberg's theory, James Rest (1979) developed the Defining Issues Test (DIT) which allows researchers to assess the moral reasoning ability of individuals. The DIT categorizes an individual's responses by Kohlberg's stages of moral reasoning. The resulting test measure, called a P score, can then be compared within and across samples of interest.

The first objective of the study is: to obtain a measure of ethical reasoning from a large sample of CMAs currently working in industry, government, or nonprofit organizations; to compare the ethical reasoning ability of CMAs with that of CPAs in public practice to address the question of whether all American accountants or only American public accountants have low ethical reasoning; and to compare ethical reasoning levels of the subsets of CMAs in industry who left public accounting firms and CMAs who were never in public practice with CPAs in public practice. The second objective is to examine characteristics of the responding CMAs that may be related to ethical reasoning including gender, social belief, and position held within an organization.

In addition to the above analysis, we take advantage of prior research using Canadian CMAs and provide an international comparison of the two peer groups. The work environment of CMAs is very much the same in the two countries, and in each country the CMA is a professional designation that does not carry with it a "license to practice." Etherington and Schulting's (1995) study of Canadian

CMAs, using Rest's DIT, showed they have a level of ethical development similar to that exhibited by Canadian public accountants (Ponemon and Gabhart 1993), which is significantly higher than that found for American CPAs. Of interest in the current study is whether different levels of moral reasoning exist between CMAs in the United States and Canada, and whether cultural differences may be a factor.

We also obtain additional data about ethical issues in the work environment. Results of these inquiries should have implications for management when devising policies that attempt to promote corporate ethical behavior, implications for corporate accountants in dealing with ethical issues, and implications for accounting educators.

## MOTIVATION

This research focuses on U.S. Certified Management Accountants (CMAs) practicing in industry, government, and nonprofit organizations. While there has been much research using Certified Public Accountants (CPAs) in public practice, there is little research which focuses on CMAs in non-public settings. Jeffrey and Weatherholt (1995) and Etherington and Schulting (1995) are among the few researchers that have examined the ethical reasoning of accountants not in public practice. Preliminary evidence from Jeffrey and Weatherholt (1995) found, in a nonrandom sample of U.S. corporate accountants from three companies, that the ethical reasoning ability of corporate accountants was not higher than that of their sample of public accountants. In contrast, Etherington and Schulting's (1995) examination of Canadian CMAs found higher ethical reasoning for these practitioners than U.S. CPAs. A focus on CMAs is necessary for two reasons. First, CMAs play a significant role in industry, government, and nonprofit organizations (referred to subsequently as "in industry"). Further, unlike public accountants, accountants in industry are not typically independent of the organizations for whom they do accounting work. De Fond and Jiambalvo (1991), Bruns and Merchant (1990), Merchant and Rockness (1994), and Mihalek et al. (1987) document that accountants in industry are subject to ethical dilemmas, particularly those where loyalty to employers and their wishes may conflict with professional standards. The outcome of ethical conflicts such as managing financial results and falsification of records may affect accountants themselves, shareholders, board members, top level management, and the reputations of their organizations.

Second, information about the ethical reasoning abilities of accountants not in public practice should provide information to support or reject explanations put forth in prior research for the low moral reasoning found for American public accountants. For example, if CMAs in industry have ethical reasoning levels similar to those of their peer groups of similar educational level outside the accounting profession (and, therefore, higher than the scores of CPAs), then further research on the practice of public accounting and the socialization of public accountants is necessary to understand and explain the lower ethical reasoning abilities found for CPAs. If, however, CMAs in industry show the same low scores as CPAs, then the low ethical reasoning abilities for all U.S. accountants can perhaps be explained by an overall examination of accounting education, the practice of accounting, or why particular students self-select into the field. We discuss the importance of this exploration below.

## BACKGROUND

While many studies have used the DIT, Armstrong (1987) was one of the first to use it in accounting research. Prior research using the DIT suggests that those with more education usually have higher levels of ethical reasoning (Rest 1986b). However, in 1984 and 1985 samples of CPAs, Armstrong (1987) found that CPAs did not have levels of ethical reasoning that were commensurate with samples having similar educational backgrounds. Instead, the CPAs showed ethical reasoning equivalent only to high school graduates. She suggested that "their college education may not have fostered moral growth" (1987, 33).

Later research has consistently produced the same results and found in addition that at the partner and manager positions, the two top positions in public accounting firms, the levels of ethical reasoning are lower than that which is found in more junior members of the firms (Ponemon 1988, 1992; Ponemon and Gabhart 1990; Shaub 1994), with partners having the lowest levels of ethical reasoning of all. It has been hypothesized that there must be something about the competitive nature of public accounting that stymies the level of ethical development normally found in persons of that educational achievement (a mean of 1.1 years of graduate education) (Ponemon 1992; Ponemon and Gabhart 1993). Considerable research has documented that those at the top tend to promote people like themselves (Chatman 1991; Maupin and Lehman 1994; McNair 1991; Ponemon and Gabhart 1993). It has been asserted that this has led those with higher levels of moral reasoning to self-select out of public accounting (Ponemon and Gabhart 1993) or to fail to be promoted and thus fall victim to the strict "up-or-out" policies found particularly in large public accounting firms (McNair 1991).

These research findings are of some concern, as CPAs in public practice have, in their responsibility for certifying financial statements, the trust of the public and, indirectly, the trust of the governments who authorize them to attest to the representativeness of financial statements and the current status of the entities they audit. Their ethics and independence from the businesses (and governments) that they audit are supposed to be beyond question. While moral reasoning is not a total predictor of ethical work behavior, ethical reasoning has been found to be related to ethical behavior of accountants by a number of researchers (Arnold and Ponemon 1991; Bernardi 1991; Ponemon 1992a, 1992b; Ponemon and Gabhart 1993; Ponemon and Glazier 1990; Rest and Narvaez 1994).

Other studies of accountants in Canada and Ireland have examined whether the lower moral development of public accountants in the United States is found in other countries. Hill et al. (1998) examined Irish public accountants and found that Irish Chartered Accountants (CAs), like American CPAs, had ethical reasoning levels significantly lower than what would be expected of people at their educational level. In contrast, Ponemon and Gabhart (1993) examined Canadian public accountants (CAs) and found the level of ethical reasoning that would be expected, given their educational level. That is, unlike American CPAs, Canadian CAs' ethical reasoning was significantly higher than that of high school graduates, and CAs at higher positions in the firms did *not* exhibit lower levels of ethical reasoning. While the Ponemon and Gabhart study was not able to ascertain causes, they suggested that cultural differences and/or differences in education may have caused the differing scores between the public accountants in the two countries. The findings are certainly of interest, especially since university accounting curricula in Canada and the United States are very similar, the Big Six firms are common to both countries, and some in-firm training in the large firms is carried out in common (Etherington and Schulting 1995).

It has been suggested that accounting may attract students with lower levels of ethical reasoning (Shaub 1994; Lampe and Finn 1992; St. Pierre et al. 1990), but other studies (Ponemon and Glazer 1990) found that students and CPAs from a liberal arts college had higher DIT P scores, and Jeffrey (1993) found accounting students had higher DIT P scores than students in other programs at the same university. Icerman et al. (1991) found that although accounting students had low DIT scores, their DIT scores were higher than other business majors. Given the somewhat conflicting results, the question of whether the field of accounting is more attractive to those with lower ethical reasoning remains unanswered.

Prior research has also identified characteristics that may positively and negatively affect DIT P scores. This study examines three variables included in prior research that show inconsistent results and may be important in this study of CMAs: gender, social belief, and position held in the organization. First, the early research on moral reasoning and gender found no significant differences between men and women subjects (Rest 1979, 1986b; Thoma 1984). Most later work in accounting, however, has found that women have significantly higher scores than their male colleagues (Etherington and Schulting 1995; Lampe and Finn 1992; St. Pierre et al. 1990; Shaub 1994). This study includes an examination of gender and ethical reasoning abilities for CMAs.

Next, there is also some evidence that those who classify themselves as having more "liberal" social views have higher moral reasoning levels than those who view themselves as "conservative" (Rest 1986b; Sweeney 1995). Sweeney posited that the results cound be attributed to conservative political doctrine emphasizing adherence to rules and authority, whereas a liberal philosophy tends to focus on balancing competing interests (1995, 218). Indeed, it has been suggested that accounting is a profession that attracts conservative rule-oriented individuals (Ponemon 1990,

1992). Eynon et al. (1996), in a recent study of Irish and American students, found that P scores decreased monotonically as self-reported conservatism in social beliefs increased. While the rule-orientation and social beliefs of U.S. public accountants as compared to those of the general population is not known, Davidson and Dalby (1993), who administered a complex personality test to Canadian public accountants, found that the Canadian CAs were significantly less rule-oriented and more expedient than the general population. They were also significantly more "liberal, critical, open to change" than the general population, as opposed to "conservative, respecting traditional ideas" (Cattell et al. 1970). The present research examines social belief and ethical reasoning for a sample of U.S. CMAs.

Finally, prior research reports inconsistencies in ethical reasoning abilities and position held within the organization. Etherington and Schulting's (1995) study of Canadian CMAs tested for differences in P scores by position held in the organization (staff, middle, or upper-level management). Findings showed that there was no significant difference between position held and P score. This is in contrast to work cited earlier, where ethical reasoning abilities for U.S. CPAs was found to be lower at advanced positions in public accounting firms. Whether American CMAs exhibit the same ethical reasoning at all levels of the organization, as do Canadian CMAs, or whether their moral reasoning varies by organization level can indicate whether theoretical reasons for lower moral reasoning levels at higher levels in CPA firms are supported.

As suggested by conflicting results in prior research, additional studies of accountants are critical. While an "accounting degree" may be considered generic, the practice of accounting is often quite specific. Analysis of moral reasoning abilities of accountants in various practices—for example, industry, "Big Six" practitioners, tax accountants, auditors in smaller firms, and so forth—may provide insights about moral reasoning levels that may better guide changes in ethical training within university curricula and/or continuing professional education. In addition, research can provide information about critical variables related positively and negatively to moral reasoning abilities. This information may also help direct changes in education and practice.

## HYPOTHESES

Prior research of DIT scores for CPAs shows lower levels of moral reasoning than expected, given age and education. That is, college-educated adults have higher DIT scores than the average DIT found for CPAs in various studies (Armstrong 1987; Ponemon 1992; Rest 1986). Some suggest, however, that accountants with higher ethical reasoning do not initially enter public accounting, leave public accounting voluntarily, or are forced out. This may explain the lower-than-average scores for CPAs. If these assertions are correct, then CMAs (who choose non-public accounting) may have higher ethical reasoning ability than the CPAs sampled in prior research. It is therefore hypothesized that:

**Hypothesis 1.** The ethical development of CMAs in industry, government, and nonprofit organizations will be significantly higher than that of U.S. CPAs in public accounting.

Following this reasoning, it is hypothesized that accountants who have left public accounting and now work in industry, government, or nonprofit organizations and those who were never in public accounting should have higher ethical reasoning than CPAs in public accounting.[1]

**Hypothesis 1a.** The ethical development of those CMAs who have left public accounting will be significantly higher than that of CPAs in public accounting.

**Hypothesis 1b.** The ethical development of CMAs who were never in public accounting will be significantly higher that that of CPAs in public practice.

Prior research, discussed above, has investigated various characteristics and their relationship to ethical reasoning ability. This study gathers data on three such variables and tests the following hypotheses. First, recent accounting research on gender and ethical reasoning has found that women have higher ethical development than their male counterparts. Hypothesis 2 tests this relationship for CMAs.

**Hypothesis 2.** The ethical development of female CMAs will be statistically higher than that of male CMAs.

Second, it was found by prior researchers that persons who have liberally oriented views on social and economic policies had higher ethical reasoning than those who classified themselves as conservatives. Therefore, the following is posited:

**Hypothesis 3.** Ethical reasoning will vary inversely with conservatism for CMAs.

Finally, as differences in ethical reasoning at different hierarchical positions in the firm appears to date to be found only in studies of U.S. CPAs, the following is hypothesized:

**Hypothesis 4.** The level of ethical development of CMAs in senior management, middle management, and staff accountant positions will be similar.

# METHOD

## The Subjects

The sample was obtained from the current membership list of the American Institute of Management Accountants, provided to the researchers by the Institute. It contained over 10,000 names, addresses, job titles, and professional designations. The sample was selected from those with the CMA designation in order to examine a sample of accountants working in industry, government, and nonprofit organizations and to enable comparisons with the prior studies of CPAs and Canadian CMAs. Using the job titles, the CMAs were preclassified into senior management, middle management, and staff accountants,[2] after which approximately 500 were randomly selected from each of the three groups. The subjects were also asked on the questionnaire to classify themselves as staff, middle, or senior management. Those who were educators, consultants, working for public accounting firms, or unemployed were not included in the sample.

To ensure anonymity, no coding that would permit identification was used. This precluded a follow-up for nonrespondents, but the response rate of 38% indicates the importance of ethics to the respondents.[3] Data about the sample is provided in Table 1. Despite the researchers' intention not to sample CMAs who worked for public accounting firms, 34 surveys in this category were returned, possibly because of inaccurate job titles on the membership list. Their scores were not included in the analysis (as noted in Table 1).

## The Instrument

The three-story version of Rest's DIT was used.[4] Rest's DIT (1986a) is an instrument based on Kohlberg's six stages of moral development and is considered to be the most prominent and reliable objective test of cognitive moral development (Gibbs and Widaman 1982). It assumes that a person can operate on more than one stage at a time and, thus, measures the comprehension and preference for a stage

**Table 1.** Sample Information

| | *Mailed* | *Received* | *Incomplete surveys* | *Work in Public Accounting* | *Inconsistent or Meaningless Scores* | *Included Responses with Valid P Scores* |
|---|---|---|---|---|---|---|
| Staff | 514 | 202 | 15 | 14 | 11 | 162 |
| Middle Management | 498 | 183 | 16 | 8 | 0 | 159 |
| Upper Management | 535 | 199 | 28 | 12 | 12 | 147 |
| *Total* | 1,547 | 584 | 59 | 34 | 23 | 468 |

**Table 2.** Demographic Profile of Sample

| Variable | n | (%) | n | (%) | n | (%) |
|---|---|---|---|---|---|---|
| *Gender* | | | | | | |
| Men | 343 | (73%) | | | | |
| Women | 125 | (27%) | | | | |
| | 468 | (100%) | | | | |
| *Age* | | | Men | | Women | |
| 24-35 years | 157 | (33%) | 108 | (31%) | 49 | (39.2%) |
| 36-45 years | 186 | (40%) | 133 | (39%) | 53 | (42.4%) |
| 45-67 years | 125 | (27%) | 102 | (30%) | 23 | (18.4%) |
| | 468 | (100%) | 343 | | 125 | |
| Mean = 40 years | | | | | | |
| *Years of accounting experience* | | | | | | |
| 2-10 years | 166 | (36%) | 113 | (33%) | 53 | (42.4%) |
| 11-20 years | 217 | (46%) | 154 | (45%) | 63 | (50.4%) |
| 21-41 years | 85 | (18%) | 76 | (22%) | 9 | (7.2%) |
| | 468 | (100%) | 343 | | 125 | |
| *Mean* = 14 total years | | | | | | |
| *Public accounting experience* | | | | | | |
| None | 350 | (75%) | 265 | (78%) | 85 | (69%) |
| 1-13 years | 115 | (25%) | 76 | (22%) | 39 | (31%) |
| | 465 | (100%) | 341 | | 124 | |
| *Highest degree earned* | | | | | | |
| H.S. diploma | 1 | (0%) | 1 | (0%) | 0 | (0%) |
| Bachelor's | 202 | (44%) | 138 | (41%) | 64 | (52%) |
| Master's | 258 | (56%) | 199 | (59%) | 59 | (48%) |
| | 461 | (100%) | 338 | | 123 | |
| *Professional certification* | | | | | | |
| CMA | 282 | (60%) | 211 | (62%) | 71 | (57%) |
| CMA and CPA | 186 | (40%) | 132 | (38%) | 54 | (43%) |
| | 461 | (100%) | 343 | | 125 | |
| *Size of organization* | | | | | | |
| Small (<100) | 89 | (19%) | 67 | (20%) | 22 | (18%) |
| Medium (101-500) | 84 | (18%) | 58 | (17%) | 26 | (21%) |
| Large (>500) | 289 | (63%) | 215 | (63%) | 74 | (61%) |
| | 462 | (100%) | 340 | | 122 | |
| *Pre-mailing position* | | | *Actual Position* | | | |
| Staff | 162 | (35%) | Staff | | 104 | (23%) |
| Middle management | 159 | (34%) | Middle management | | 219 | (47%) |
| Upper management | 147 | (31%) | Upper management | | 140 | (30%) |
| | 468 | (100%) | | | 463 | (100%) |
| | | | (5 people did not classify themselves) | | | |

of moral reasoning. The most used index of the DIT is the "P" score, with P standing for principled morality. The DIT has been validated in over 1,000 studies on hundreds of thousands of subjects in over 40 countries (Rest and Narvaez 1994). The study questionnaire also asked for demographic and other data, some data about the respondents' organizations, and some further questions about ethical dilemmas they had encountered in the workplace. The DIT instruments were scored by the Center for the Study of Ethical Development at the University of Minnesota. The scoring included an "M" score (identifying those with meaningless responses) and identified those who had unacceptably high inconsistency scores. These were eliminated. Table 2 contains a demographic profile of the subjects, together with additional data about the respondents' organizations, and both the preliminary and current/actual subjects' positions in their organizations.

## RESULTS AND DISCUSSION

Table 3 presents the mean P scores for the CMA sample of accountants in industry and for the subsets of CMAs who once worked in public accounting and who have no public accounting experience. For comparison, Table 3 also reports mean P score results from prior research for U.S. CPAs in public accounting, and for Canadian CMAs in industry. The U.S. CMA sample, contrary to expectation, shows a mean level of ethical reasoning that is not significantly different from that of U.S. CPAs ($t = 0.45$, $p = 0.63$). The mean P score of 39.3 is lower than would be expected, given the respondents' educational level and, like that of U.S. CPAs, reaches only the level of adults in the general population. Hypotheses 1 cannot be accepted. The mean P score of CMAs who left public accounting was also compared to the mean P score of CPAs in public accounting. Again, the difference in means was not significant ($t = 1.52$, $p = 0.1545$) and Hypothesis 1A cannot be accepted. The DIT P scores for those without public accounting experience was also not significantly different from the scores of CPAs, also failing to provide support for Hypothesis 1B. As no significant differences in moral reasoning have been identified, the findings cast some doubt on the contention that those in public accounting with higher levels of moral reasoning left voluntarily or were forced out because their ethical reasoning was incompatible with their firms. The findings also do not support the contention that the lower moral reasoning of CPAs occurs because those with higher moral reasoning levels initially choose to pursue accounting careers other than public accounting.

Table 3 also reports the mean P scores of Canadian CMAs as reported in prior research. The significant difference ($p = 0.035$) between moral development of U.S. and Canadian CMAs (39.3 versus 43.5, respectively) concurs with the difference observed in the 1993 Ponemon and Gabhart study between American CPAs and Canadian CAs.

***Table 3.*** Comparison of DIT P Scores

| | *n* | $\bar{x}$ | σ |
|---|---|---|---|
| U.S. CMAs | 468 | 39.3 | (16.5) |
| Left public accounting | 115 | 41.38 | (16.6) |
| Never in public accounting | 350 | 38.7 | (16.5) |
| U.S. CPAs | 313[1] | 38.9 | (8.95) |
| Canadian CMAs | 76[2] | 43.5 | (15.7)* |

***Notes:*** [1]Weighted average from Ponemon (1992) and Ponemon and Gabbart (1993).
[2]From Etherington and Schulting (1995).
*Significant difference at $p < .05$.

***Table 4.*** ANOVA of Scores by Gender, Social Beliefs, Employment Category, and Public Accounting Experience

| | *F* | P |
|---|---|---|
| *Main Effects* | | |
| Gender | 9.45 | 0.0022 |
| Liberal/Conservative | 3.24 | 0.0401 |
| Actual Position | 2.08 | 0.1262 |
| Public Accounting (yes/no) | 0.00 | 0.9895 |
| *Interactions* | | |
| Gender*Liberal/Conservative | 1.68 | 0.1880 |
| Gender*Position | 2.89 | 0.0565 |
| Gender*Public Accountant | 0.13 | 0.7168 |
| Liberal/Conservative*Position | 0.31 | 0.8714 |
| Liberal/Conservative*Public Accountant | 0.10 | 0.9053 |
| Actual Position*Public Accountant | 1.66 | 0.1916 |
| Overall | 3.32 | 0.0001 |

Table 4 presents the results of an ANOVA with P score as the dependent variable, and gender, social belief classification, and hierarchical employment level as independent variables. Experience in public accounting is included as a control variable. Table 5 presents mean P scores for the various variable categories. As shown in Tables 4 and 5, gender has a strong significant effect on moral reasoning levels, with women's P scores significantly higher than those of men. This finding supports Hypothesis 2 and other recent research showing gender differences in ethical reasoning among accountants. Further, the difference found here is not explainable by education differences, as a higher proportion of men (59%) than women (48%) have graduate degrees.

**Table 5.** DIT P Score Means and Test of Mean Difference

| *Variable* | | $n =$ | P Score Mean | Std. Dev. |
|---|---|---|---|---|
| *Overall* | | 468 | 39.3 | (16.5) |
| *Gender** | | | | |
| Men | | 343 | 36.6 | (15.5) |
| Women | | 125 | 46.7 | (16.9) |
| *Education** | | | | |
| Bachelor's | | 202 | 41.1 | (16.9) |
| Master's | | 258 | 37.9 | (15.9) |
| *Education by Gender* | | | | |
| Men: | Bachelor's | 138 | 38.4 | (15.9) |
| | Master's | 199 | 35.3 | (14.9) |
| Women: | Bachelor's | 64 | 46.9 | (17.9) |
| | Master's | 59 | 46.7 | (16.2) |
| *Public accounting experience* | | | | |
| None | | 350 | 38.7 | (16.5) |
| 1-13 years | | 115 | 41.4 | (16.6) |
| *Pre-mailing position* | | | | |
| Staff | | 162 | 38.8 | (16.4) |
| Middle | | 159 | 40.5 | (16.8) |
| Senior | | 147 | 38.7 | (16.2) |
| *Actual position* | | | | |
| Staff | | 104 | 38.9 | (17.1) |
| Middle | | 219 | 40.1 | (16.9) |
| Senior | | 140 | 38.3 | (15.4) |
| *Social beliefs* | | | | |
| Liberal (1) | | 7 | 47.6 | (18.2) |
| Liberal (2) | | 57 | 46.6 | (14.2) |
| Moderate (3) | | 99 | 40.1 | (16.7) |
| Conservative (4) | | 224 | 39.7 | (15.7) |
| Conservative (5) | | 77 | 31.6 | (17.4) |
| *Employment* | | | | |
| Business Enterprise | | 410 | 40.0 | (16.4) |
| Not-for-profit | | 29 | 34.4 | (13.4) |
| Government or public | | 29 | 35.3 | (18.7) |

***Note:*** *Significant difference in a *t*-test $p < .05$.

Hypothesis 3, that moral reasoning varies inversely with conservatism, is also strongly supported, with ANOVA results indicating the social belief variable is statistically significant. Results in Table 5 show monotonically lower mean P scores as conservatism increases. The results indicate that those who report liberal social beliefs have significantly higher levels of moral reasoning. This is of interest as only 14% of the sample describe themselves as liberal, while 66% describe themselves as espousing conservative views.

The ANOVA main effect for position as classified by the respondents reported in Table 4 cannot be rejected at conventional levels of 0.05. Table 5 reports P score means for each employment position. Hypothesis 4, that those in different positions would be similar in ethical reasoning was supported.[5]

## Other Findings

Table 5 also reports P scores for other variables of interest for CMAs. Surprisingly, the results of a *t*-test show lower levels of ethical reasoning for those with master's degrees (statistically significant at $p = 0.04$). The result appears when all subjects are grouped together, but not in same-gender tests. The results are driven by the predominance of male subjects, whose ethical reasoning is lower both for bachelor degree-holders and for those with master's degrees. Contrary to Rest's findings (1986b), those with more education do not exhibit higher ethical reasoning for either sex. Men's mean P scores are lower ($p = 0.07$) for those with graduate degrees while women's were not significantly different. It appears, at least for accountants in industry, that increased education does not result in higher moral reasoning.

The questionnaire also asked the respondents to identify their current employment as part of a business enterprise, government/public sector or a nonprofit organization. Data are presented in Table 5. We analyzed current employment as an independent variable in an ANOVA together with the variables previously mentioned (The ANOVA is not reported here). Those in different places of employment were not found to have statistically different P scores, but this may be driven by the comparatively small number of respondents working in nonprofit and government organizations.

Through the survey instrument, we also asked CMAs how often in the past three years they had been involved in or witnessed an ethical dilemma in the workplace. As shown in Table 6, a strong majority (79%) had recently faced such dilemmas. When asked whom they would consult regarding an ethical dilemma, 58% indicated they would consult a senior colleague at work who was not involved, and 50% would consult other colleagues at work. These results do not imply that only approximately 50% of the respondents consult a work colleague. Since respondents could indicate more than one person with whom to consult regarding an ethical dilemma, there is some overlap of responses. In fact, 27% of respondents indicated they would consult both a colleague and senior level colleague.

**Table 6.** Ethics-Related Data

| | n | U.S. CMAs | n | Canadian CMAs |
|---|---|---|---|---|
| ***How often have you been involved in or witnessed an ethical dilemma in the last 3 years?*** | | | | |
| Never | 99 | (21%) | 26 | (34.2%) |
| Once | 82 | (18%) | 15 | (19.7%) |
| More than once | 280 | (61%) | 35 | (46.1%) |
| | 461 | (100%) | 76 | (100%) |
| ***Whom would you consult regarding an ethical conflict?*** | | | | |
| Senior colleague at work who is not involved | 273 | (58%) | 60 | (79%) |
| Family member | 234 | (50%) | 34 | (45%) |
| Other colleague at work | 236 | (50%) | 33 | (43%) |
| Other | 84 | (18%) | 16 | (21%) |
| IMA/SMAC | 68 | (15%) | 16 | (21%) |
| No one | 16 | (3%) | 3 | (4%) |

***Notes:*** Results sum to >100% as subjects could answer in more than one category. The percentages are out of 468 respondents for U.S. and 76 for Canadian.

Only 19% indicated they would not consult a work colleague. It is important to note, however, that although respondents may choose not to consult a colleague at work, that would not necessarily indicate that ethical values in the workplace do not play an important role.

Table 6 also shows that half of the CMA respondents would consult a family member. This suggests that for many CMAs, discussing a workplace ethical dilemma with a family member apparently is not considered to violate workplace confidentiality (although it is quite possible that some ethical dilemmas faced by management accountants would not involve confidential information). We note also that only 15% indicated that they would consult the IMA, although this could be affected by the nature of the ethical dilemma that was encountered. Classifying ethical dilemmas into categories was beyond the scope of this paper, so it is possible that some may have involved such aspects as, for example, sexual or racial discrimination, making the IMA not the most applicable source for consultation.

Finally, Table 6 includes some results from Etherington and Schulting's (1995) study of Canadian CMAs.[6] First, it is noteworthy that more Americans than Canadians had faced recent ethical dilemmas at work. This may imply that the Canadian business environment may operate in a more ethical climate. If so, it would be compatible with the significantly higher moral reasoning found for Canadian accountants.

It is also noted that a much greater percentage of Canadian CMAs than American CMAs (79% versus 58%) would consult a senior colleague at work who was not involved in the conflict. It appears that American accountants are less willing than Canadians to turn to upper-level management for help in addressing ethical conflicts. This could imply a greater confidence by Canadians in finding ethical managers in higher positions in their organizations. The U.S. CMAs' rank orders

**Table 7.** Multinomial Logit Model

| *Variable* | *1 Ethical Dilemma* | | | *More than 1 Ethical Dilemma* | | |
|---|---|---|---|---|---|---|
| P score | 0.0072 | (0.0096) | $p = 0.4515$ | 0.0148 | (0.0076) | $p = 0.0508$ |
| Gender | 0.4913 | (0.3461) | $p = 0.1558$ | −0.1269 | (0.2866) | $p = 0.6579$ |
| Middle management | 0.3681 | (0.3706) | $p = 0.3206$ | 0.3056 | (0.2840) | $p = 0.2819$ |
| Senior management | 0.7417 | (0.4284) | $p = 0.0834$ | 0.8347 | (0.3343) | $p = 0.0125$ |
| Intercept | −0.9766 | (0.4615) | $p = 0.0343$ | 0.1324 | (0.3566) | $p = 0.7103$ |
| $n = 461$; $df = 10$ | | | | | | |
| *chi-square* = 252.22 | | | | | | |

***Notes:*** The baseline categorical dependent variable is "0" ethical dilemmas, Level 1 is "1" ethical dilemma, and Level 2 is "more than 1" ethical dilemma.

Hierarchical employment is included as a dummy variable with staff position as the omitted group.

for "Whom would you consult?" are the same as the Canadian CMAs, and other responses have similar percentages for both countries.

Table 7 reports results of a multinomial logit model with frequency of ethical dilemma as the dependent variable, and P score, gender, and current position as independent variables. Results show that those with higher P scores are more likely to have been involved in more than one ethical dilemma ($p = 0.05$), suggesting that those with higher moral reasoning may either pay more notice to ethical issues, or possibly may consider some issues as ethical dilemmas that those with lower ethical reasoning may not. Individuals with lower moral reasoning tend to refer to authority figures (Trevino 1986) and may not perceive a specific issue as a moral dilemma.

Table 7 also shows that accountants in senior management positions are more likely to have been involved in more than one ethical dilemma in the past three years ($p = 0.01$) than those in staff positions (the omitted group). This suggests, logically, that decisions concerning many ethical dilemmas such as managing financial results and/or falsifying records are more likely to be encountered by an upper level manager than by a staff accountant. There is no difference by gender, with men and women being equally likely not to have encountered a workplace ethical dilemma, to have encountered only one, or to have encountered ethical dilemmas more than once in the past three years.

## IMPLICATIONS AND CONCLUSIONS

The study findings that U.S. CMAs have relatively low ethical reasoning levels are of considerable concern when taken together with the prior findings of low ethical reasoning levels for U.S. CPAs. The accounting profession has had strong codes of ethics for a considerable time and believes that the public has confidence in the professional integrity of its practitioners, whether in industry or public accounting. While scores on a widely validated and widely used test of moral reasoning are not a total predictor of work behavior, there is evidence from accounting studies of a significant relationship between them (Ponemon and Gabhart 1993; Arnold and Ponemon 1991). When considering the significant differences between P scores of U.S. and Canadian accountants it is important to note that low moral reasoning and low P scores are not characteristic of all Americans (Rest 1979, 1986b). While cultural differences have been cited by Ponemon and Gabhart (1993) and Etherington and Schulting (1995) as possible reasons for the differences between accountants in the two countries, national cultural differences do not explain accountants' P scores being lower than those found in the general U.S. population for people with a college education. It is possible that an overriding emphasis on "the bottom line" may be a stronger characteristic of the American business environment than of Canadian. It is also possible that following the "economic man" model is seen as more of a virtue in the United States, which may in itself lead to less emphasis on ethical concerns.[7]

The finding that the more conservative a person's social beliefs, the lower their ethical reasoning, is also disturbing, since the majority (66%) of U.S. CMAs state that they espouse conservative social beliefs. It has been suggested by Ponemon and Gabhart (1993) and Lampe and Finn (1992) that one of the factors may be that accounting attracts those with a strong rule-orientation. Whether this is true for American accountants has not been determined, but, as noted earlier, Davidson and Dalby (1993) found Canadian CAs were more expedient and disregarding of rules than the average population as well as less conservative. Conservatism and rule-orientation may be related. Further study to explore whether there are such links could be fruitful. Whether rule-orientation is related to moral judgment and behavior could be an important area for study, as suggested both by the findings of this study and by Gaa (1992). Jeffrey and Weatherholt (1995) have also examined accountants' rule orientation and their ethical reasoning using the DIT, but their analysis did not directly examine whether there was a relationship between the two nor was it able to compare the rule orientation of accountants with that of the general population.

The finding that mean P scores at different hierarchical positions in companies were not significantly different (see Table 5) indicates that those with high ethical reasoning *do* reach top levels in industry, government, and nonprofit organizations. In contrast, those with high levels of ethical reasoning in U.S. public accounting firms are found infrequently at the partner level (Ponemon 1992; Ponemon and Gabhart 1993; Shaub 1994). Wider diversity likely exists in the social conditioning and climate of American business than in public accounting, which tends toward isomorphism (DiMaggio and Powell 1983).

A limitation of the study is one common to survey methodology—that the respondents were those who "volunteered," and the difference between them and nonrespondents cannot be determined. The large sample size and that subjects were randomly selected from all parts of the country, however, makes it more likely that the sample is representative of U.S. CMAs in industry. A further limitation is that delineating the concepts of liberal and conservative was beyond the scope of the paper. It is possible that respondents classified themselves according to differing interpretations of those terms. Further examination, possibly by administering sophisticated well-validated psychological tests, could prove fruitful in determining whether U.S. accountants and accounting students have characteristics significantly different from the general population that may explain their lower scores on the DIT. Such knowledge could assist in changing recruitment appeals to young students making career choices. It is also possible that the results of such inquiries may indicate changes in the introductory financial accounting course. Emphasis on the more creative and challenging aspects of accounting work may attract students who are less rule-oriented and traditional—those who may currently be "turned off" accounting as a career choice.

Education at both university and professional levels in the United States ought to be further explored as means of raising ethical reasoning in accountants. For the international differences discussed earlier (U.S. versus Canadian), it is possible that differences in education have affected ethical reasoning. One difference, although not specifically an ethics-related difference, is in the qualifying examinations for Canadian accountants, where both the CA and CMA examinations include case studies and require, in addition, a substantial portion of written answers and explanations. It is possible that written examinations test a different kind of thinking, perhaps requiring more thinking that involves fewer "rules" and more judgment, with Canadian examinations thus having proportionately less of an overriding technical emphasis.

In sum, the results of this study indicate that U.S. CMAs reason ethically at a lower level than Canadian CMAs and college students in general. However, as Rest (1986b, 60) states, ethical reasoning is only part of an individual's overall capacity to frame and resolve ethical issues. What the findings do suggest is that accountants in the United States tend to think at the conventional level (Kohlberg's stage 3), while Canadian accountants are more likely to reason at higher levels (Stages 4 and 5) (Kohlberg 1976, 1984). This implies that the ethical reasoning of U.S. accountants is primarily influenced by their need to affiliate with peers or a referent group such as the client, the firm's management, or office colleagues when attempting to resolve ethical issues. Ponemon and Gabhart noted the same (1993, 106) when discussing managers and partners of CPA firms in the United States. Trevino (1986, 612) also proposed that for individuals not at the principled level of cognitive moral development, their "ethical behavior will be influenced significantly by the demands of authority figures."

Research is needed to determine the type of intervention that may effect changes in these behaviors, from education in business ethics to development of corporate ethical codes. Perhaps top manager commitment to their observance also needs monitoring. It may be that accounting in the United States is attracting an excess of rule-oriented conservative individuals who are too willing to accept guidance from authority figures. The accounting bodies themselves have set an example by showing great concern and interest in ethical issues, by placing ethics topics on their conference programs, in assisting researchers through funding, by making their membership lists available to researchers, and by encouraging participation in such research. Hopefully, such cooperation between practitioners and academics may result in the development of useful interventions that will raise accountants' ethical reasoning to that found in those of their educational level in the general population.

## ACKNOWLEDGMENTS

The authors acknowledge with appreciation funding from the Social Sciences and Humanities Research Council of Canada and the British Columbia Institute of Chartered Accountants.

## NOTES

1. Of the 115 CMAs who reported having public accounting experience (1 to 13 years), 89 are also CPAs.

2. Staff accountants were those with the job titles of Chief Accountant or Accountant (Senior, Tax, Staff, Plant). Middle management were those with titles of Assistant Corporate Controller, Plant Controller, Manager, and General Manager. Upper management titles were CFO, Ceo/Owner, Corporate Controller, Vice-President.

3. As noted by Allreck and Settle (1985), for unsolicited surveys "mail surveys with response rates of over 30 percent are quite rare" (45).

4. From a sample of 1080 subjects, the P score from the short version correlates .91 with the P score based on the six-story version (Rest 1986a).

5. Tests were done using both the pre-mailing classification and the respondents' actual positions. As neither showed significance, we only report the anlysis for actual positions.

6. The researchers had access to the Canadian data.

7. Much of the accounting research on differences between countries has focused on differential effects of disclosure practices and policy formulation (i.e., Gray 1988; Nobes and Parker 1991; Salter and Niswander 1995). In other research, the United States and Canada are often grouped together when cultural differences are analyzed (i.e., Hofstede 1980, 1991).

## REFERENCES

Alreck, P., and R. Settle. 1985. *The Survey Research Handbook.* Homewood, IL: Irwin.

Armstrong, M. 1987. Moral development and accounting education. *Journal of Accounting Education* (Spring): 27-43.

Arnold, D.F., Sr., and L.A. Ponemon. 1991. Internal auditors' perceptions of whistle-blowing and the Influence of moral reasoning: An experiment. *Auditing: A Journal of Practice & Theory* 10(2): 1-15.

Bernardi, R. 1991. Fraud detection: An experiment testing differences in perceived client integrity and competence, individual auditor cognitive style and experience, and accounting firms. Unpublished Ph.D. dissertation, Union College, Schenectady, NY.

Bruns, W.J., and K.A. Merchant. 1990. The dangerous morality of managing earnings. *Management Accounting* 72(2): 22-25.

Catell, R.B., H.W. Eber, and B. Tatsuoka. 1970. *Handbook for the Sixteen Personality Questionnaire.* Champaign, IL: Institute for Personality and Ability Testing.

Chatman, J.A. 1991. Matching people and organizations: Selection and socialization in public accounting firms. *Administrative Science Quarterly* 36: 459-494.

Cohen, J.R., and L. Pant. 1989. Accounting educators' perceptions of ethics in the curriculum. *Issues in Accounting Education* (Spring): 70-81.

Coreless, J., and L. Parker. 1987. The impact of MAS on auditor independence: An experimental study. *Accounting Horizons* (September): 25-30.

Davidson, R.A., and J.T. Dalby. 1993. Personality profiles of Canadian public accountants. *International Journal of Selection and Assessment* (April): 107-116.

DeFond, M.L., and J. Jiambalvo. 1991. Incidence and circumstances of accounting errors. *The Accounting Review* (July): 643-655.

DiMaggio, P.J., and W. Powell. 1983. The iron cage revisited: Institutional isomorphism and collective rationality in organizational fields. *American Sociological Review* 48: 147-160.

Etherington, L.D., and L. Schulting. 1995. Ethical development of management accountants: The case of Canadian CMAs. In *Research on Accounting Ethics,* Vol. 1, ed. L.A. Ponemon, 237-253. Greenwich, CT: JAI Press.

Eynon, G., N. Hill, K. Stevens, and P. Clarke. 1996. An international comparison of ethical reasoning abilities: Accounting students from Ireland and the United States. Journal of Accounting Education (Winter): 477-492.

Farmer, T., L. Rittenberg, and G. Trompeter. 1987. An investigation of the impact of economics and organizational factors on auditor independence. *Auditing: A Journal of Theory and Practice* (Fall): 1-14.

Finn, D.W., and J.C. Lampe. 1992. A study of whistleblowing among auditors. *Professional Ethics: A Multidisciplinary Journal* 1(3/4): 137-168.

Gaa, J. 1992. Discussion of a model of auditors' ethical decision processes. *Auditing: A Journal of Practice and Theory* 11(Supplement): 60-66.

Gibbs, J.C., and K.F. Widaman. 1982. *Social Intelligence: Measuring the Development of Sociomoral Reflection.* Englewood Cliffs, NJ: Prentice-Hall.

Gray, S.J. 1988. Toward a theory of cultural influence on the development of accounting systems internationally. *Abacus* 24(1, March): 1-15.

Hill, N.T., K.S. Stevens, and P. Clarke. 1998. Factors that affect ethical reasoning abilities of U.S. and Irish small-firm accounting practitioners. In *Research on Accounting Ethics*, Vol. 4, ed. L.A. Ponemon. Greenwich, CT: JAI Press.

Hofstede, G. 1980. *Culture's Consequences: International Differences in Work Related Values.* Beverly Hills, CA: Sage.

Hofstede, G. 1991. *Culture and Organizations: Software of the Mind.* Maidenhead, UK: McGraw-Hill.

Icerman, R., J. Karcher, and M. Kennelly. 1991. A baseline assessment of moral development: accounting, other business and non-business students. *Accounting Educator's Journal* (Winter): 46-62.

Jeffrey, C. 1993. Ethical development of accounting students, business students, and liberal arts students. *Issues in Accounting Education* (Spring): 86-96.

Jeffrey, C., and N. Weatherholt. 1996. Ethical development, professional commitment, and rule observance attitudes: A study of CPAs and corporate accountants. *Behavioral Research in Accounting* 8: 8-31.

Kohlberg, L. 1976. Moral stages and moralization: The cognitive-developmental approach. In *Moral Development and Behavior,* ed. T. Likona, 31-53. New York: Holt, Rinehart and Winston.

Kohlberg, L. 1984. *The Psychology of Moral Development.* San Francisco, CA: Harper and Rowe.

Kohlberg, L. 1989. Stages and Sequences: The cognitive developmental approach to socialization. In *Handbook of Socialization Theory and Research*, ed. D. Goslin, 347-480. Chicago, IL: Rand McNally,

Lampe, J.C., and D.W. Finn. 1992. A model of auditors' ethical processes. *Auditing: A Journal of Practice and Theory* 11(Supplement): 33-59.

Loeb, S.E. 1971. A survey of ethical behavior in the accounting profession. *Journal of Accounting Research* 9: 287-306.

Loeb, S.E. 1984. Codes of ethics and self-regulation for non-public accountants: A public policy perspective. *Journal of Accounting and Public Policy* (Spring): 1-8.

Loeb, S.E. 1988. Teaching students accounting ethics: Some crucial issues. *Issues in Accounting Education* (Fall): 281-94.

Loeb, S.E. 1990. Whistleblowing and accounting education. *Issues in Accounting Education* (Fall): 316-329.

Maupin, R.J., and C.R. Lehman. 1994. Talking heads: Stereotypes, status, sex-roles and satisfaction of female and male auditors. *Accounting, Organizations and Society* (19): 427-437.

McNair, C.J. 1991. Proper compromises: The management control dilemma in public accounting and its impact on auditor behavior. *Accounting, Organizations and Society* 16(7): 635-653.

Merchant, K.A., and J. Rockness. 1994. The ethics of managing earnings. *Journal of Accounting and Public Policy* 13(Spring): 79-94.

Mihalek, P.H., A.J. Rich, and C.S. Smith 1987. Ethics and management accountants. *Management Accounting* 69(6): 34-36.

Nobes, C., and R. Parker. 1991. *Comparative International Accounting,* 3rd edition. New York: Prentice Hall.

Ponemon, L. 1988. A Cognitive-Developmental Approach to the Analysis of Certified Public Accountants' Ethical Judgments. Unpublished Ph.D. dissertation, Union College of Union University.

Ponemon, L. 1990. Ethical judgements in accounting: A cognitive-developmental perspective. *Critical Perspectives on Accounting* (1): 191-215.

Ponemon, L. 1992a. Auditor underreporting of time and moral reasoning: An experimental-lab study. *Contemporary Accounting Research* (9): 171-189.

Ponemon, L. 1992b. Ethical reasoning and selection-socialization in accounting. *Accounting, Organisations and Society* (17): 239-258.

Ponemon, L. 1993. *Ethical Reasoning in Accounting and Auditing.* Vancouver: CGA Canada Research Foundation.

Ponemon, L., and D. Gabhart. 1990. Auditor independence judgments: A cognitive-developmental model and experimental evidence. *Contemporary Accounting Research* (Fall): 227-251.

Ponemon, L., and A. Glazer. 1990. Accounting education and ethical development: The influence of liberal learning on students and alumni in accounting practice. *Issues in Accounting Education* 5(2): 21-34.

Rest, J. 1979. *Development in Judging Moral Issues.* Minneapolis, MN: The University of Minnesota Press.

Rest, J. 1986a. *Manual for the Defining Issues Test.* Minneapolis, MN: The University of Minnesota Press.

Rest, J. 1986b. *Moral Development: Advances in Theory and Research.* New York: Praeger,

Rest, J., and D. Narvaez. 1994. *Moral Development in the Professions.* Hillsdale, NJ: Lawrence Erlbaum Associates.

Salter, S.B., and F. Niswander. 1995. Cultural influence on the development of accounting systems internationally: A test of Gray's [1988] theory. *Journal of International Business Studies* (Second Quarter): 379-397.

Shaub, M.K. 1994. An analysis of the association of traditional demoraphic variables with the moral reasoning of auditing students and auditors. *Journal of Accounting Education* 12(1): 1-26.

Shaub, M.K., D.W. Finn, and P. Munter. 1993. The effects of auditors' ethical orientation on commitment and ethical sensitivity. *Behavioral Research in Accounting* (5): 145-169.

St. Pierre, K., E. Nelson, and A. Gabbin. 1990. A study of the ethical development of accounting majors in relation to other business and nonbusiness disciplines. *The Accounting Educators Journal* (Summer): 23-35.

Sweeney, J.T. 1995. The moral reasoning of auditors: An exploratory analysis. In *Research on Accounting Ethics,* Vol. 1, ed. L.A. Ponemon, 213-234. Greenwich, CT: JAI Press.

Thoma, S.J. 1984. Estimating Gender Differences in the Comprehension and Reference of Moral Issues. Unpublished manuscript, University of Minnesota, Minneapolis.

Trevino, L. 1986. Ethical decision making in organizations: A person-situation interactionist model. *Academy of Management Review* 11(3): 601-617.

# THE EMPIRICAL DEVELOPMENT OF A FINANCIAL REPORTING ETHICS DECISION MODEL: A FACTOR ANALYSIS APPROACH

Thomas G. Hodge, Dale L. Flesher, and James H. Thompson

## ABSTRACT

Interest in ethical behavior in business is possibly at an all-time high. Interest in ethics exists in both the private and public sectors. The development of numerous codes of ethics by professional organizations and individual firms and the presence of congressional hearings on the functioning of the accounting profession's self-regulation process provide evidence of a heightened awareness regarding ethics. This study investigates the decision-making behavior of management accountants who are faced with an ethical dilemma involving financial reporting. A two-phased questionnaire is utilized in this study to develop an ethics decision model involving financial reporting. Subjects include management accountants with supervisory experience. The study represents one of the first efforts in accounting research to use protocol analysis in decision situations involving ethics.

**Research on Accounting Ethics, Volume 4, pages 247-266.**

**ISBN: 0-7623-0339-5**

## INTRODUCTION

Interest in ethical behavior in business is possibly at an all-time high. One evidence of interest in ethics is the development of codes of ethics. Codes of ethics are frequently adopted in an effort to promote ethical behavior. While many accounting and other business and nonbusiness organizations adopted codes of ethics/conduct many years ago, many business firms are now adopting codes of ethics and instituting ethics training courses. Murphy (1995) reports that 91% of the responding firms in his study have adopted a formal code of ethics. In addition, one-half have published values statements, and one-third have a corporate credo.

A heightened awareness of ethics is not limited to the private sector. Congressional hearings have recently addressed ethical questions in business. The Dingle Committee hearings were conducted to investigate the functioning of the accounting profession's self-regulation process and the SEC's oversight responsibilities. These hearings resulted from a growing number of corporate failures and abuses, financial institution collapses, alleged audit failures, and lawsuits linked with the names of prominent accounting firms. The Dingell Committee's investigation revealed that management accountants had participated in the preparation of their company's fraudulent financial reports (Shultis and Williams 1985). The purpose of this study is to investigate the decision-making behavior of management accountants faced with an ethical dilemma involving financial reporting.

## BACKGROUND

In recent years, several models have been presented in an effort to describe the ethical decision-making process in business (Ferrell and Gresham 1985; Trevino 1986; Bommer et. al. 1987; Ferrell et. al. 1989; Cooper and Frank 1992). These models and other papers suggest a number of factors that may influence a manager's decisions when confronted by ethical dilemmas. Fritzsche (1987) identified a number of factors related to personal values, corporate policy, and organizational climate which can significantly affect the ethical behavior of managers. Bommer et. al. (1987) report that factors related to one's professional environment can be important in the ethical behavior of managers who are members of professions. For example, the behavior of professionals may be affected by the profession's code of ethics.

Ethical dilemmas are common for subordinates in a business. One condition for the existence of a dilemma is that there is no clear basis on which one can settle it. A subordinate cannot acquiesce in a superior's decision just because it is a superior's decision. Yet, a subordinate might reason that since no one can be sure what the correct resolution to a problem is, one answer is as good as another; there is no good reason for not doing what the superior decides ought to be done (Robison, 1991).

Stanford (1991) identifies three questions may help a manager verify that a reasoned solution is acceptable. First, is the situation a legal issue or one that violates company policy? If the answer is yes, one has an ethical issue, probably black or white. More likely, the situation is gray and an ethical dilemma exists. Second, does the outcome treat the stakeholders equally? The caution here is that there may not be one right answer; rather, one chooses between right and right within contingency parameters. Third, would my family be proud of me if they were aware of my decision or would I care if my decision were printed in the newspaper? This litmus test may help resolve the issue ethically when one has resorted to "splitting hairs" to find justification, or has succumbed to the temptation of thinking that, in this case, rules do not apply.

Research studies report that management accountants frequently encounter ethical dilemmas involving financial reporting. Indeed, Merz and Groebner (1981) found that pressure to manipulate profits was a common ethical dilemma faced by management accountants. In addition, Mihalek, Rich, and Smith (1987) found that approximately 50% of the middle- and lower-level management accountants surveyed had experienced pressure to materially alter reported net income. Leung and Cooper (1995) study of Australian Society of Certified Practicing Accountants revealed that approximately 50% of respondents have little or no knowledge of the Society's Code of professional Conduct. The survey also found that respondents appeared to have faced an average of four ethical problems relating to financial reporting and professionalism.

## METHODOLOGY

A two-phased questionnaire was utilized in this study to develop an ethics decision model involving financial reporting. Phase 1 of the questionnaire was developed from reports of ethical dilemmas in practice involving financial reporting. From these anecdotes, a financial reporting vignette was developed. Due to the length of time required for each subject to respond to the vignette (approximately two hours), a convenience sample of 30 management accountants employed by 15 organizations was selected.

Concurrent verbal protocol was used to gather data in Phase 1. Each subject was asked to "think aloud" as he or she made decisions. The protocol sessions were tape-recorded, transcribed, and verified. These data were then utilized in developing the questionnaire used in Phase 2 of the study. Phase 2 of the study utilized a shorter version of the vignette used in Phase 1. Subjects were asked to read the vignette and to indicate which action(s) they would take to resolve the dilemma. The action choices were based on the action(s) taken by the subjects in the Phase 1 verbal protocols.

After selecting his or her action choice(s), each subject was asked to indicate his or her level of agreement with statements taken from the verbal protocols obtained

***Table 1.*** List of Factors

| *Factor Number* | *Factor Name* |
|---|---|
| Factor I | Professional/Personal Ethics and Values |
| Factor II | Personal Consequences From Unethical Behavior |
| Factor III | Obligations to Those Other Than Local Management, Employees, and Family |
| Factor IV | Personal Benefits From Ethical Behavior |
| Factor V | Internal Control/Deterrent |
| Factor VI | Mistrust of Management |
| Factor VII | Obedience to Authority/Loss of Employment |
| Factor VIII | Characteristics of Entries: Current and Prior |
| Factor IX | Obligations to Local Management and Employees |
| Factor X | Internal Controls/Protection |
| Factor XI | Aggressive Behavior/Pressure From Parent to Increase Profits |
| Factor XII | Open Communication Channels |
| Factor XIII | Obligations to One's Family |
| Factor XIV | Rationalizations For Committing Unethical Act |

in Phase 1 of the study. These statements represented the reasoning statements verbalized by the subjects as they were making their decisions in Phase 1. A six-point Likert scale was used for this part of the questionnaire. The agreement choices were strongly agree, agree, slightly agree, slightly disagree, disagree, and strongly disagree. To determine the underlying dimensions of the subjects' decision processes, factor analysis was performed on the responses to these statements. Based on the objectives of this study, oblique factor extraction was used. The scree test criterion was used to determine the optimum number of factors. In addition, variables with a factor loading of .40 or greater were considered significant (Hair et al. 1979). In Phase 2, questionnaires were mailed to a random sample of 485 CMAs.[1]

## RESULTS

A total of 144 usable responses were received in the Phase 2 questionnaire, representing a 30% response rate. Bartlett's test of sphericity and the Kaiser-Meyer-Olkin (KMO) measure of sampling adequacy were conducted to determine if the data were suitable for factor analysis. Bartlett's test of sphericity was significant to .00001 and the KMO was .69753.[2] Both tests indicated that the variables belong together and were appropriate for factor analysis.

As shown in Table 1, the factor analysis results produced 14 factors based on the scree test. All factors had an eigen-value of 1.0 or above. These 14 factors account for a total of 68.2% of the total variance in the data. This percentage is relatively

**Table 2.** Eigenvalues and Percent of Variance Extracted for Unrotated Principal Components Factor Solution

| *Factor* | *Eigenvalues* | *Percent of Variance* |
|---|---|---|
| Factor I | 9.59 | 17.8 |
| Factor II | 5.02 | 9.3 |
| Factor III | 3.53 | 6.5 |
| Factor IV | 2.38 | 4.4 |
| Factor V | 2.22 | 4.1 |
| Factor VI | 2.17 | 4.0 |
| Factor VII | 1.99 | 3.7 |
| Factor VIII | 1.82 | 3.4 |
| Factor IX | 1.76 | 3.3 |
| Factor X | 1.50 | 2.8 |
| Factor XI | 1.29 | 2.4 |
| Factor XII | 1.25 | 2.3 |
| Factor XIII | 1.16 | 2.2 |
| Factor XIV | 1.12 | 2.1 |
| Total | | 68.2 |

high considering the imprecise nature of the data found in social science research. A factor solution that accounts for less than 60% of the variance is often considered satisfactory (Hair et al. 1979). Table 2 gives a breakdown of the eigenvalues and the percent of variance for each factor.

After the initial factor analysis was completed, six variables loaded as single-variable factors. As suggested by Sharpe and Smith, these variables were omitted from subsequent factor analysis (Hair et al. 1979). Table 3 lists these factors and their respective loadings.

One advantage of using factor analysis is the possibility of discovering underlying dimensions that may not be apparent from simply observing the individual variables. The following discussion emphasizes the underlying dimensions that seem to be reflected in the data.

## Professional/Personal Ethics and Values

Factor I was named professional responsibilities and personal values (see Table 4). The discovery that professional and personal values loaded along the same dimension and on the same factor is not surprising. Many professional accounting organizations, such as the IMA, the AICPA, and the IIA, have worked to assist members in their professional development. These organizations have provided

***Table 3.*** Single Loading Factors

| Item Number | Item | Factor Loadings |
|---|---|---|
| 1. | If I refuse to make the entry I might receive support from the president | .7649 |
| 6. | If you report this to corporate those involved will only be lightly reprimanded | .7822 |
| 17. | If I made the entry jobs might be saved | .7836 |
| 24. | Knowing the parent company had an internal audit department might affect your behavior | .7723 |
| 28. | The entry will not solve the problem | .8185 |
| 45. | If I refuse to make the entry the controller might be replaced | .6396 |

***Table 4.*** Factor I: Professional/Personal Ethics and Values

| Item Number | Item | Factor Loadings |
|---|---|---|
| 14. | A professional should have integrity | .6637 |
| 26. | This entry would violate generally accepted accounting principles | .4915 |
| 27. | You have a responsibility to follow professional accounting standards | .7357 |
| 31. | Your own ethical standards should guide you | .6817 |
| 33. | I couldn't sleep at night if I did this | .5739 |
| 42. | You must always follow professional ethics | .6525 |
| 48. | If I made the entry my conscience would bother me | .5092 |
| 53. | You have a responsibility to keep accurate accounting records | .6466 |
| 56. | I wouldn't feel good about myself if I made the entry | .6034 |
| 58. | Preparing this type of entry is unethical | .6633 |

members with professional development programs, various publications, and codes of professional ethics.

Personal values and professional values are very difficult to separate. Without proper personal values, an individual may have difficulty maintaining proper professional values. One of the variables states that your own ethical standards should

***Table 5.*** Factor II: Personal Consequences from Unethical Behavior

| *Item Number* | *Item* | *Factor Loadings* |
|---|---|---|
| 22. | If I made the entry I could go to prison | .7031 |
| 32. | If I made the entry I might lose my professional certification (i.e., CPA, CMA) | .6577 |
| 37. | Preparing this type of entry is illegal | .6918 |
| 41. | If I made the entry I could lose my job | .4669 |
| 46. | If I made the entry I could be held personally liable | .6099 |
| 52. | If I made the entry I might get into trouble with government agencies | .7221 |
| 59. | If I made the entry I could be asked to commit similar acts in the future | .4536 |

guide you. Perhaps deterrents can be used to help prevent unethical behavior for individuals who do not choose to conduct themselves in an ethical manner.

## Cost/Benefit Tradeoffs in Predicting Outcomes

Jones (1982) defines this category of ethical dilemmas in terms of the costs and benefits of each alternative course of action. In the financial reporting vignette, subjects stressed that costs and benefits that would affect them personally. The two factors were Factor II, Consequences From Unethical Behavior, and Factor IV, Personal Benefits From Ethical Behavior.

### *Personal Consequences From Unethical Behavior*

Trevino's (1986) person-situation interactionist ethics decision model includes responsibility for consequences as a moderating variable. Individuals generally exhibit more ethical behavior when they are responsible for the consequences of their actions. In the financial reporting vignette used in Phase 1 of the study, the management accountant was responsible for preparing and approving the entry. These subjects were more apt to ignore the entry and take no action when the controller prepared and approved the entry.

As evidenced by the high factor loadings, most of the consequences resulted from the illegality of preparing the entry (see Table 5). A subject's illegal action could result in a prison term, trouble with government agencies, loss of professional certification(s), personal liability, or loss of employment, and he or she could be asked to commit similar acts in the future. Awareness of the consequences could serve as a deterrent to unethical behavior.

**Table 6.** Factor IV: Personal Benefits from Ethical Action

| Item Number | Item | Factor Loadings |
|---|---|---|
| 8. | If I refuse to make the entry I might receive support from the parent company | .7905 |
| 10. | Talking to corporate will not help | <.7370> |
| 18. | If I refuse to make the entry the controller might reconsider and not want the entry to be made | .5301 |
| 54. | If I refuse to make the entry I will gain respect from others | .6146 |

### *Personal Benefits from Ethical Action*

The most significant loadings on this factor involve assistance from the parent company (see Table 6). Another variable indicates that if the subject refuses to make the entry, the controller might reconsider and not want the entry made. The controller might reconsider because of fear that the subject would receive assistance from the parent company. Gaining respect from others for refusing to make the entry could also include gaining respect from the parent company management, fellow workers, employees, and family.

## Conflicting Obligations

Conflicting obligations were included in Jones' (1982) model to classify ethical dilemmas involving the social interdependence of people in our society. This social interdependence results in conflicting obligations and ultimately to ethical dilemmas. The factor analysis identified three dimensions of conflicting obligations: Factor IV, Obligations to Those Other than Local Management, Employees, and Family; Factor IX, Obligations to Local Management and Employees; and Factor XIII, Obligations to Family.

### *Obligations to Those Other than Local Management, Employees, and Family*

Except for the parent company, the parties included in "those other than local management, employees, and family" represent many of the outside users of an organization's financial statements (see Table 7). The most significant loadings on this factor represent obligations to groups not directly associated with the organization. The strong obligation to government agencies could result from the possible consequences of providing these groups with misleading financial statements. People in the community could also represent a significant group of financial statement users. Local suppliers and bankers clearly have an interest in the financial state-

***Table 7.*** Factor III: Obligations to Those Other Than Local Management, Employees, and Family

| *Item Number* | *Item* | *Factor Loadings* |
|---|---|---|
| 36. | I would feel a strong obligation to the stockholders | .5830 |
| 39. | I would feel a strong obligation to the parent company | .5064 |
| 44. | I would feel a strong obligation to government agengies (i.e., SEC, IRS, FTC) | .7104 |
| 50. | I would feel a strong obligation to people in the community | .6116 |
| 57. | I would feel a strong obligation to other outside users of the financial statements | .7105 |

ments of local organizations to which they extend credit. In addition, outside users of financial statements represent potential third-party claimants. The parent company and stockholders should be regarded differently from other users of the financial statements. These groups are in a position to exert influence and control over the activities of the organization. They are also in a position to demand information that other groups would find difficult or impossible to obtain. Perhaps the most significant feature of this factor is that the parent company and stockholders are included in the same group as outside financial statement user groups.

### *Obligation to Local Management and Employees*

Table 8 presents the three variables that have significant loading for Factor IX, Obligation to Local Management and Employees. Although management accountants should be loyal to company management, they also have a responsibility to let superiors know about wrongdoings. In fact, the IMA's Standards of Ethical Con-

***Table 8.*** Factor IX: Obligation to Local Management and Employees

| *Item Number* | *Item* | *Factor Loadings* |
|---|---|---|
| 11. | I would feel a strong obligation to local management of the company | .7141 |
| 16. | You have a responsibility to let superiors know when they are doing wrong | <.5468> |
| 21. | I would feel a strong obligation to other company employees | .6371 |

duct for Management Accountants states, "Management accountants shall not commit acts contrary to these standards nor shall they condone the commission of such acts by others within their organization."

Perhaps the loading of variables relating together the obligations to management and employees is not unusual. The financial well-being of both groups is dependent on the success of the organization. The temptation to alter financial results could stem from a management accountant's feeling of obligation to these two groups.

### *Obligations to One's Family*

An individual's obligations to his or her family can take many forms. When an ethical dilemma is job-related, loss of employment can cause the person's family to suffer financially. The desire to fulfill the financial obligation to one's family can result in an otherwise ethical person committing an unethical act (see Table 9). As a result of committing an unethical act, a person's career could be damaged in many ways. Perhaps two of the most serious consequences could be the loss of professional certifications, and legal action resulting in prosecution. Finding employment in the future might prove difficult if a management accountant's career is damaged as the result of committing an unethical act.

## Tone at the Top

The Treadway Commission determined that an ethical "tone at the top" was one of the best deterrents to unethical behavior. Four factors were found to affect a management accountant's relationship with an organization's management. The factors are Factor VI, Mistrust of Management; Factor VII, Obedience to Authority/Loss of Employment; Factor XI, Aggressive Behavior/Pressure From Parent to Increase Profits; and Factor XII, Open Communication Channels.

### *Mistrust of Management*

Mistrust between workers and management is not a new phenomenon. In fact, certain labor unions got started and have flourished by exploiting this mistrust (Wren 1979). Although management accountants are actually considered a part of

**Table 9.** Factor XIII: Obligation to One's Family

| *Item Number* | *Item* | *Factor Loadings* |
|---|---|---|
| 5. | If I made the entry this could damage my career | .4072 |
| 7. | I would feel a strong obligation to my family | .7376 |

**Table 10.** Factor VI: Mistrust of Management

| Item Number | Item | Factor Loadings |
|---|---|---|
| 12. | Sometimes superiors set up employees to take the blame | .7839 |
| 25. | Lower-level employees often get blamed and dismissed in cases like this | .7161 |
| 35. | If I refuse to make the entry my future with the company would not be good | .6075 |

the management team, the mistrust expressed in Factor VI is directed toward superiors in the management structure (see Table 10). The first two variables actually express this mistrust. The first variable expresses concern that superiors sometimes set employees up to take the blame, while variable two expresses an additional concern that an employee will be both blamed and dismissed. The third variable expresses a sense of mistrust from a different perspective. One might infer from this variable that management can not be trusted to provide support if the employee refuses to make the entry.

### Obedience to Authority/Loss of Employment

The underlying dimension of Factor VII relates to obedience to authority (see Table 11). When a management accountant is faced with a superior who demands that he or she commit an unethical act, the individual is faced with the decision to either commit the act or refuse to commit the act. For some individuals, these situations create an ethical conflict with their personal value system. This conflict can become so severe that individuals no longer want to be associated with the organization. This action could indicate that an individual believes superiors will not change their position and take the ethical action.

**Table 11.** Factor VII: Obedience to Authority/Loss of Employment

| Item Number | Item | Factor Loadings |
|---|---|---|
| 2. | Most presidents want you to agree with them | .5434 |
| 3. | Questioning your boss could jeopardize your future with the company | .5466 |
| 40. | This is a company that I would no longer want to be associated with | .7169 |
| 60. | If I refuse to make the entry I might lose my job | .6278 |

The fear of losing his or her employment could be a very significant factor in a person's decision to commit an unethical act. One subject in the current study stated:

> In today's work place ethical behavior and the code of ethics take a back seat to doing what your boss tells you to do. I had rather take my chances with either internal or external auditors on the subject of unethical behavior than to disagree with my boss, who I have to work with every day.

Organizations must have open communication channels to encourage employees to come forward and report the unethical behavior of their superiors.

### *Aggressive Behavior/Pressure from Parent to Increase Profits*

One of the variables loading on Factor XI involves pressure to manipulate profits (see Table 12). Consistent with prior research findings, this study found that eight (27%) of the 30 subjects in Phase 1 had either experienced or witnessed an ethical dilemma involving the manipulation of reported profits. Wilson (1983) suggests that one way to help prevent this type of behavior is to structure management incentive programs in a way that does not encourage the manipulation of results. Carroll (1975) found that 78% of the managers responding to his survey expressed varying levels of agreement with the following statement, "I can conceive of a situation where you have sound ethics running from top to bottom but, because of pressures from the top to achieve results, the person down the line compromises."

Another variable loading on this factor involves management accountants "stretching a point" in favor of their company. The process of selecting and applying accounting principles to enhance profits or other financial results is not a new phenomenon in the accounting profession. This process, often called "cooking the books," has resulted in the issuance of a number of pronouncements by accounting standard-setting bodies. For example, Accounting Principles Board Opinion No. 20, "Accounting Changes," requires a company to establish that a new accounting principle is preferable to the old principle before a change to the new principle can

***Table 12.*** Factor XI: Aggressive Behavior/Pressure from Parent to Increase Profits

| *Item Number* | *Item* | *Factor Loadings* |
|---|---|---|
| 9. | Local management could be under pressure from the parent to increase profits | .6713 |
| 34. | Management accountants often have to stretch a point in favor of their company | .4708 |
| 43. | Most presidents are aggressive | .7306 |

be made. In addition, the cumulative effect of a change in accounting principle must be disclosed in the income statement when a new accounting method is adopted. Obviously, all of the opportunities for "cooking the books" have not been eliminated. Indeed, one subject in the current study stated:

> I would suggest another means, entirely legal, that would accomplish the same result. It is not necessary to commit illegal actions. Any good financial VP should be able to use grey areas and opinions to create profits for many months even if a company is actually losing money.

Altering the financial results by the use of grey areas and stretching points in favor of the company can lead to a violation of professional ethics. When the solution to an accounting problem is not "black or white," the accountant must exercise professional judgment in deciding how to solve the problem. The accountant must always maintain the professional responsibility to communicate information fairly and objectively.

### *Open Communication Channels*

The importance of having open communication channels through which employees can report the unethical behavior of their superiors can not be overemphasized. In fact, Waters (1978) suggests that the reporting of unethical behavior through a regular chain of command is inadequate. He believes that employees may feel blocked by the chain of command unless they are provided with direction. Waters suggests that perhaps an ombudsman could serve as a receiving point through which individuals could blow the whistle. South Central Bell's corporate code of conduct provides employees with a choice of departments through which they can report unethical behavior. These departments include the security department, the legal department, or the internal audit department (Harrison 1988).

Two important points are described by the variables loading on this factor (see Table 13). Unless prospective whistle-blowers have someone at the corporate office they would feel comfortable talking to, they are not apt to report unethical behavior. The second variable indicates that an individual's decision to talk with the president would be dependent on their relationship. Top management must set

**Table 13.** Factor XII: Open Communication Channels

| *Item Number* | *Item* | *Factor Loadings* |
|---|---|---|
| 4. | I would be willing to report this to corporate only if I had contact(s) there | .7122 |
| 51. | My relationship with the president would determine whether I would talk to him | .5586 |

the proper tone to encourage ethical behavior, and they must set the proper tone to convince prospective whistle blowers that they will be protected from reprisal.

## Internal Controls

One of the 49 recommendations made by the National Commission on Fraudulent Financial Reporting specifically addressed the role of administrative internal controls in the prevention of fraudulent financial reporting:

> Public companies should maintain internal controls that are adequate to prevent and detect fraudulent financial reporting (National Commission 1987, 29).

Two factors in the current study provide two underlying dimensions of internal control. The factors are Factor V, Internal Control/Deterrent, and Factor X, Internal Control/Protection. Factor V represents the deterrent aspect of internal control, and Factor X represents the role that internal control can play in protecting individuals caught in the middle of a dilemma.

### *Internal Controls/Deterrent*

Realizing the importance of both the internal and external audit function, the Treadway Commission issued several recommendations involving internal audit departments, board of director audit committees, and the selection and review process for external auditors. Obviously, a strong internal control system, an effective internal audit department, and competent external auditors all serve as deterrents to certain types of unethical behavior. Factor V indicates that both internal and external auditors would discover the entry (see Table 14).

### *Internal Controls/Protection*

Three variables loaded on Factor X (see Table 15). One variable indicated that superiors might request that a subordinate do things to protect the superior's job. The other two variables provide ways in which protection might be provided to those asked to commit unethical acts: having proper controls in place to prevent

***Table 14.*** Factor V: Internal Control/Deterrent

| *Item Number* | *Item* | *Factor Loadings* |
|---|---|---|
| 13. | If I made the entry auditors would find the entry | .8853 |
| 20. | Internal auditors would find the entry | .7954 |
| 55. | External auditors would find the entry | .8659 |

**Table 15.** Factor X: Internal Controls/Protection

| Item Number | Item | Factor Loadings |
|---|---|---|
| 15. | Superiors might ask you to do things to protect their job | .6244 |
| 19. | Companies should have controls to prevent this from occurring or to catch it when it does occur | .6000 |
| 23. | I wouldn't lie to the auditors if they asked me about the entry | .6892 |

and catch unethical acts and being truthful with the internal audit department concerning the entry.

The Treadway Commission emphasized that one way to protect a company from the fraudulent acts of its employees is to organize and maintain a strong internal audit department. In some companies the internal audit department conducts an annual ethics audit. The main emphasis of this audit is to measure compliance with the company's code of conduct (Harrison 1988). Perhaps the audit should include questions directed to employees concerning their knowledge of unethical acts that have been committed within the organization.

## Characteristics of Entries: Current and Prior

Although the financial reporting vignette involves the preparation of misleading journal entries, the variables loading on Factor VII could relate to the types of information an individual would need when confronted with other ethical dilemmas. First, has this situation occurred in the past or is this the first occurrence? If it has occurred in the past, how similar is this occurrence to those in the past? Second, what is the intent of this action? Third, what is the impact of this action on other people and organizations (i.e., employees, creditors, other users of financial statements, and so forth)?

**Table 16.** Factor VIII: Characteristics of Entries: Current and Prior

| Item Number | Item | Factor Loadings |
|---|---|---|
| 29. | I would want to know the amounts of any similar entries made in prior years | .7838 |
| 38. | I would want to know the intent of the entry | .7381 |
| 47. | I would want to know whether the amount of the entry was material | .5693 |

***Table 17.*** Factor XIV: Rationalizations for Committing Unethical Act

| *Item Number* | *Item* | *Factor Loadings* |
|---|---|---|
| 30. | Even if I made the entry the controller would be ultimately responsible | .5252 |
| 49. | If I refuse to make the entry someone else will make the entry | .5728 |

Of all the variables loading on Factor VII (see Table 16), the intent of the entry is possibly the most important. Perhaps similar entries have been made in the past, and the amounts might have been considered immaterial. However, if the intent of the entry is to manipulate the financial statements, then the preparation of the entries is unethical.

## Rationalizations for Committing Unethical Act

The process of rationalization is a psychological mechanism driven by the need for self-protection. Rationalizing often occurs in situations where an individual might suffer significant social and material losses. When an individual makes a decision that he or she knows is not the proper decision, that person attempts to redefine the problem and inserts inappropriate motives and goals. Redefining the problem represents an effort to mask the unpleasant aspects (Toffler 1986). Both variables loading on Factor XIV represent examples of rationalizing (see Table 17).

These variables represent an individual's attempt to shift responsibility for the unethical act to someone else. After individuals commit unethical acts, they often suffer guilt and regret. This process is known in the psychology literature as cognitive dissonance. Rationalizing serves to relieve this guilt and regret (dissonance) so the person can live with the decision (Toffler 1988).

Variable III represents how individuals redefine a problem. The original problem was to decide whether to make the entry. Now, the problem has been redefined to determine who will make the entry. Although the process of rationalization can never be completely eliminated, individuals and organizations need to be constantly aware that it does exist. Individuals must learn how to recognize and deal with the impact of rationalization on the decision process. Individuals are often forced to rationalize in decision situations in which they are not allowed to make an ethical decision. Indeed, organizations that provide employees with an ethical decision-making environment should help reduce the impact of rationalization.

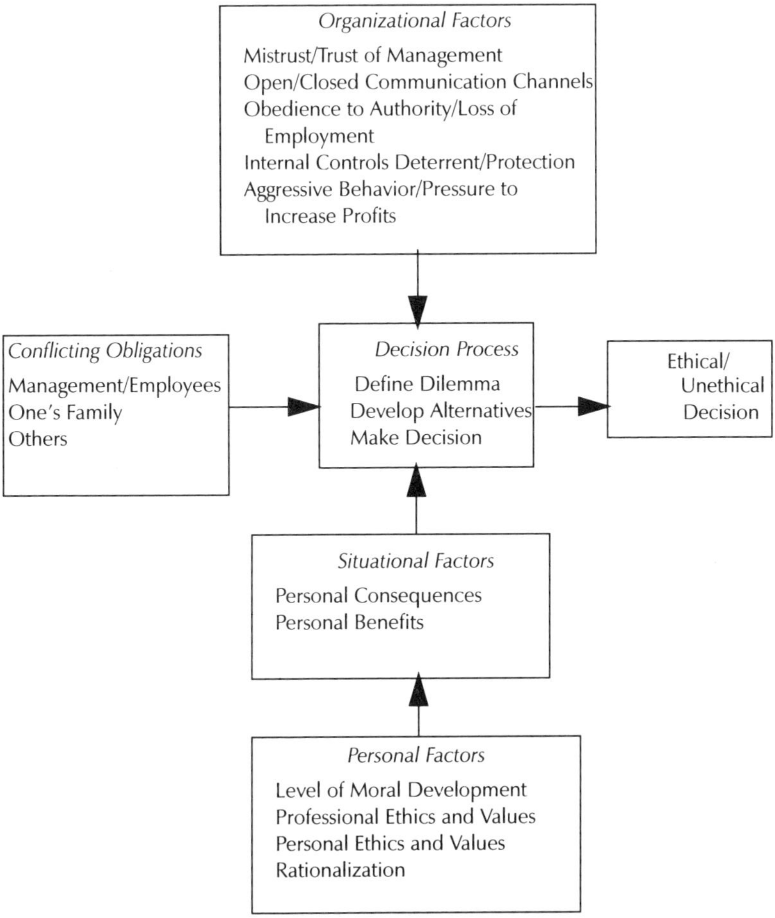


***Figure 1.*** Financial Reporting Decision Model

## DEVELOPMENT OF THE DECISION MODEL

The ultimate goal of Phase 2 was the development of a ethics decision model involving financial reporting. Development of this model was accomplished by utilizing the factors produced from the factor analysis (see Figure 1).

The model represents a decision environment which includes the decision maker, the organization, and other parties to whom the decision maker is obligated. The model includes personal factors, organizational factors, conflicting obligations, and situational factors. Personal factors include a person's level of moral development, professional ethics and values, personal ethics and values, and rationalization.

The findings from Phase 1 of this study provide justification for including a factor for level of moral development.[3] Professional/personal ethics and values were important factors in the ethics decision process. However, situational factors might prove to be important moderating variables that help explain why an otherwise ethical person commits an unethical act. Since most decision alternatives involve consequences (costs) and/or benefits for both the organization and the decision maker, these factors can serve to modify the behavior of the decision maker and other employees within the organization. Research has found that individuals are generally more ethical when they are responsible for the consequences of their actions. Organizations must make employees aware that they will be responsible for their actions. In addition, employees must be made aware that they will be reprimanded for unethical acts, regardless of their position within the organization.

Conflicting obligations can also affect the decision process. Obligations to management/employees, one's family, and others were included in the decision model. Although all obligations can influence the decision process, the financial obligation to one's family might create sufficient pressure to cause an otherwise ethical person to commit an unethical act.

Organizational factors play a critical role in influencing the behavior of individuals within the organization. These factors contribute to the ethical/unethical environment within an organization. The factors identified were mistrust/trust of management, open/closed communication channels, obedience to authority/loss of employment, internal controls deterrents/protection, and aggressive behavior/pressure to increase profits. Perhaps the ability to terminate an employee is one of the most powerful weapons a manager can use to influence an employee's behavior. Loss of employment is a possible consequence for not following orders. This weapon should not be available for managers to use as a method of encouraging unethical behavior. Building trust between employees within an organization, providing open communication channels, installing proper internal controls, and setting realistic profit goals can all serve to deter unethical behavior. Top management must instill in employees that reaching goals "at all costs" will not be acceptable when the costs include unethical behavior.

## CONCLUSION

The complex environment in which ethical decisions are made can never be recreated with the use of hypothetical vignettes. Though this limitation restricts the ability to generalize the findings, this study produced some significant results. The study represents one of the first efforts in accounting research to use protocol analysis in decision situations involving ethics. All of the protocol data represent the thoughts of subjects during the decision-making process. These thoughts were freely expressed by the subjects without intervention by the researcher. In addition, the data were used to prepare the Phase 2 questionnaire. Since the Phase 2 questionnaire was empirically developed, this should help to increase the validity of the findings in Phase 2. Though the study is exploratory, the factors that it identifies provide a foundation for future study of ethical dilemmas involving financial reporting by management accountants.

## NOTES

1. This sample was taken from the Institute of Management Accountants CMA database.
2. According to Hair et al. (1979), a KMO test result of more than 0.5000 is needed to conclude that factor analysis is appropriate.
3. Certain results from Phase 1 of the study were not discussed in this paper. These results provided significant evidence that an individual's level of moral development influences the variables considered by that individual in a decision situation involving ethics.

## REFERENCES


American Institute of Certified Public Accountants (AICPA). 1988. News report. *Journal of Accountancy* (April): 18-19.

Bommer, M. 1987. A behavioral model of ethical and unethical decision making. *Journal of Business Ethics* 6: 265-280.

Carroll, A.B. 1975. Managerial ethics: A post Watergate view. *Business Horizons* (April).

Cooper, R.W., and G.L. Frank. 1992. Professionals in business: Where do they look for help in dealing with ethical issues? *Business and Professional Ethics Journal* 11(2): 41-56.

Ferrell, O.C., and L. Gresham. 1985. A contingency framework for understanding ethical decision making in marketing. *Journal of Marketing* 49: 87-95.

Ferrell, O.C., L.A. Gresham, and J. Fredrick. 1989. A synthesis of ethical decision models for marketing. *Journal of Macromarketing* 9: 55-64.

Green, P.E., and D.S. Tull. 1978. *Research for Marketing Decisions*. Englewood Cliff, NJ: Prentice Hall.

Guerrette, R.H. 1988. Corporate ethical consulting: Developing management strategies for corporate ethics. *Journal of Business Ethics* (May): 373-380.

Hair, J.H., Jr., R. Anderson, R. Tatham, and B. Grablowsky. 1979. *Multivariate Data Analysis*. Tulsa, OK: Petroleum Publishing Co.

Harrison, S.R. 1988. South Central Bell and the Treadway Commission Report. *Management Accounting* (August): 21-27.

Jones, D.G. 1982. *Doing Ethics in Business*. Cambridge, MA: Oelgeschlager, Gunn & Hain.

Leung, P., and B.J. Cooper. 1995. Ethical dilemma in accountancy practice. *Australian Accountant* (May): 28-33.

Mihalek, P.H., A. Rich, and C. Smith. 1987. Ethics and management accountants. *Management Accounting* (December): 34-36.

Murphy, P.E. 1995. Corporate ethics statements: Current status and future prospects. *Journal of Business Ethics* (September): 727-740.

National Commission on Fraudulent Financial Reporting. 1987. *Report of The National Commission on Fraudulent Financial Reporting*, Washington, D.C.

Robison, W.L. 1991. Subordinates and Moral Dilemmas. *Business & Professional Ethics Journal* 10(4): 3-21.

Shultis, R.L., and K. Williams. 1985. Accountants must clean up their act. *Management Accounting* (May): 21-23, 53-56.

Stanford, S.E. 1991. Ethics: The first fifty years. *Internal Auditor* (June): 102-104.

Toffler, B. 1986. *Tough choices-Managers talk ethics*. New York: John Wiley & Sons.

Trevino, L.K. 1986. Ethical decision making in organizations: A person-situation interactionist model. *Academy of Management Review* (July): 601-618.

Waters, J.A. 1978. Catch 20.5: Corporate morality as an organizational phenomenon. *Organizational Dynamics* (Spring): 3-19.

Wilson, G.T. 1983. Solving ethical problems and saving your career. *Business Horizons* (November-December): 16-20.

Wren, D.A. 1979. *The Evolution of Management Thought*. New York: John Wiley & Sons.

# ACCOUNTING INFORMATION SYSTEMS:
## BUSINESS ETHICS CASE—EMPLOYEE MONITORING

Alfred R. Michenzi


### ABSTRACT

Accounting and business curricula focus on ethics by using techniques such as, discussions, presentation or case work analysis. In auditing, textbooks often present a chapter on ethics with approaches to resolving audit-related ethical dilemmas. In financial accounting, textbooks present illustrations or vignettes that ask the student to consider the ethical issues related to false or misleading presentation of accounting transactions. However, most current texts in Accounting Information Systems do not present ethics either as a separate chapter or as cases for student discussion. This paper presents a brief discussion of possible Accounting Information Systems ethics issues and develops a case for use in the course.


**Research on Accounting Ethics, Volume 4, pages 267-273.**

**ISBN: 0-7623-0339-5**

## INTRODUCTION

With the widespread use of computers and network technology in the processing of accounting information, it becomes possible for employers to monitor the activities of their employees. Monitoring of employees' activities presents management with an opportunity to develop employees' skills or to intrude on their privacy. This raises a serious question concerning the lack of discussion of ethics in the accounting information systems course as a means of resolving such issues.

Accounting Information Systems courses need to consider the inclusion of ethics topics as an element of its content. Ethical issues cover the entire spectrum of business and accounting education. The curriculum must integrate the treatment of ethics into all major areas that shape the thinking of students. A brief survey of the popular textbook showed no substantial treatment of ethics issues. This paper presents an approach to involving students in an ethics case that would fit into the Accounting Information Systems' course. Many ethics issues are generally addressed in all courses taken in the accounting curriculum. These frequently deal with traditional reporting, recording, auditing, and tax issues. The Accounting Information Systems course offers a unique opportunity to integrate the technology of computers, accounting and employee relations as part of ethics cases. The power of the computer, along with its uses in recording and processing information, gives management a tool that promotes efficiency and profitability of the business. The computer can monitor the interactions and activities of all the computer system users. This monitoring can result in a positive or negative impact on the employee.

The positive effects of monitoring employee interactions result from the use of the computer as a training device. Training elements consist of monitoring error rates, keystrokes, time required to process a transaction, and so forth. Management can develop personalized training programs to improve employee efficiency. The negative effects of monitoring employee interaction result from the control that this technology can exercise over the individual and the potential for invasion of privacy. Examples of control consist of monitoring keystrokes and continuously requiring increased speed beyond the standard required by the job description, or monitoring idle and personal time. The violation of privacy can also occur when management reads e-mail files or employee-created private files without the employee's knowledge. The positive aspects of monitoring can result in improved skills and opportunities for advancement of the employee. The negative aspects result from the control over the employee. This gives rise to low morale and suppression of individual rights and the potential for litigation by the employee against the business.

The ethical issues that an Accounting Information Systems course might consider relate to the use of monitoring by management that violates a stated or implied rule or the violation of privacy of an employee. A brief review of the popular literature describes how management's monitoring of employees has resulted in litigation and poor morale. The basis of the litigation involved the violation of the employees' rights to privacy. Federal and many state statues do not clearly address employer and employee rights, duties, or obligations relating to electronic monitoring.

## APPROACH

Since the legal aspects of employee monitoring results in a vague set of possibilities, this topic seems appropriate for the presentation of some ethical issues in a case format. The automated information system possesses many options which management can use to monitor activities, such as: the keystokes of clerks; the idle time of the computer and, therefore, the operator; the e-mail files of employees; and error rates associated with data entry and processing. Management's purpose for monitoring these activities generally relates to management's responsibility concerning the effective and efficient use of the company's resources. Also, management evaluates the employees' knowledge of the tasks and assesses the training needs of the employees.

The potential exists for management to use the monitoring capabilities of the network computer systems to exceed its legitimate purpose. Management could possibly single out one or more employees and use this information for harassment and the invasion of privacy. The harassment could involve reading the e-mail correspondence of employees without informing them. Management could maintain a log of idle time or error rates and bring action against the employee without giving the employee an opportunity for retraining.

## CASE

Able Company manufactures and sells sports clothing. It sells the sports clothing by catalog to individuals nationwide. It experienced increased growth and the order-entry function required a total revision. The firm of Macc & Hensey, accounting and computer consultants, received the contract to develop and install much-needed software and hardware to revise the order-entry function. The revision went smoothly and was completed on time and only slightly over budget. Management and employees received training in the use of the new system and all agreed that it would increase efficiency, reduce order-entry errors, and increase customer satisfaction. Macc & Hensey gave management all the system's specifications and showed them that the system contained various management control overrides and monitoring capabilities. These capabilities gave management tools that could help them improve current employee skills and training new employees. Management did not inform any of the order-entry personnel about the system's monitoring capabilities. No one then believed it was important to do so.

Eight months passed and the system worked well. One employee, Sue Frost, a bright and energetic person, became active in the unionizing of the order-entry personnel. Generally, employee and management relations were satisfactory. Management's attitude toward a union was not very understanding and in a short time became more hostile. The vice president, Bruce Young, came to you. He said that you, as the controller and the person in charge of order-entry personnel, must stop this union activity. You pointed out that the unionization activity had not taken place on company time and you believed that Sue Frost had the right to pursue this action. Several days passed and Bruce Young came to you again and said that he understood that the order-entry system had a monitoring capability. He suggested that you used this capability to monitor Sue Frost and gather information to build a case to fire her. He was emphatic in his desire to prevent a union in Able Company's organization. Bruce gave you a one-week time limit to develop a plan to fire Sue Frost based on the monitoring capabilities of the computer system. You did not give him an answer and asked for time to think through all possible approaches available to Able Company.

You went home that evening and reflected on the day's event. You recalled from your college accounting ethics discussions that this issue could be resolved using some or all the techniques noted in these discussions. After a dinner with your family, you searched for your old college notes. You found these and reviewed them, hoping to find an answer to this difficult problem. You realized that your actions could result in the firing of a qualified employee, potential litigation by the union and the employee, yourself being fired for not complying with the vice president's requests, your loss of his support in the future for raises and promotions and other possible outcomes.

Your notes set out six steps in the resolving of ethical dilemmas. These six steps follow:

1. Obtain the relevant facts.
2. Identify the ethical issues from the facts.
3. Determine who is affected by the outcome of the dilemma and how each person or group is affected.
4. Identify the alternatives available to the person who must resolve the dilemma.
5. Identify the likely consequences of each alterative.
6. Decide the appropriate action (Arens and Loebbecke, 1994, 69, 70).

## TEACHING NOTES

The relevant facts are:

1. The employees are not aware of the monitoring capabilities of the computer system.

2. The employees are not aware of any "standard" error or efficiency rates that the company expects, once they pass their probationary period with Able Company. This period is three months.
3. Sue Frost appears to be a very competent worker.
4. Sue Frost wants to bring a union in and organize her co-workers.
5. The vice president does not want a union to organize any Able Company workers.

The ethical issues are:

1. Is it ethical for Able Company to use the monitoring capabilities of the computer system without informing all order entry employees?
2. Is it ethical for Able Company to single out one employee for monitoring based on the employee's interest in establishing a union?

Those affected and the impact on each are:

1. Sue Frost may be fired.
2. Sue Frost may be retained as an employee.
3. The controller may be fired.
4. The controller may be retained.
5. The controller may be retained but denied future promotions and pay raises.
6. The controller may be involved in a lawsuit brought by Sue Frost and the union.
7. Bruce Young, the vice president, may be rewarded for his actions.
8. Bruce Young may be involved in a lawsuit brought by Sue Frost and the union.
9. Able Company may be involved in a lawsuit brought by Sue Frost and the union.
10. Able Company may be successful in discouraging other unionizing activities.
11. Able Company may succeed but the result may be the lowering of employee morale and efficiency.
12. Able Company employees may become suspicious of all actions of management.
13. Able Company employees may continue the union organizing activities.

The controller's alternatives are:

1. Refuse to use the monitoring capabilities to fire Sue Frost.

2. Inform Sue Frost of management's intentions.
3. Quit working for Able Company.
4. Inform the President and Board of Directors of Bruce Young's request to gain an understanding their intentions.
5. Do as Bruce Young requests and gather evidence to fire Sue Frost.
6. Inform Sue Frost that you will be monitoring her computer terminal for errors and idle time but do not inform her of the company's anti union intentions.

The above alternatives have consequences that result in both short-term and long-term effects. The short-term effects include the firing of the controller for the refusal to gather evidence to support the firing of Sue Frost. Another possible short-term effect is that the controller monitors the employee and gathers the evidence that supports her firing since she is performing below the standard for entry-level/probationary employees. Long-term effects could range from the establishment of a union and improved management and employee relations to lower morale of all employees. This could result in loss of sales and customers. Another long-term effect could be that the controller quits and finds a better job. The company might find itself in a long and expensive lawsuit filed by the employee and the union with the possibility of winning or losing the court battle. Other alternatives and consequences do exist.

## SUMMARY

To avoid finding Able Company in this circumstance, management could have informed all employees of the system's monitoring capabilities and communicated this in writing to all employees. Management should disclose that it has the right to examine e-mail files and other peronal files residing on company equipment. Management should also set standards for the uses of the information gathered. This should be documented, and open discussion should take place between employees and management. Management could let its intention regarding the unionization efforts of any employee be known; also, management must follow the laws governing employees' rights. Management should explain how it used the monitoring information to improve the productivity of employees and the consequences of the poor productivity. If management clearly communicates to employees, in writing, all that the company expects, then there will be reduced potential for misuse of monitoring information and for discord between employer and employees.

## REFERENCE

Arens, A.A., and J.K. Loebbecke. 1994. *Auditing: An Integrated Approach.* Upper Saddle River, NJ: Prentice Hall.

# ACADEMIC ACCOUNTING ETHICS:
## A CODE IS NOT NEEDED

Allen G. Schick, Lawrence A. Ponemon,
Sharon H. Fettus, and Robert J. Nagoda

### ABSTRACT

This paper challenges Loeb's contention that a code of ethics is needed for accounting educators. We provide a critical analysis of Loeb's arguments for and against a code. We conclude that his arguments for an ethics code do not indicate that the ethical issues faced by academic accountants are so profound or unique that they warrant a separate code for individuals who teach accounting. His arguments against a code of ethics suggest that the existence of a code may have unintended negative consequences. Specifically, having a code of ethics may impede rather than promote ethical thought, analysis, and judgment and, hence, may lead to less rather than more ethical behavior. This paper is organized into three main parts. The first section provides an introduction on the role and purpose of ethics codes in organizations and professional associations. The second provides an analysis of Loeb's six arguments for an ethics code applicable to academic accountants. The third part assesses the effectiveness of ethics codes in promoting ethical behavior.

**Research on Accounting Ethics, Volume 4, pages 275-290.**

**ISBN: 0-7623-0339-5**

## INTRODUCTION

Recently, the accounting profession in the United States has been under close scrutiny by regulatory bodies, the financial press, and the legal profession for alleged instances of ethical lapses. Perhaps the most notable instance of ethical breaches concerns the issue of auditor negligence and possible fraudulent acts when attesting to the financial condition of savings and loan institutions (Bacon and Berton 1992). A less serious but more common practice involves a lack of independence in the auditor-client relationship, where auditors permit client management to take an overly aggressive and deleterious stance on financial reporting issues (Merchant and Rockness 1994).

In light of alleged ethical problems, the accounting and financial community, governmental regulatory bodies, and accounting educators have intensified their interest in seeking to understand the nature and extent of these problems. Their purpose is to find ways to prevent or minimize the occurrence of ethical problems or, if they do occur, to find ethical resolutions to these problems. The reason for their interest is based, at least in part, on the general belief or perception that unethical behavior can be extremely costly to the accounting profession by virtue of lost reputation and increased litigation risk.

Understanding ethical issues in any professional context requires a normative framework to guide in the analysis of determining "appropriate" ethical behavior (Scribner and Dillaway 1989). One framework that has been advanced by cognitive developmental psychologists pertains to the moral development of individuals. This framework is concerned with the underlying reasoning and decision-making processes of individuals when facing an ethical dilemma. According to Kohlberg (1969), Rest (1979), and others, there is a deliberate sequence of steps or stages of development in the way people frame, process, and resolve ethical conflicts.

The lowest stages of the moral development model state that an individual is concerned about maximizing self-interest when deciding whether or not to act ethically. For instance, an auditor with lower moral reasoning skills might exercise a cost/benefits approach when choosing whether or not to report an incident of wrongdoing in the organization. An individual with moderate moral reasoning skills is less concerned with self-interest and more with affiliation or being part of a social group, organization, or institution. Thus, an auditor with moderate moral reasoning skills would be guided by his or her perception of the organization's best interest when deciding whether or not to act ethically. The highest stages of moral reasoning state that the individual has achieved an ability to separate principle from rule when making a judgment about an ethical problem. Hence, when the principle conflicts with mere rules, the high moral reasoning individual may violate rules in order to follow his or her ethical beliefs and values. For instance, an auditor might decide to walk away from a profitable and prestigious client that is believed to be too aggressive in its financial reporting practices, even though the client is in conformance with Generally Accepted Accounting Principles.

A second normative framework that has been advanced by theologians, philosophers, and a number of ethicists asserts that there are universal ethical principles or virtues that transcend time and situational context. For example, the Judeo-Christian tradition specifies the Ten Holy Commandments which should guide all human behavior. The value of having universal ethical principles is that they are absolute—that is, there should be no exceptions. Hence, they foster consistent and unequivocal behavior simply because everyone faced with the same ethical challenge is supposed to act in exactly the same manner. While there are always exceptions to universal ethical principles, any possible violation of these principles, such as stealing bread to feed a starving child, should be thoughtfully analyzed and examined for its appropriateness by a learned authority figure (e.g., member of clergy, attorney, or law enforcement official) to sanction or legitimize behavior.

Universal ethical principles are appealing to all organizations and professional associations because, in promoting consistency and unequivocality, they reduce uncertainty. Conversely, the implication of moral development is that an individual's principles should take precedence over the mere rules of any organization in guiding ethical behavior, thus increasing the possibility of uncertainty. In extreme cases, this could lead to organizational chaos and turmoil. Consequently, to reduce uncertainty, organizations and professional associations seek to invoke universal ethical principles by establishing codes of conduct (Beach 1984). Like clearly defined routines and standard operating procedures for all other organizational processes and functions, ethical codes seek to ensure that individuals will behave in conformity to relevant group norms or organizational dictates.

Interestingly, this appears to be the approach of two of the three pillars supporting the accounting profession in the United States. The first pillar is the American Institute of Certified Public Accountants (AICPA), which requires CPAs, who are members of the Institute, to uphold a code of ethical conduct. The AICPA's code contains a general statement of ideal conduct as well as specific rules delineating the profession's responsibility to the public. The second pillar pertains to professional accountants in organizations who are members of the Institute of Management Accountants (IMA). The IMA also provides its members with a clearly defined code of principles pertaining to competence, integrity, and confidentiality that should guide ethical conduct in everyday practice.

The last pillar pertains to the community of college and university teachers in accounting across the country. At present, there is no particular code of ethics that solely encompasses accounting academics. Rather, the community is guided by the individual university and college statements of mission and definitions of faculty duties within the institution. Further, the American Association of University Professors has developed a *Statement of Professional Ethics*, which clearly requires professors to (Academe 1987, 49) "devote their energies to developing and improving scholarly competence, . . . demonstrate respect for students . . . [and recognize that obligations] derive from common membership in the community of scholars." Despite these organizational policies and a general code of professional ethics for professors, and despite the view of accounting practitioners that accounting faculty are honest and highly competent and have a high degree of personal integrity (Carver and King 1986), a call nevertheless has been made for a specific code applicable to accounting educators (Loeb 1990, 1994).

Loeb (1990, 123) posed the question of "a code of ethics for academic accountants?" concluding that a code of ethics for academic accountants should be considered. More recently, Loeb (1994, 191) argued not only that a code is necessary but that the need "is so pressing." We believe differently. Instead, we think that the conclusion of the American Accounting Association (AAA) Committee on Academic Independence, which considered the idea of a code of ethics for academic accountants in 1981, is still valid. The Committee's conclusion was that "there is no need for the development by the AAA of a code of ethics for academic accountants. No strong support appears to exist for such a code nor does there appear to be any impelling evidence to suggest that such a code is needed" (AAA 1981, 42). In this paper, we argue that a code of ethics is not needed for accounting educators. Our arguments are presented in the following sections.

## LOEB'S SIX ARGUMENTS FOR A CODE

1. The AICPA has a code of ethics for its members.

In our opinion, the primary reason why Loeb (1990, 1994) believes a code of ethics is needed for accounting academics is that the AICPA has a code of ethics for its members. If this reason is valid, then it would be an example of mimetic isomorphism (Dimaggio and Powell 1983) or, in more colloquial terms, what McKinley, Sanchez, and Schick (1995) call "cloning." Cloning is a social force that institutional theorists claim helps explain why organizations in the same industry or market become similar over time. According to McKinley et al. (1995, 34), cloning is a response to uncertainty, which occurs when "organizations mimic the actions of the most prestigious, visible members of their industry." The power of cloning is not based on hard evidence, for example, that organizational codes of ethics improve the ethical behaviors of organizational members. Instead, cloning derives its appeal from similarity. As an illustration, the AAA would establish a code to imitate the AICPA to show that the AAA is "with it" when it comes to ethics and that the AAA is actively doing something to address "purported" ethical problems. However, unless it can be shown that academic accountants operate in an ethical vacuum, which a code of ethics would overcome, the fact that the AICPA has a code for its members is not a valid reason why the AAA also should have one. Further, according to Beach (1984, 323), establishing a code "simply to put up a good façade . . . can be more dangerous than not having a code at all."

2. There is the potential for some kind of schism.

Loeb (1994) has noted that the AICPA has begun to clarify its ethical standards for their accounting educator members. However, he is concerned that such a clarification of AICPA ethical standards would not affect a large number of accounting faculty. He believes that this would create the potential for some kind of schism between those accounting faculty who are subject to the AICPA code and those who are not—hence, his call for the AAA to establish a code of ethics applicable to all accounting educators. While the potential for a schism may be a valid concern, we would be more comfortable with Loeb's argument if it were based on logical analysis or empirical evidence. Loeb (1994), however, provides no examples of possible ethical schisms nor a statement of conditions under which ethical schisms might occur. Since we are unaware of any ethical schisms between accounting faculty who are subject to the AICPA code and those who are not, we do not see the need for the AAA to establish its own code of ethics.

Indeed, there are more likely to be ethical schisms between accounting practitioners and accounting educators, than between different groups of accounting academics (Carver and King 1986). For example, one such schism may involve the external fund-raising efforts of many university accounting departments (Carver, Hirsch, and Strickland 1993). They state that ethical issues abound in fund raising. One reason is that "fund raising is directly related to an institution's values and priorities [whereby] . . . each accepted gift is a statement about what the institution is willing to become" (Carver et al. 1993, 301). A second reason is that large amounts of monies may be involved. If some of those who contribute large sums believe in the golden rule—that those who have the gold, rule—then they may pressure accounting faculty to conform to the norms and values they advocate. If these contributors can dictate or have considerable influence over the development of accounting curriculum, the type of courses to be taught, and course content, then at issue could be the academic independence and integrity of accounting faculty. Will accounting academics become *advocates* for the accounting profession and *trainers* of their future employees? Or will accounting faculty be able to preserve their independence to teach, and criticize constructively and publicly, not only the technical aspects of accounting but also the institutional and organizational aspects of the accounting profession?

3. One cannot teach accounting ethics without being subject to a code.

We have two difficulties with Loeb's idea. First, no logical arguments are made nor empirical evidence provided as to why being subject to a code of ethics would enhance accounting education or pedagogy. Since no connection is made between having a code and effectively teaching accounting ethics, there is no reason why a code should be a prerequisite to teaching ethics. Indeed, academic philosophers, who are often the faculty in universities who teach business ethics, are unlikely to adhere to a specific organizational or professional code. They teach ethics because they have been educated to do so. This suggests that educating accounting academics about ethics may be more relevant to their teaching ethics than subjecting them to an ethics code.

Our second difficulty is Loeb's implicit assumption underlying his call for a code. Specifically, he seems to assume that academic accountants, who are not professionally certified, need a code of ethics to correct their ethical failings that arise either because they are aethical—do not know anything about ethics—or they are unethical. If this assumption is valid, then as Loeb (1990, 1994) suggests, a code of ethics could act as a set of standards that would encompass societal expectations, enhance the reputation of accounting educators as well as educate them, and be used to train accounting doctoral students on how to resolve ethical dilemmas.

Suppose, however, that ethical behavior has two components. The first is knowing what is right or wrong behavior. In essence, this component represents how people ought to behave, and most likely would behave, if they functioned in a vacuum. The second component is the actual behavior. In our opinion, individuals can act unethically not because they do not know the difference between right and wrong but because they are afraod to follow their ethical convictions. That is, acting ethically may lead to bad consequences, whereas acting unethically may lead to good consequences. In essence, individuals have to choose between acting ethically or in their own self-interests. Under such circumstances, and because individuals generally cannot control most of the consequences associated with their actions, improved ethical behavior would require not a code of ethics but rather a mechanism that would tighten the fit between ethically desirable behavior and the occurrence of good consequences. In other words, external reinforcements would need to be changed such that acting ethically and acting in one's self-interest would become compatible, rather than remain in conflict.

An example of possibly unethical behavior that conflicts with one's self-interest is the "inattention to teaching and students" cited by Loeb (1994, 125) that he seems to associate with increased pressure on faculty to do research and publish. Implied by the publication pressure that universities place on their faculty is that research is more important than teaching. This generally held perception is reinforced through annual merit salary increases as well as promotion and tenure decisions. For example, it is our observation that in some research schools, only those faculty who have published in the applicable academic year receive annual merit salary increases. Similarly, faculty who publish in the so-called "leading" academic journals tend to receive favorable tenure or promotion decisions. If our perception is accurate, then teaching and service, the other two components of a university's tripartite mission, are rewarded less or not at all. Given such an environment, it seems somewhat ingenious for Loeb (1994, 125) to suggest or imply that "inattention to teaching and students" is an ethical lapse of accounting faculty.

Indeed, an ethical analysis of this issue might conclude that ethical improprieties are committed by academic administrators who lure undergraduate students to their schools with the promise of an outstanding education and excellent teaching, knowing full well that faculty are encouraged to pay attention to research and publishing and not to teaching and students. It seems that the inattention to teaching issue could easily be resolved if university administrators would acknowledge the importance of teaching and reward it accordingly. If a code of ethics really is needed, then maybe it is needed for university administrators, who set the ethical tone of the university and its respective subunits, rather than for the academic accountants who labor largely within preset ethical boundaries.

4. To serve as a guide to accounting faculty who unintentionally engage in unethical behavior.

Loeb (1994, 195-196) provides examples of behavior that accounting faculty may do, without realizing that such behavior may be viewed as unethical. These examples are as follows:

- Unintentionally violating someone's copyright in preparing a manuscript;
- Accepting some form of compensation in return for adopting a textbook;
- Failure to treat students with disabilities differently from students without disabilities; and
- Using someone else's ideas without giving proper credit to the source of the ideas.

In suggesting that a code of ethics serves as a guide to ethical behavior, Loeb (1990, 1994) implicitly is saying one of two things about the ethical issues faced by accounting educators. First, the ethical issues are unique and, hence, there is a pressing need for accounting educators to have a code. Or second, even if the ethical issues are not unique, no harm occurs if the AAA establishes an ethics code.

The above examples clearly are not unique to accounting faculty but are relevant to all university faculty. Thus, if a code is needed, it should be university-wide. No reason seems to exist for singling out accounting faculty and suggesting that they have a greater need for a written ethics code than faculty in other disciplines. But even if there is agreement that the ethical issues faced by academic accountants are not unique, then what harm occurs anyway in having a code of ethics? In our opinion, the harm comes in the imposition of one person's ethical beliefs upon another and, with it, the possible infringement of a faculty member's academic freedom.

For it is not at all clear that Loeb's (1994) examples of possible unethical behavior are indeed examples of unethical behavior. For instance, consider an accounting educator accepting some form of compensation in return for adopting a textbook. Support for this behavior as being unethical is provided by Engle and Smith (1990). In their study of ethical standards of accounting academics, they found that 62% of their 245 full, associate, and assistant professor respondents believed that (1990, 15) "a severe reprimand or dismissal . . . were appropriate in the case of a faculty member adopting a textbook in return for assets donated to the accounting department by the publisher."

Suppose, however, that one is a faculty member in a university that enrolls a large number of minority students (e.g., African-American, Hispanic). On average, such students typically have fewer financial resources available to attend college and need to work more hours to get their resources than students who attend nonminority private or public universities (e.g., Babson, Vermont). Under such circumstances, it may make sense for accounting faculty and department heads to negotiate with textbook publishers to contribute needed resources to an accounting department (e.g., textbooks, practice sets, computer software) in exchange for adopting a textbook. This is particularly so for core courses, where a large number of texts are involved and mostly common material is found across texts. Negotiating behavior appears a creative way for accounting faculty to scale down the financial impediments that make it difficult for minority students to successfully complete their education. Such faculty should be considered heroes and heroines, rather than unethical people subject to reprimand or dismissal.

5. Research pressures may result in questionable ethical behavior.

Research pressures clearly exist for accounting academics. These pressures, nonetheless, also exist for academicians in most other disciplines. Thus, if research pressures may result in questionable ethical behavior, then these pressures should affect not only accounting faculty but other faculty as well. Indeed, it was the finding of possible errors and/or fraud in the hard sciences (e.g., chemistry, physics, biomedical science) and economics that raised concern about errors, fraud, and plagiarism in accounting research (see Davis and Ketz 1991; Loeb 1990). Although we do not know if fraud has occurred in accounting research, we do know that Davis and Ketz (1991, 109) state that "no accusations of fraud in accounting research have been published." Hence, we wonder why identifiable instances of research fraud in other disciplines require a code of ethics for the accounting discipline? Since there is nothing unique about accounting research nor does there appear to be any indication that accounting faculty are doing research in unethical ways, we do not see the need for a separate code of ethics for accounting academics.

6. Substance abuse.

Loeb (1990, 1994) provides no explanation for why the problem of substance abuse (i.e., alcoholism and drug abuse) in society requires a code of ethics to be established for academic accountants. Nevertheless, we can think of two reasons that might underlie Loeb's suggestion. The first is that alcoholism and drug abuse not only exist but occur more frequently among academic accountants than among other academics. The second is that alcoholism and drug abuse represent unethical behaviors.

Loeb, however, presents no evidence, nor are we familiar with any, that indicates that alcoholism and drug abuse are more of a problem among academic accountants than other faculty. Nevertheless, if one assumes that the population of academic accountants is representative of society, then it seems reasonable to believe that some accounting faculty may abuse alcohol and/or drugs. Since this assumption can be made about the population of academics in other disciplines, too, it also seems reasonable to believe that a number of nonaccounting faculty may abuse alcohol and/or drugs. However, without evidence to the contrary, it is unreasonable to believe that alcoholism and drug abuse are more prevalent for academic accountants than for any other faculty members. Once again, as discussed above in point 4, Loeb provides no apparent reason for singling out accounting faculty and suggesting that they have a greater need for a written ethics code than faculty in other disciplines.

With respect to the second implicit reason, do alcoholism and drug abuse represent unethical behaviors? Yes, say many respondents queried by Engle and Smith (1990, 14-15). In their study, they reported that 41% of their 245 respondents thought that a faculty member was extremely unethical and should be dismissed if he or she performed university responsibilities under the influence of drugs or alcohol. An additional 44% thought that the above behavior was moderately to extremely unethical and that a severe reprimand was in order. This finding would seem to be compatible with the view that a code of ethics is necessary to address the substance abuse issue.

It may be appropriate, however, to give more study and thought to substance abuse as an ethical issue. Do faculty in other disciplines agree with the academic accountants cited in Engle and Smith (1990) that performing university responsibilities under the influence of drugs and alcohol is an ethical problem? If they do, and if they think that substance abuse might affect a wider range of academics than just accounting faculty, then it would seem that a code of ethics concerning substance abuse should be applicable to all faculty, as well as all other university personnel.

Perhaps more importantly, however, is the possibility that substance abuse could be an illness. For example, "alcoholism may be viewed as a disease . . . [or] a symptom of an underlying psychological or physical disorder." Its cause may be "a hereditary defect, a physical malfunction, a psychological disorder, [or] a response to economic or social stress" (*New Encyclopedia Britannica* 1989, Vol. 1, 229). Indeed, a number of approaches to the treatment of alcoholism assumes alcoholism is an illness (e.g., the use of drugs to promote abstinence). Therefore, if alcoholism is viewed as a disease, like diabetes or epilepsy, then it may be unethical to treat alcoholics differently than diabetics or epileptics or faculty members with other diseases. Consequently, to fire or reprimand an academic accountant for performing university responsibilities under the influence of alcohol or drugs, as recommended by 85% of the accounting faculty respondents in the Engle and Smith (1990) study, may be an unethical act.

We think that to call for a code of ethics in response to substance abuse reflects the attitude that drug and alcohol abuse occur because individuals are weak of character and do not have the discipline to say no to drugs and alcohol. In essence, the user is blamed for his or her abuse of drugs or alcohol. Yet, to blame the user for being dependent on drugs or alcohol seems to ignore the reality that large quantities of drugs (e.g., sedatives, tranquilizers, sleeping remedies) and alcohol are readily available for consumption. Dependence on prescribed drugs can occur, particularly with respect to widely used tranquilizers such as Valium and Librium. Nevertheless, physicians issue millions of prescriptions every year for these drugs, despite their potential for abuse and dependence (*New Encyclopedia Britannica* 1989, Vol. 4, 233). Concerning alcohol, "the alcohol beverage industry produces countless millions of gallons of wine and spirits and countless millions of barrels of beer each year" (*New Encyclopedia Britannica* 1989, Vol. 13, 240.). Enough drugs and alcohol are prescribed and manufactured such that "one might conclude that there is a whole drug culture . . . [and] that existing attitudes are at least inconsistent, possibly hypocritical" (*New Encyclopedia Britannica* 1989, Vol. 13, 240). This suggests that those concerned with substance abuse may be viewing the problem too narrowly when they focus on the user. Instead, their focus should be broadened to include the societal environment and culture within which the excessive use of drugs and alcohol takes place.

In summary, we have discussed Loeb's arguments for a code of ethics for academic accountants. Our conclusion, is that these arguments are inadequate to justify implementing a code for academic accountants. A number of the issues raised transcend accounting academics and are common to all faculty, while Loeb's arguments are either nonexistent or narrow in focus or ignore the contextual element. Despite our conclusion, further analysis of Loeb's call for a code is warranted because many organizations, associations, and professions have a code. Thus, irrespective of need, it is not unreasonable to assume that a code of ethics could also be developed for academic accountants. Unanswered, however, is the question of whether codes of ethics promote ethical thought, analysis, and judgment ultimately leading to more ethical behavior? The purpose of the next section is to address this question.

## LOEB'S THREE ARGUMENTS AGAINST A CODE

Loeb identified three arguments against a code of ethics for academic accountants. They are (1994, 194): "(a) the existence of one or more codes of ethics to which academic accountants are already subject, (b) the difficulty in finding agreement on what should be in such a code of ethics, and (c) the low probability that such a code of ethics would be effective." As noted by Loeb, the belief that academic accountants are already subject to the codes of the accounting profession is not a valid argument against a code for academic accountants. The reason is that many accounting faculty are not professionally certified and, hence, would not be subject to professional standards. His second argument—namely, the difficulty of reaching consensus on a code of ethics—is also not valid in our opinion. The reason is that diverse professional groups, such as the AICPA and the IMA, have been successful in developing a code of ethics for their members. The existence of these codes implies one of two things. First, consensus can be achieved among diverse groups of practitioners in various settings such as business organizations, governmental entities and professional firms. Or second, codes can be developed and imposed without consensus. Conceptually, therefore, consensus or the lack of it does not prevent a code of ethics from being established. However, the existence or nonexistence of consensus may affect how well a code of ethics works. Thus, we think that the only valid argument against a code of ethics—to which we now direct our analysis—is that codes of ethics may not be effective at mitigating unethical behaviors. Our analysis is comprised of four parts, as follows.

1. Top management's shunning of codes.

Ethical climate is defined as that "part of an organization's culture or informal philosophy that deals with ethical problems and the resolution of moral conflict" (Ponemon 1994, 124). Awareness of a corporation's ethical climate is important, since a weak ethical climate has been identified as one cause of fraudulent financial reporting (National Commission on Fraudulent Financial Reporting 1987). Conversely, it is perceived that companies with strong ethical climates are less likely to engage in fraudulent financial reporting. Thus, written codes of ethics have been proposed as one way to foster a stronger ethical climate and, hence, reduce the incidence of fraudulent financial reporting (National Commission on Fraudulent Financial Reporting 1987).

Nevertheless, codes of ethics may not improve the ethical climates of corporations. For example, in a study of 119 fraudulent financial reporting actions, it was found that management in those companies repeatedly had been able to override systems of internal accounting control (National Commission on Fraudulent Financial Reporting 1987). Management apparently continues to have this control overriding ability, since management override of existing controls still is believed to be one of the most common causes of financial fraud (Hooks, Kaplan, and Schultz 1994). This suggests that unless top managements are willing to follow their own corporate codes of ethics, these codes may have a fine appearance but, as noted by Treviño and Nelson (1995, 205), may be of little value in strengthening ethical climates in corporations.

2. Whistle-blowing as an unethical act.

A possible reason for why codes of ethics do not appear to improve a corporation's ethical climate and, hence, discourage fraud, may be that the corporation has not yet learned how to make its top management adhere to the same ethical principles and rules as the rest of its employees. One approach that can be used to encourage a top management to conform to its corporate code of ethics is whistle-blowing (Hooks et al. 1994; Ponemon 1994). Whistle-blowing involves individuals disclosing their perceptions of unethical practices to other individuals who are outside the organization's normal reporting structure or chain of command. Disclosures can be either internal or external, depending upon whether they are made to individuals inside or outside the organization. Both types of whistle-blowing are considered an inherent part of an organization's internal control environment. Thus, they are advocated as ways that could strengthen an organization's ethical climate, assisting in fraud prevention and detection.

Implicit in the advocacy of whistle-blowing as a mechanism to report organizational wrongdoing is the assumption that whistle-blowing is an ethically appropriate practice. Yet, the standards of ethical conduct for both management accountants (IMA 1983) and internal auditors (IIA 1985) expressly forbid the disclosure of wrongdoing to parties outside the organization. In short, according to these standards, external whistle-blowing is deemed to be an ethical violation.

Codes of ethics are enacted with the purpose of helping individuals through the thought processes of deciding what is or is not ethical or unethical behavior (Finn 1995). However, rather than helping, the professional standards of management accountants and internal auditors have created an ethical dilemma for them by stating that external whistle-blowing is unethical. The reasons for this dilemma are two-fold. First, some consider whistle-blowing not only to be permissible but even obligatory under certain circumstances (DeGeorge 1982; Finn 1995). Second, the types of behaviors and values that the aforementioned professional standards appear to sanction as ethical, assuming that all internal alternatives to stopping the wrongdoing have been exhausted, can be seen as ethically questionnable. One apparently sanctioned behavior is to go along with the wrongdoing, although you disagree with it, and keep your mouth shut. The value promoted is "my group, right or wrong." Or, if you cannot go along with the wrongdoing, the second professional behavior is to express righteous indignation, get out, and keep your mouth shut. The value endorsed will be familiar to those who lived through the era of the Vietnam War: "love it or leave it." Thus, the professional standards appear to view external whistle-blowing as comparable to "tattling," "snitching," "ratting," "betraying," or "informing."

In our opinion, the behaviors and values promoted by the professional standards are simplistic. They harken back to a time when management accountants and internal auditors were kids and were not supposed to tell on each other to adults. The standards fail to take into account the increased interdependence between organizations and societies, leading to more frequent and complex ethical issues. They do not appear to increase ethical awareness, thought, and analysis among management accountants and internal auditors and, quite probably, do not assist them to behave more ethically. Instead, if management accountants and internal auditors agree that external whistle-blowing is inappropriate, then the standards may do more harm than good to ethical behavior.

3. Ethics codes stymie ethical reasoning.

Why might professional codes be detrimental to greater ethical awareness and limit the development of ethical reasoning? One reason may be that ethics education in accounting "has emphasized conformance to regulations rather than the underlying ethical issues and ethical behavior" (Langenderfer and Rockness 1989, 63). Thus, most teaching of ethics in accounting classes follows the "teach-the-code" approach (Langenderfer and Rockness 1989, 63). The consequence is that students may begin to view the professional codes as the source of all ethical wisdom. They "may never think about the implications of these rules, other ethical situations that are not covered by rules, or whether the rules themselves are ethical" (Langenderfer and Rockness 1989, 63). In short, students may not think in a critical sense. In addition, they may be being taught that it is okay not to think critically.

Ethics training for auditors also, apparently, follows the "teach-the-code" approach, since auditors generally are directed to adhere to the AICPA Code of Professional Conduct (Dreike and Moeckel 1994). As with the students, auditors would be expected to show little ethical reasoning while focusing on the code as the basis for their ethical decision making. Support for this expectation is provided in an empirical study by Dreike and Moeckel (1994). Their study sought to determine: (1) what audit seniors thought were ethical problems or issues, and (2) what their responses were to them. They reported two findings. First, in general, "auditors have a very narrow definition of what constitutes an ethical issue and what the appropriate course of action is when faced with an ethical issue." Second, "both definition and actions are strongly conditioned by a literal interpretation of the Code" (Dreike and Moeckel 1994, 261). Thus, it should not be surprising that the audit senior respondents in their study felt that client confidentiality was more important than responsibility to shareholders or the public, and that it was inappropriate for them to apprise the public of client wrongdoing.

Teaching ethics as conformity to professional codes is tantamount to treating the codes as "standard operating procedures" for ethics. Standard operating procedures are created for people to follow without thinking critically. Thus, one effect of having professional codes is that individuals will conform to them without thought, as apparently did the audit senior subjects reported on by Dreike and Moeckel (1994). Or, conversely, individuals will assume that all unethical actions are prohibited by the codes and that, therefore, anything not explicitly prohibited is okay to do (Armstrong 1993). Relying on codes to determine what are or are not ethical issues or ethically appropriate behaviors transfers responsibility for ethical decision making from the individual to the codes. As Gerboth (1987, 98) states, "it is surely harmful to teach professionals that they are not responsible for the results of their decisions." Whether or not accountants conform to or act in ways not expressly forbidden by professional codes, we are skeptical that such codes can create greater ethical awareness and a stronger ethical climate. Codes, rather than making it easier, seem to make it more difficult to develop the abilities needed to think about and resolve ethical conflicts and to cope with the uncertainties of the accounting profession (Scribner and Dillaway 1989).

4. The efficacy of ethics codes.

Finally, if professional codes of ethics are desirable, then one might expect people subject to them to be more ethical beings, both in intellect and behavior, than those who are not subject to such codes. However, the results of some studies suggest that this may not be the case. In one study, Armstrong (1987) found that the moral development of her CPA subjects was lower than the moral development of college students. In another study of solely CPAs, Ponemon (1990) found that the ethical capacity of CPAs decreased as they moved toward the partner ranks. In essence, those in firms with the most exposure to the code were the ones whose ethical capacities decreased the most. In still another study, Engle and Smith (1990) reported that faculty who were not CPAs had stricter ethical standards than academic CPAs in virtually all the 29 activities they examined, although in only three activities was the difference in standards statistically significant. Taken together, these studies suggest that the existence of a code is not associated with the disposition of CPAs to act more ethically. Thus, academic accountants not subject to a code should ask the following salient questions. First, why is a code of ethics needed? And second, what good would it do?

In summary, we have identified Loeb's (1994) arguments against a code of ethics for academic accountants. We agreed with Loeb that academic accountants, who are not professionally certified, are not subject to various professional codes of ethics. We also agreed that consensus on a code probably would be difficult to attain. Nevertheless, since there are codes of ethics, we addressed the argument that a code of ethics would not be effective. Our thought was to evaluate the effectiveness of codes of ethics in other settings to determine whether or not a code of ethics would work for academic accountants.

Based upon our analysis, we conclude that codes of ethics are ineffective and their ineffectiveness represents a valid argument against having a code of ethics for academic accountants. Codes do not appear to work for the following four reasons. First, they do not have a mechanism to prevent top managements from violating their own codes and acting unethically. Second, some professional codes expressly forbid external whistle-blowing, placing particularistic interests above the common good. Third, professional codes do not appear to increase ethical awareness or stimulate the development of ethical reasoning. Last, groups subject to professional codes of ethics do not appear to be more ethical than groups or individuals not under the aegis of a code.

## CONCLUSION

In this paper, we have examined Loeb's arguments for and against a code of ethics for academic accountants. We concluded that there are no valid arguments for a code of ethics for academic accountants. The ethical issues posed by Loeb are not unique to academic accountants. In addition, they are more rich in their ethical implications than Loeb suggests and, as a result, require greater ethical insights than a code of ethics would provide. We also concluded that codes of ethics are ineffective and that their ineffectiveness is an appropriate argument against a code of ethics for academic accountants. In our opinion, nothing has changed since 1981 that would invalidate the conclusion of the American Accounting Association (AAA) Committee on Academic Independence that there is not a need for a code of ethics for academic accountants. Expressed in our words, their conclusion and ours is simply that no code is warranted, none is needed.

## ACKNOWLEDGMENTS

The authors gratefully acknowledge the helpful comments provided by Bernie Bobal and Joan Grossman on an earlier version of this manuscript.

## REFERENCES

*Academe*. 1987. Statement on professional ethics (July-August): 49.

American Accounting Association (AAA), Committee on Academic Independence. 1981. *Report of the Committee on Academic Independence of the American Accounting Association*. Sarasota, FL: AAA, June.

Armstrong, M.B. 1987. Moral development and accounting education. *Journal of Accounting Education* (Spring): 27-43.

Armstrong, M.B. 1993. Ethics and professionalism in accounting education: A sample course. *Journal of Accounting Education* (Spring): 77-92.

Bacon, K.H., and L. Berton. 1992. Ernst to pay $400 million over audit of 4 big thrifts. *The Wall Street Journal* (November 24): A3.

Beach, J.E. 1984. Code of ethics: The professional catch 22. *Journal of Accounting and Public Policy* (Fall): 311-323.

Carver, M.R., Jr., M.L. Hirsch, Jr., and D.E. Strickland. 1993. The responses of accounting administrators to ethically ambiguous situations: The case of fund raising. *Issues in Accounting Education* (Fall): 300-319.

Carver, M.R., Jr., and T.E. King. 1986. Attitudes of accounting practitioners towards accounting faculty and accounting education. *Journal of Accounting Education* (Spring): 31-43.

Davis, S.W., and J.E. Ketz. 1991. Fraud and accounting research. *Accounting Horizons* (September): 106-109.

DeGeorge, R. 1982. *Business Ethics.* New York: Macmillan.

Dimaggio, P.J., and W.W. Powell. 1983. The iron age revisited: Institutional isomorphism and collective rationality in organizational fields. *American Sociological Review* (April): 147-160.

Dreike, E.M., and C. Moeckel. 1994. Audit seniors' responses to scenarios containing ethical issues and factors affecting actions. In *Proceedings of the Ernst & Young Research on Accounting Ethics Symposium,* June, 257-283.

Engle, T.J., and J.L. Smith. 1990. The ethical standards of accounting academics. *Issues in Accounting Education* (Spring): 7-29.

Finn, D.W. 1995. Ethical decision making in organizations: An employee-organization whistleblowing model. In *Research on Accounting Ethics,* Volume 1, ed. L.A. Ponemon, 291-313. Greenwich, CT: JAI Press.

Gerboth, D. 1987. The accounting game. *Accounting Horizons* (December): 96-99.

Hooks, K.L., S.E. Kaplan, and J.J. Schultz, Jr. 1994. Enhancing communication to assist in fraud prevention and detection. *Auditing: A Journal of Practice and Theory* (Fall): 86-117.

Institute of Internal Auditors (IIA). 1985. *Statement on Internal Auditing Standards No. 3: Deterrence, Detection, Investigation, and Reporting of Fraud.* Altamonte Springs, FL: IIA.

Institute of Management Accountants (IMA) [formerly, the National Association of Accountants]. 1983. *Standards of Ethical Conduct.* Montvale, NJ: IMA.

Kohlberg, L. 1969. Stages and sequences: The cognitive developmental approach to socialization. In *Handbook of Socialization Theory and Research,* edited by D. Goskin, Chicago: Rand McNally.

Langenderfer, H.Q., and J.W. Rockness. 1989. Integrating ethics into the accounting curriculum: Issues, problems and solutions. *Issues in Accounting Education* (Spring): 58-69.

Loeb, S.E. 1990. A code of ethics for academic accountants? *Issues in Accounting Education* (Spring): 123-128.

Loeb, S.E. 1994. Accounting academic ethics: A code is needed. *Issues in Accounting Education* (Spring): 191-200.

McKinley, W., C.M. Sanchez, and A.G. Schick. 1995. Organizational downsizing: Constraining, cloning, learning. *Academy of Management Executive* (August): 32-42.

Merchant, K.A., and J. Rockness. 1994. The ethics of managing earnings: An empirical investigation. *Journal of Accounting and Public Policy* (Spring): 79-94.

National Commission on Fraudulent Financial Reporting (The Treadway Commission). 1987. *Report of the National Commission on Fraudulent Financial Reporting.* New York: AICPA, October.

*New Encyclopedia Britannica.* 1989. Chicago, IL: Encyclopedia Britannica.

Ponemon, L.A. 1990. Ethical judgments in accounting: A cognitive-developmental perspective. *Critical Perspectives in Accounting*, 191-215.

Ponemon, L.A. 1994. Whistle-blowing as an internal control mechanism: Internal and organizational considerations. *Auditing: A Journal of Practice and Theory* (Fall): 118-130.

Rest, J. 1979. *Development in Judging Moral Issues.* Minneapolis, MN: University of Minnesota Press.

Scribner, E., and M.P. Dillaway. 1989. Strengthening the ethics content of accounting courses. *Journal of Accounting Education* (Spring): 41-55.

Treviño, L.K., and K.A. Nelson. 1995. *Managing Business Ethics: Straight Talk About How to Do it Right.* New York: John Wiley & Sons.

# THE ROLE OF VIRTUE IN AUDITORS' ETHICAL DECISION MAKING:

## AN INTEGRATION OF COGNITIVE-DEVELOPMENTAL AND VIRTUE-ETHICS PERSPECTIVES

Linda Thorne

## ABSTRACT

Recently, accounting researchers have drawn attention to the importance of "virtue" to auditors' ethical decision making (Dobson and Armstrong 1995; Francis 1990; Mintz 1995; Moizer 1995). Virtue describes the characteristics and motivation of the decision maker, the possession and exercise of which tends to increase his or her propensity to act ethically (MacIntrye 1984). However, the exact role and influence of "virtue" on auditors' ethical decision processes remain to be explained. Accordingly, the objective of this paper is to enrich our understanding of the ethical judgments and actions of auditors by describing the role and influence of virtue in auditors' ethical decision making. This will be achieved through the development of a model of auditors' ethical decision processes which integrates virtue-ethics and cognitive-developmental perspectives. Using this model, significant findings from the existing

**Research on Accounting Ethics, Volume 4, pages 291-308.**

**ISBN: 0-7623-0339-5**

research in auditors' ethical judgments and behaviors are examined and fruitful areas for future research are suggested.

## INTRODUCTION

The usefulness and importance of accounting ethics research lay in its potential to mitigate the unethical acts of accountants and auditors. The realization of this end, however, requires an understanding of the ways in which ethical dilemmas manifest themselves and are resolved in practice (Ponemon 1995). Ethical dilemmas are complex, unpredictable, and not amenable to resolution through the application of concrete rules (Ladd 1991; Dienhart 1995). An auditor's resolution of an ethical dilemma involves a complex decision process which includes the following components (Rest 1983, 1994): (1) the *ethical judgment* of "the ideal solution" to a particular dilemma; (2) the *ethical intention* of whether to comply or not to comply with the *ethical judgment*; and (3) the *action* of carrying out the *ethical intention*. Definitions of "the ethical solution" to a particular ethical dilemma can vary from auditor to auditor. Hence, an understanding of auditors' ethical decision processes is integral to the evaluation of the appropriateness of their ethical actions (Gaa 1992a).

Accounting ethics researchers have initiated an investigation of the antecedents and factors associated with auditors' ethical behaviors (e.g., Lampe and Finn 1992; Ponemon 1992; Ponemon and Gabhart 1993; Tsui and Gul 1996; Windsor and Ashkanasy 1995). Generally, this work has relied heavily upon a cognitive-developmental perspective which focuses on an individual's level of moral development (Kohlberg 1958). For example, several empirical models of auditors' ethical decision processes using a cognitive-developmental approach have been used to *predict* the effect of *situational* factors on auditors' exercise of professional judgments and ethical actions (e.g., Lampe and Finn 1992; Ponemon and Gabhart 1993). Nevertheless, a theoretical model which describes the influence of personal characteristics, other than the level of moral development, on auditors' ethical decision processes has yet to be developed.

Recently, accounting researchers have drawn attention to the importance of "virtue" to the exercise of auditors' ethical judgment (Dobson and Armstrong 1995; Francis 1990; Mintz 1995; Moizer 1995). Virtue describes the characteristics and motivation of the decision maker, the possession and exercise of which tends to increase his or her propensity to exercise sound ethical judgment (MacIntrye 1984). However, the influence of "virtue" in auditors' ethical decision making remains to be explained. Accordingly, the objective of this paper is to enrich our understanding of the ethical actions and judgments of auditors by describing the role and influence of "virtue" in auditors' ethical decision-making processes. This is achieved through the development of a model of ethical decision making which integrates virtue-ethics theory with a cognitive-developmental perspective. From this model, a theoretical framework useful for understanding the ethical aspect of auditors' professional judgments is proposed.

This paper is organized as follows. First, the cognitive-developmental perspective to ethical decision making is reviewed. Second, a brief overview of the concept of "virtue" and virtue-ethics theory is presented. Next, an integrated model of the ethical decision process is developed. From this model, a framework describing the ethical aspect of auditors' professional judgments and actions is proposed. Using this framework, significant findings from the existing research in auditors' ethical judgments and behaviors are examined and fruitful areas for future research are suggested.

This paper aims to make two contributions to accounting-ethics research. First, it hopes to contribute to the development of a comprehensive framework useful for understanding auditors' ethical decisions and actions. Second, it hopes to add to recent efforts of accounting-ethics researchers by drawing attention to the significance of virtue to the exercise of auditors' professional judgments (Dobson and Armstrong 1995; Francis 1990; Mintz 1995; Moizer 1995). Furthermore, this paper hopes to contribute to the general domain of ethics research by offering a theoretically grounded framework useful for understanding the role of virtue in ethical decision making.

## AUDITORS' ETHICAL DECISION PROCESSES FROM A COGNITIVE-DEVELOPMENTAL PERSPECTIVE

Accounting researchers recently have made significant inroads into understanding the ethical actions and judgments of professional accountants by adopting a cognitive-developmental perspective (e.g., Armstrong 1987; Ponemon 1988, 1990, 1992; Ponemon and Gabhart 1993, 1994). According to this perspective, an individual's moral sophistication progresses through a series of developmental levels in reaching maturity (Kohlberg 1958). As applied to the domain of auditing, research using this perspective has demonstrated the existence of an association among an auditor's level of moral development, and his or her ability to exercise professional judgment in a superior way (Gaa and Ponemon 1994; Sweeney 1995), and his or her propensity to act ethically (Ponemon 1992). Furthermore, empirical findings also show that situational (Ponemon 1992; Ponemon and Gabhart 1993) and personal variables (Tsui and Gul 1996; Windsor and Ashkanasy 1995) interact with an auditor's level of moral development to affect his or her ethical decision processes.

Cognitive-developmental researchers also have attempted to understand the *process* of ethical decision making. In particular, Rest (1986) asserts that ethical actions are not the outcome of a single, unitary decision process but result from a combination of various cognitive structures and psychological processes. Rest (1983, 1994) proposes a model of ethical action based upon the presumption that

***Table 1.*** Rest's (1994) Four-Component Model

| | *Process* | *Outcome* |
|---|---|---|
| 1. | Ethical sensitivity | Identification of a dilemma |
| 2. | Prescriptive reasoning | Ethical judgment of "ideal" |
| 3. | Ethical motivation | Ethical intention |
| 4. | Ethical character | Ethical Action |

an individual's ethical behavior is related to his or her level of moral development. Rest's model of ethical action distinguishes four components inherent to the ethical decision making process. Each component is described according to a psychological process and an outcome as identified in Table 1.

## Component One: Ethical Sensitivity

*Ethical sensitivity* initiates the ethical decision making process through the *identification of an ethical dilemma. Ethical sensitivity* is the awareness that the resolution of a particular dilemma may affect the welfare of others (Rest 1994). It involves the perception and interpretation of the cognitive aspects of a situation and an evaluation of the effects of the potential alternatives on the welfare of others (Rest 1983).

## Component Two: Prescriptive Reasoning

An individual's ethical cognition of what "ideally" ought to be done to resolve an ethical dilemma is called *prescriptive reasoning* (Rest 1979). The outcome of an individual's prescriptive reasoning is his or her *ethical judgment* of the ideal solution to an ethical dilemma. Generally, an individual's prescriptive reasoning reflects his or her cognitive understanding of an ethical situation as measured by his or her level of moral development (Kohlberg 1976, 1979).

## Component Three: Ethical Motivation

The importance of ethical motivation to ethical decision making is suggested by Rest's (1994) contention that the "notoriously evil people in the world" are not cognitively limited but lack ethical motivation (e.g., Hitler, Stalin). *Ethical motivation* reflects an individual's willingness to place ethical values (e.g., honesty, integrity, sincerity, and truthfulness) ahead of nonethical values (e.g., wealth, power, and fame) which relate to self-interest. An individual's *ethical motivations* influence his or her *ethical intention* to comply or not comply with his or her *ethical judgment* in the resolution of an ethical dilemma.

### Component Four: Ethical Character

Individuals do not always behave in accordance with their *ethical intention.* Nisan and Kohlberg (1982) attribute the deviation between an individual's *intention* to act ethically and his or her *ethical actions* to a lack of *ethical character.* Trevino (1986) suggests three aspects of an individual's character which affects his or her propensity to act ethically: ego strength, field dependence, and locus of control.[1] Individuals with a strong *ethical character* will be more likely to carry out their *ethical intentions* than individuals with a weak ethical character (Rest 1983).

## A COMPARISON OF VIRTUE-ETHICS' AND INTERACTIONISTS' PERSPECTIVES TO ETHICAL DECISION MAKING

Rest's (1983, 1994) model of ethical action has been used as a foundation for "interactionist" frameworks which attempt to *predict* the influence of key individual and situational factors on ethical judgments (e.g., Dubinsky and Loken 1989; Ferrell and Gresham 1985; Hunt and Vitell 1986; Jones 1991; Trevino 1986). The interactionist perspective primarily has been adopted by accounting-ethics researchers to explain how situational factors influence the ethical behavior of auditors (e.g., Lampe and Finn 1992; Ponemon and Gabhart 1990). Although useful insights into auditors' ethical decision processes have been obtained from interactionists' frameworks, these frameworks fail to provide a theoretical description of the role of personal characteristics, excepting level of moral development, in auditors' ethical decision processes. In order to fill this gap, this paper uses an integrated perspective to present a theoretically grounded model of individuals' ethical decision processes upon which a better understanding of auditors' actions and professional judgments may be developed.

The proposed integrated model of ethical action is consistent with interactionists' frameworks which postulate that personal characteristics moderate individuals' levels of moral development to affect their ethical decision processes (e.g., Ferrell and Gresham 1985; Trevino 1986). The "essential difference" between the cognitive-developmental and the virtue-ethics perspective to ethical decision making is mainly emphasis (Dobson and Armstrong 1995). While cognitive-development theory focuses on understanding the role and development of moral cognition, virtue-ethics theory focuses on understanding the role of one's character in influencing one's ethical judgments and ethical actions (Duncan 1995).

Virtue-ethics theory provides a theoretical foundation based upon Aristotelian moral philosophy which describes the role and importance of personal *virtue* to ethical decision making. Virtues are the characteristics of the decision maker, the possession and exercise of which tends to increase his or her propensity to exercise sound ethical judgment (MacIntyre 1984). Falk (1995) describes the four basic attributes of virtue ethics theory: (1) the notion of virtue, (2) ethical judgment, (3) nurturing community, and (4) moral exemplars. The first two attributes are concerned with the *personal dimension* of ethical decision making and the second two attributes are concerned with the *community* from which an individual's definition of virtue is based (Solomon 1993). It is the personal dimension of virtue-ethics theory which describes the role of personal characteristics in ethical decision making. Accordingly, the proposed integrated model of ethical decision making incorporates the personal dimension of virtue-ethics theory into Rest's (1983, 1994) model of ethical action.

## The Notion of Virtue Defined

According to the Aristotelian virtue-ethics perspective, virtues are dispositions that describe an individual's concern to act in the interests of others, even when self-interest is involved. Virtue is the inner strength or "good" in an individual which is reflected by his or her tendency to do the "good" thing for its own sake (Wallace 1978). For example, a virtuous person is one who, despite personal risks, performs actions considered to be for the good of humanity (e.g., Gandhi, Florence Nightingale, Sister Mary-Theresa). Furthermore, virtue is not simply a tendency to perform "good" actions, a virtuous individual must *knowingly* choose the actions and choose them for *the sake of goodness* (Pincoffs 1986; Wallace 1978). This suggests that a virtuous individual must understand what is "good" in addition to possessing the desire and capability to affect "good" outcomes.

Pincoffs (1986) provides a categorization of virtues to clarify our understanding of the fundamental character traits of a virtuous person: instrumental virtue and moral virtue. Moral virtue is the inherent "goodness" within a person. Individuals which are especially concerned to take others' interests into account are described as possessing moral virtue. According to Pincoffs (1986), morally virtuous individuals "are kind, benevolent, altruistic, generous, charitable, selfless, forgiving, sensitive, helpful, and understanding people" (p. 91). For example, Pincoffs describes justice as the quintessential moral virtue. Also classified as moral virtues are honesty, integrity, sincerity, truthfulness (Mintz 1995).

In contrast to moral virtues, instrumental virtues describe the properties of individuals' characters essential to the accomplishment of their ethical intentions (e.g., persistence, courage, carefulness, prudence and determination). In Pincoffs' own words:

> Persons, as individuals, are more likely to succeed in undertaking of any difficulty if they are persistent enough not to be easily discouraged, courageous enough to face daunting challenges, alert enough to perceive pitfalls and opportunities, careful enough not to make needless errors, resourceful enough to devise alternative strategies, prudent enough to plan ahead for eventualities that are likely to be encountered, energetic and strong enough to carry though what they have planned, coolheaded enough to meet emergencies without panicking, and confident and determined enough not to give way to evanescent feeling and desires that would lead them away from their tasks (1986, 84).

Thus, instrumental virtues are characteristics of an individual which relate to his or her capability to *carry out* his or her "good" intentions.

Although the focus of virtue-ethics theory is on the examination of the personal characteristics which result in "good" or ethical action, its conception of the ethical decision process is similar to the approach advocated by the cognitive-developmental perspective in three important ways. First, both perspectives suggest that *ethical action is the end result of a rational, ethical decision-making process.* Second, both perspectives are concerned with an *individual's* ethical decision process. Third, both perspectives acknowledge the critical role of *cognition* to individuals' ethical decision making. Nevertheless, a model of the ethical decision process as advocated by the virtue-ethics perspective has yet to be developed.

## AN INTEGRATED MODEL OF THE ETHICAL DECISION PROCESS

Given the similarities between the virtue-ethics and cognitive-developmental perspectives, an enriched understanding of ethical decision making results from the integration of their "essential differences." Their "essential differences" reflect the distinct focus taken by the respective theories (Dobson and Armstrong 1995). On the one hand, cognitive-developmental theory focuses on ethical cognition and provides a framework for understanding the relationship between significant components of the ethical decision process (Rest 1983, 1994). On the other hand, virtue-ethics theory provides a theoretical foundation which examines the character and motivation of the decision maker when faced with ethical dilemmas (Dobson and Armstrong 1995; Klein 1989); however, virtue-ethics theory lacks a framework with which to develop a comprehensive understanding of the role of virtue in ethical decision making. In order to advance understanding of ethical decision making, an integrated model is proposed which consolidates these two theoretical perspectives by explicitly identifying the influence of virtue on the different components of an individual's ethical decision process. Figure 1 presents the integrated model of the ethical decision-making process.

Figure 1 indicates that in addition to level of moral development, an individual's virtue is important to the resolution of ethical dilemmas and integral to the determination of ethical behavior. More specifically, an integrated model extends our understanding of the ethical decision-making process in three ways. First, the model identifies the influence of *moral virtue* in the determination of *ethical motivation.* Second, the model also identifies the influence of *instrumental virtue* in the determination of one's *ethical character.* Third, the model recognizes an association between an individual's *moral virtue and prescriptive understanding.* The significance of each of these factors is described below.

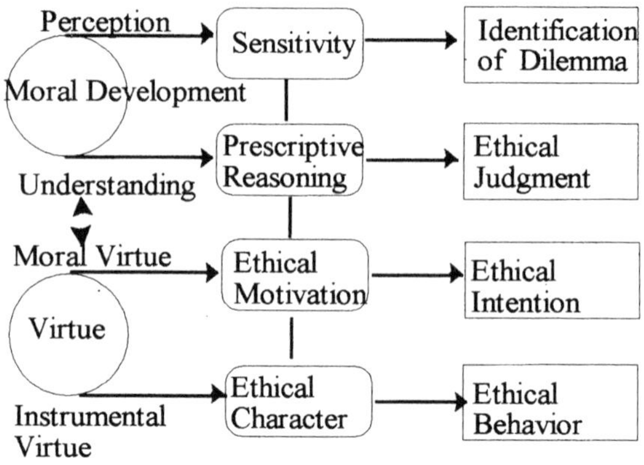


***Figure 1.*** An Integrated Model of Ethical Decision Making

## Moral Virtue and Ethical Motivation

*Moral virtue* and *ethical motivation* are each concerned with an individual's intention to act in accordance with his or her *ethical judgment*. As discussed earlier, *moral virtue* is the positive attribute of character which describes an individual's direct concern for the interests of others despite personal risks (Pincoffs 1986), and *ethical motivation* describes an individual's willingness to place the interests of others ahead of his or her own (Rest 1994).

Although the similarity between these two concepts is evident, an enriched understanding of the factors which influence an individual's *ethical intention* is obtained from integrating the cognitive focus of the cognitive-development perspective with the dispositional focus of the virtue-ethics perspective. Hence, an integrated view suggests that an individual's *ethical motivation* is a reflection of his or her *moral virtue.* This, in turn, suggests that individuals which are *morally virtuous* are *more motivated* in their *intention* to act ethically than less *morally virtuous* individuals.

## Instrumental Virtue and Ethical Character

Similarly, an enriched understanding of the factors which influence one's *ethical intentions* also may be obtained through the integration of the virtue-ethics concept of *instrumental virtue* with the cognitive-developmental concept of *ethical character.* According to the cognitive-developmental perspective, an individual's *ethical character* affects his or her willingness and ability to act in accordance with his or her *ethical intention* (Nisan and Kohlberg 1982; Rest 1994; Trevino 1986). According to virtue-ethics theory, the property of character critical to an individual's ability to implement his or her ethical intention is *instrumental virtue*

(Pincoffs 1986). Hence, the integrative perspective suggests that an individual's *ethical character* is a reflection of his or her *instrumental virtue*.

### Association between Moral Virtue and Prescriptive Understanding

The two-headed arrow connecting *prescriptive understanding* and *moral virtue* in Figure 1 is used to depict the reflexive nature of the association between these two concepts. Although not inconsistent with cognitive-development theory, the nature of this association largely reflects a virtue-ethics emphasis which accepts that virtuous individuals possess both the understanding of what is "good" and the desire to be "good." As MacIntyre (1984) describes it:

> The immediate outcome of the exercise of a virtue is a choice which issues in right action. . . . The educated moral agent must of course know what he is doing when he judges or acts virtuously. Thus he does what is virtuous *because* it is virtuous. . . . The genuinely virtuous agent acts on the basis of a true and rational judgment (149-150).

Hence, the integrated perspective explicitly acknowledges that an individual's *prescriptive understanding* of an ethical dilemma is integral to his or her desire and ability act *virtuously,* and that an individual's *ethical character* is integral to his or her *prescriptive understanding* of an ethical dilemma.

### Summary

The development of an integrated model of ethical decision making is achieved through the incorporation of the virtue-ethics concepts of moral virtue and instrumental virtue into the framework provided by Rest's (1983, 1994) model of moral action. The integrated model provides an enriched understanding of the roles of virtue and moral development in individuals' ethical decision making process by identifying the critical influence of virtue in individuals' ethical decision making. It may be inferred from the integrated model that both virtue and moral development are integral to individuals' ethical decision making processes and jointly influence individuals' ethical actions.

## THEORETICAL FRAMEWORK OF ETHICAL COMPONENTS OF AUDITORS' PROFESSIONAL JUDGMENT

"Professional judgment" is judgment exercised with due care, objectivity, and integrity within the framework provided by applicable professional standards, by experienced and knowledgable people (Gibbins and Mason 1988, 5). Support for the ethical dimension to auditors' professional judgment has been provided by numerous accounting researchers (i.e., DeAngelo 1981; Lampe and Finn 1992; Ponemon and Gabhart 1993; Watts and Zimmerman 1983). The exercise of

**Table 2.** The Issuing of a Qualified Audit Opinion According to the Four Components of the Integrated Model

| | |
|---|---|
| 1. | The first step in an auditor's evaluation of whether or not to issue a qualified audit opinion is his or her identification of the existence of a client's transgression. This is analogous to the first component of the integrated model: *identification of an ethical dilemma.* |
| 2. | The second step requires the auditor to formulate his or her professional opinion of what the client ought to report in this particular situation. This is analogous to the second component of the model: *prescriptive reasoning.* |
| 3. | Given that a transgression exists, the auditor is required to deliberate on whether or not he or she intends to qualify the audit opinion. This is analogous to the third component of the model: ethical intention. |
| 4. | Finally, the auditor's action of issuing the qualified audit opinion is analogous to the fourth component of the model: ethical action. |

professional judgment requires more than just following rules (Gaa 1992b). It involves the application of "professional standards" to situations where the rules are unclear or when explicit standards do not yet exist (Gaa 1994, 1995). It also requires that auditors possess the moral fortitude "to withstand client pressures to disclose selectively in the event a breach is discovered" (DeAngelo 1981, 115).

An examination of the components of the integrated model of the ethical decision process suggests that it appears to capture the essential aspects of an auditor's professional judgment process. This is illustrated through an example, included in Table 2, which draws an analogy between the steps involved in the issuing of a qualified audit opinion and the four components of the integrated model.

Accordingly, the integrated model of ethical decision making is used as a framework to consider the existing accounting-ethics research which examines the ethical aspect of auditors' professional judgments. The scope of this review is limited to accounting-ethics research investigating the professional judgment of auditors from a cognitive-developmental perspective. This examination will be used to assess whether the integrated model is useful for describing the ethical aspect of auditors' professional judgment and to provide insight into opportunities for future research. Table 3 identifies the accounting-ethics research studies classified according to the component of the integrated model that each examines.

## Identification of Transgressions

The ethical decision process is initiated by the recognition that a particular situation will affect the welfare of others (Rest 1983). In the audit context, this is analogous to the identification that a client has violated the rules. Several studies (Bernardi 1994; Ponemon and Gabhart 1993; Ponemon 1993; Shaub, Finn, and

***Table 3.*** Literature Review of Auditors' Ethical Decision Making as Categorized by Integrated Framework of Auditors' Professional Judgment Process

| | *Component of Ethical Decision Process* | *Component of Professional Judgment Process* | *Research Studies* |
|---|---|---|---|
| 1. | Identification of ethical dilemma | Identification of client transgressions | Bernardi (1994)<br>Ponemon (1993)<br>Ponemon & Gabhart (1993)<br>Shaub, Finn and Munter (1993) |
| 2. | Prescriptive judgment | Formulation of professional judgment | Gaa and Ponemon (1994) |
| 3. | Ethical intention | Intention to exercise professional judgment | Ponemon and Gabhart (1990) |
| 4. | Ethical action | Exercise of professional judgment | Ponemon (1992)<br>Tsui and Gul (1996)<br>Windsor and Ashkanasy (1995) |

Munter 1993) have examined the association between an auditor's ethical sensitivity and his or her ability to detect a client's transgressions, intentional or otherwise. Each is examined in turn.

Bernardi (1994) examined the relationship between auditors' level of moral development and their ability to detect a client's fraudulent financial statement information. This study required auditors with different levels of experience to review a realistic set of financial information that contained a seeded error which strongly pointed to the possibility of material fraud. The findings of this study suggest that auditors with higher levels of moral development and experience were more likely than other auditors (lower levels of moral development and/or inexperienced) to discover fraud in clients' financial statements. Two other studies, Ponemon and Gabhart (1993) and Ponemon (1993), examined the association between an auditor's level of moral development and his or her sensitivity to contextual cues which may signify the existence of client transgressions. These studies indicate that for a given level of technical competence, auditors at higher levels of moral development are more sensitive to contextual cues that a transgression has occurred than are auditors at lower levels of moral development. It may be inferred from these studies that auditors at higher levels of moral development may be better able to perceive the existence of material misstatement of financial statements. This, in turn, suggests that an auditor's level of ethical sensitivity is related to his or her capability to perform tasks integral to the auditors' role and an auditor's level of moral development is integral to his or her capability to formulate professional judgments.

Another study using a cognitive-developmental approach also considers an auditor's ethical sensitivity to the existence of a client's transgression. Shaub et al.

(1993) developed a measure of auditors' ethical sensitivity and examined the association between an auditor's *ethical sensitivity* and *commitment.* Commitment was measured as an auditor's willingness to exert effort on behalf of the profession or audit firm. Shaub et al. (1993) did not find evidence of a significant association between commitment and ethical sensitivity as they anticipated. Notwithstanding, their findings are not unexpected, given the framework provided by the integrated model of ethical decision making. Although it is difficult to ascertain the nature of "commitment" as operationalized in this study, it appears that it may be analogous to instrumental virtue as described by Pincoffs (1986). The integrated model suggests that instrumental virtue is related to a person's tendency to comply with his or her ethical intention and not to that individual's ethical sensitivity. Additional research is needed to evaluate the association between commitment (as operationalized in Shaub's et al.'s study) and an auditor's ethical intention to exercise professional judgment. Nevertheless, the potential usefulness of the integrated model as a guide to empirical accounting-ethics research is suggested by this result.

## Formulation of Professional Judgment

The formulation of professional judgment requires more than just following the rules (Moizer 1995). It involves the application of the rules to situations where they are unclear or where explicit standards have yet to be specified. The formulation of professional judgment compels an auditor to prescriptively assess what *should* be done if the financial statements are discovered to be materially misstated (Moizer 1995). Although numerous studies have described the prescriptive reasoning capability of auditors (e.g., Lampe and Finn 1992; Ponemon 1990, 1992; Ponemon and Gabhart 1993, 1994; Shaub 1994), very little work has considered the association between prescriptive reasoning and the professional judgment of auditors.

An extensive review of the literature revealed one unpublished study which considers this association. Gaa and Ponemon (1994) examined the way in which auditors with higher and lower levels of moral development and higher and lower levels of technical expertise, resolved a realistic audit dilemma which required the trade off between two conflicting accounting principles. Their analysis involved an examination of concurrent verbal protocols of auditors. The findings show that both *a higher level of moral development* and *a higher level of technical expertise* are necessary for an auditor to display forward reasoning in the prescriptive resolution of a realistic auditing case. The study suggests that for auditors at higher levels of technical expertise, an auditor's level of moral development is associated with his or her capability to formulate professional judgment at an expert level. This, in turn, suggests that moral expertise, defined as the moral understanding held by individuals with higher levels of moral development, is necessary to the formulation of expert professional judgment by an auditor. Furthermore, this study provides support for the relationship described in the proposed integrated model that the second component of the integrated model, prescriptive reasoning, is analogous to the formulation of an auditor's professional judgment.

## Intention to Exercise Professional Judgment

The *intention to exercise professional judgment* requires an auditor to deliberate on whether or not to comply with the "ideal" professional judgment as established by codified rules or as determined by his or her prescriptive reasoning process. Thus, an auditor's intention to exercise professional judgment reflects his or her ethical motivation. Ponemon and Gabhart (1990) examined the association between auditors' *intentions to exercise professional judgment,* the influence of situational consequences attached to alternative ethical choices, and ethical cognition as measured by level of moral development. Ponemon and Gabhart's study shows that auditors' *intention to exercise professional judgment* appears to be a joint function of their ethical cognition and the situational consequences. Furthermore, this study also suggests that situational consequences may *differentially* influence auditors' propensity to exercise professional judgment depending upon their level of moral development.

Although Ponemon and Gabhart's study provides some support for the integrated model as applied to auditors' professional judgments, the usefulness of the integrated perspective is difficult to assess given the lack of research investigating the association between *moral virtue* and auditors' *intentions to exercise professional judgment.* However, this gap in knowledge also suggests a future research opportunity which has great potential to enhance our understanding of the ethical decision process and the professional judgment process of auditors.

## Exercise of Professional Judgment

The exercise of professional judgment describes the action of complying with one's intended professional judgment. The integrated framework suggests that auditors' *intentions to exercise professional judgment* are moderated by the strength of their *instrumental virtue* to affect *their exercise of professional judgment.* This, in turn, suggests that the personal characteristics of *moral virtue and instrumental virtue* influence auditors' exercise of professional judgment, as well as *moral development.*

Only one study using a cognitive-development approach has investigated the association between the ethical actions of auditors and their levels of moral development: Ponemon (1992) examined staff-level auditors' tendencies to underreport the time taken to complete a simulated audit task. The findings of this study show that auditors at lower levels of moral development underreport their own time to a greater extent than auditors at higher levels of moral development. Thus, this study provides support for the association between an auditor's ethical cognition and ethical action as suggested in the integrated model.

In addition, two other studies have examined the association between auditors' *instrumental virtue* as captured by Rotter's (1966) measure of "locus of control" and auditors' responses to realistic ethical dilemmas encountered in the workplace (Tsui and Gul 1996; Windsor and Ashkanasy 1995). Trevino (1986) contends that individuals who are designated as "internals" on Rotter's scale possess more instrumental virtue than individuals designated as "externals." Hence, auditors designated as "internals" are more likely to exercise "ethical" professional judgment than auditors designated as "externals." Support for Trevino's contention is found in both Tsui and Gul (1996) and Windsor and Ashkanasy (1995). These studies show that for a given level of moral development, auditors designated as "internals" were more likely to make "ethical" judgments than auditors designated as "externals." In addition, consistent with the integrated framework, these findings also show that auditors' levels of moral development and locus of control jointly affect their ethical behavior. Individuals of higher levels of moral development and "internals" were more likely to act ethically than individuals of lower levels of moral development and "externals."

The empirical evidence indicates that an auditor's likelihood of complying with his or her professional judgment is a joint function of his or her virtue and moral development. Thus, these studies provide support for the validity and the usefulness of the integrated framework for understanding the integral roles of virtue and moral development in the exercise of auditors' professional judgments.

## CONCLUSIONS

Although empirical validation is required, the integrated framework provides theoretical support to accounting-ethics researchers who maintain that the exercise of ethical professional judgment demands that auditors demonstrate certain "virtues" of character as well as possess sufficient levels of technical and moral expertise (e.g., Dobson and Armstrong 1995; Doucet and Ruland 1994; Gaa and Ruland 1995; Mintz 1995). Rather than viewing virtue or moral development as a peripheral constraint on the execution of an auditor's professional judgment, the adoption of an integrated approach to ethical decision making makes both virtue and moral development central to the essence of what is required to be a good professional (Dobson and Armstrong 1995). Consistent with a virtue-ethics perspective, the integrated framework suggests that the possession of both technical and moral expertise are necessary but not sufficient for auditors to meet their professional obligations to society (Mintz 1995). To be a good auditor, one must be virtuous as well as technically and morally competent (Dobson and Armstrong 1995).

## Future Research Opportunities

The objective of developing an integrated model of ethical decision making is to provide a theoretical framework for understanding the respective roles of virtue and moral development in the professional judgment process of auditors. This framework may be used as a guide to understanding the existing empirical accounting-ethics research and to inform accounting-ethics researchers about future research opportunities regarding the ethical decision process of auditors.

Several future research opportunities are evident from the resulting synthesis of the existing accounting-ethics research. Although an association between auditors' moral virtue and professional judgment may be inferred from two research studies (Tsui and Gul 1996; Windsor and Ashkanasy 1995), the existence of empirical research which explicitly and comprehensively examines auditors' virtue and its influence on professional judgment appears to be lacking. The scarcity of empirical research on auditors' virtue suggests the existence of an opportunity and a need for the development of a research program which defines and measures the personal characteristics of a virtuous auditor and empirically investigates the influence of auditors' virtues in the formulation and exercise of professional judgment.

## Limitations

The scope of the integrated model developed in this paper is limited to considering the integration of the personal dimension as described by virtue-ethics theory into a cognitive-developmental framework. Thus, the proposed integrated model does not specifically incorporate theoretical aspects suggested by the community dimension of virtue-ethics theory. Nevertheless, the role and nature of *community* and *practice* as suggested by the virtue-ethics perspective may provide useful insights into the effect of contextual factors on the ethical decision process. For example, the integrated framework may provide a theoretical foundation for understanding the effect of the professional code of ethics on auditors' professional judgments and ethical actions. Additionally, the integrated framework may be used to provide a theoretical foundation for understanding the effect of ethical culture, as characterized by the virtue of those in positions of authority, on an organization's internal control. Thus, the integrated framework may provide the launching pad for an investigation of the effect of the "tone at the top" on an organization's internal control. This investigation would address the concern expressed by the Treadway Commission (1987) and the CICA's Criteria of Control Committee (1994) regarding the influence of people in positions of authority on an organization's internal control (Gaa and Ruland 1995).

The model presented in this paper limits its scope to the integration of virtue-ethics theory and the cognitive-developmental perspective. However, additional insights into the ethical aspect of auditors' professional judgment may be gained through the consideration of alternative theoretical approaches. For example,

Cohen, Pant, and Sharp (1995) suggest a multidimensional approach to accounting-ethics research to examine the first component of the integrated model: ethical sensitivity (e.g., Flory et al. 1992; Cohen, Pant, and Sharp, 1992, 1995). Accordingly, an integration of the findings of other streams of research into the integrated model may provide an enriched understanding of auditors' ethical decision processes. Furthermore, additional investigation of the similarities and differences between various theoretical paradigms used by accounting-ethics researchers may facilitate the development of a more comprehensive understanding of the professional judgment of auditors.

## ACKNOWLEDGMENTS

The author would like to acknowledge the insightful comments of several anonymous reviewers, Mary Beth Armstrong, Susan Bartholomew, Louis Culumovic, Dawn Massey, and participants of the American Accounting Association Ethics Symposium, Chicago, August 1996. Funding for this project was provided by the York-CGA Research Program.

## NOTE

1. Ego strength describes one's tendency to resist impulse and follow his or her own convictions. Field dependence describes the degree to which one uses social referents as a guide for behavior (Witkin and Goodenough 1977). Locus of control reflects the extent to which one perceives that one has control over his or her own fate (MacDonald 1976).

## REFERENCES

Armstrong, M. 1987. Moral development and accounting education. *Journal of Accounting Education* 5: 27-43.

Bernardi, R. 1994. Fraud Detection: The effect of client integrity and competence and auditor cognitive style. *Auditing: A Journal of Practice and Theory* 14(Supplement): 68-97.

Cohen, J., L. Pant, and D. Sharp. 1992. Cultural and socioeconomic constraints on international codes of ethics: Lessons from accounting, *Journal of Business Ethics.* (11): 687-700.

Cohen, J., L. Pant, and D. Sharp. 1995. Towards a model of moral decision making: A multidimensional study of Canadian accounting students and accountants. Working paper, Western Ontario.

Criteria of Control Committee (CICA). 1994. *Exposure Draft on the Guidance on Criteria of Control.* Toronto: CICA.

DeAngelo, L. 1981. Auditor independence, low balling and disclosure regulation. *Journal of Accounting and Economics* 3: 113-127.

Dienhart, J. 1995. Rationality, ethical codes and an egalitarian justification for ethical expertise. *Business Ethics Quarterly* 5(July): 419-450.

Dobson , J., and M. Armstrong. 1995. Application of virtue-ethics theory. In *Research on Accounting Ethics,* Vol. l, ed. L.A. Ponemon, 187-202. Greenwich, CT: JAI Press.

Doucet, M., and R. Ruland. 1994. Exploring the necessary virtues for professional accountants. Working paper, Bowling Green State University.

Dubinsky, A., and B. Loken. 1989. Analysing ethical decision making in market. *Journal of Business Research* 19(2): 83-107.

Duncan, S. 1995. *A Primer of Modern Virtue Ethics.* Lanham, MD: University Press of America.

Falk, H. 1995. Professional services and ethical behavior. In *Research on Accounting Ethics,* Vol. 1, ed. L.A. Ponemon, 203-212. Greenwich, CT: JAI Press.

Ferrell, O., and L. Gresham. 1985. A contingency framework for understanding ethical decision making in marketing. *Journal of Marketing* 29(3): 87-96.

Flory, S., T. Phillips, R. Reidenbach, and D. Robin. 1993. A multidimensional analysis of selected issues in accounting. *Accounting Review* 67(2, April): 284-302.

Francis, J. 1990. After virtue? Accounting as a moral and discursive practice. *Accounting, Auditing and Accountability Journal* 3(3): 5-17.

Gaa, J. 1992a. The auditor's role: The philosophy and psychology of independence and objectivity. In *Proceeding of the 1992 Deloitte & Touche/University of Kansas Auditing Symposium,* eds. R. Srivastava and A. Ford, 7-43, Lawrence, KS: University of Kansas Press.

Gaa, J. 1992b. Discussion of auditors' ethical decision processes. *Auditing: A Journal of Practice and Theory* 11: 60-67.

Gaa, J. 1994. *The Ethical Foundations of Public Accounting.* Vancouver: Canadian Certified General Accountants' Research Foundation.

Gaa, J. 1995. Moral judgment and moral cognition: A comment. In *Research on Accounting Ethics,* Vol. 1, ed. L.A. Ponemon, 253-266. Greenwich, CT: JAI Press.

Gaa, J., and L. Ponemon. 1994. Towards a theory of moral expertise: A verbal protocol study of public accounting professionals. Working paper, Macmasters University, Canada.

Gaa, J., and R. Ruland. 1995. Ethics in accounting: An overview of issues, concepts and principles. Working paper.

Gibbins, M., and A. Mason. 1988. *Professional Judgment in Financial reporting.* CICA Research Study. Toronto: CICA.

Hunt, S., and S. Vital. 1986. A general theory of marketing ethics. *Journal of Macro Marketing* 6(1): 5-16.

Jones, T. 1991. Ethical decision making by individuals in organization: An issue-contingent model. *Academy of Management Review* 16(2): 366-395.

Klein, S. 1989. Platonic virtue theory and business ethics. *Business and Professional Ethics Journal* 8(4): 59-81.

Kohlberg, L. 1958. The development of modes of moral thinking and choice in the years ten to sixteen. Ph.D. dissertation, University of Chicago.

Kohlberg, L. 1976. Moral stages and moralization: The cognitive-developmental approach to socialization. In *Handbook of Socialization Theory and Research,* ed. D. Goskin, 347-480. Chicago: Rand McNally.

Kohlberg, L. 1979. *The Meaning and Measurement of Moral Development.* Worcester, MA: Clark University Press.

Ladd, J. 1991. Bhopal: An essay on moral responsibility and civic virtue. *Journal of Social Philosophy* 22(Spring): 241-263.

Lampe, J., and D. Finn. 1992. A model of auditors' ethical decision processes. *Auditing: A Journal of Practice and Theory* 11: 33-59.

MacDonald, A. 1976. Internal/external locus of control. In *Measure of Social Psychological Attitudes,* eds. J.P. Robinson and P.R. Shaver, 413-582. Ann Arbor, MI: Institute for Social Research.

MacIntyre, A. 1984. *After Virtue*, 2nd edition. Notre Dame, IN: University of Notre Dame Press.

Mintz, S. 1995. Virtue ethics and accounting education. *Issues in Accounting Education* 10(2, Fall): 247-267.

Moizer, P. 1995. An ethical approach to the choices faced by auditors. *Critical Perspectives on Accounting* 6: 415-431.

Nisan, M., and L. Kohlberg. 1982. Universality and cross-cultural variation in moral development: A longitudinal and cross-sectional study in Turkey. *Child development* 53: 359-369.

Pincoffs, E. 1986. *Quandaries and Virtues.* Lawrence, KS: University Press of Kansas.

Ponemon, L. 1988. A cognitive-developmental approach to the analysis of certified public accountants' ethical judgments. Ph.D. dissertation, Union College.

Ponemon, L. 1990. Ethical judgments in accounting: A cognitive-developmental perspective. *Critical Perspectives on Accounting* 1: 191-215.

Ponemon, L. 1992. Auditor under reporting of time and moral reasoning: A lab study. *Contemporary Accounting Research* 9(1, Fall): 171-189.

Ponemon, L., ed. 1995. *Research on Accounting Ethics,* Vol. 1. Greenwich, CT: JAI Press.

Ponemon, L., and D. Gabhart. 1990. Auditor independence judgments: A cognitive-developmental model and experimental evidence. *Contemporary Accounting Research* 7: 227-251.

Ponemon, L., and D. Gabhart. 1993. *Ethical Reasoning in Accounting and Auditing.* Vancouver: Canadian Certified General Accountants' Research Foundation.

Ponemon, L., and D. Gabhart. 1994. Ethical reasoning research in accounting and auditing professions. In *Moral Development in the Professions: Psychology and Applied Ethics,* eds. J. Rest and D. Narvaez, 101-121. Hillsdale, NJ: Erlbaum Associates.

Ponemon, L., and A. Glazer. 1990. Accounting education and ethical development: The influence of liberal learning on students and alumni in accounting practice. *Issues in Accounting Education* 5: 195-208.

Rest, J. 1979. *Development in judging moral issues.* Minneapolis, MN: University of Minnesota Press.

Rest, J. 1983. Morality. In *Manual of Child Psychology: Cognitive Development,* Vol. 3, eds. J. Flavell and E. Markman, 556-629. New York: Wiley.

Rest, J. 1986. *Moral Development: Advances in Research and Theory.* New York: Praeger.

Rest, J. 1994. Background theory and research. In *Moral Development in the Professions,* eds. J. Rest and D. Narvaez, 1-26. Hillsdale, NJ: Erlbaum & Associates.

Rotter, J. 1966. Generalized expectancies for internal versus external control of reinforcement. *Psychological Monographs* 609.

Shaub, M. 1994. An analysis of Factors affecting the Cognitive moral development of auditors and auditing students. *Journal of Accounting Education* (Fall): 1-26.

Shaub, M., D. Finn, and P. Munter. 1993. The effects of auditors' ethical orientation on commitment and ethical sensitivity, *Behavioural Research in Accounting* 5: 145-169.

Solomon, R. 1992. Corporate role, personal virtues: An Aristotelian approach to business ethics. *Business Ethics Quarterly* 2: 317-339.

Solomon, R. 1993. *Ethics and Excellence: Cooperation and Integrity in Business.* Oxford, UK: Oxford University Press.

Sweeney, J. 1995. The moral expertise of auditors: An exploratory analysis. In *Research on Accounting Ethics,* Vol. l, 213-234, ed. L.A. Ponemon. Greenwich, CT: JAI Press.

Treadway Commission. 1987. *Report of the National Committee on Fraudulent Financial Reporting.*

Trevino, L. 1986. Ethical decision making in organizations: A person-situation interactionist model. *Academy of Management Review* 11: 601-617.

Tsui, J., and F. Gul. 1996. Auditors' behaviour in an audit conflict situation: A research note on the role of locus of control and ethical reasoning. *Accounting, Organizations and Society* 21(1, January): 41-54.

Wallace, J. 1978. *Virtues and Vices.*, Ithaca, NY: Cornell University Press.

Watts, R., and J. Zimmerman. 1986. *Positive Accounting Theory.* Englewood Cliffs, NJ: Prentice-Hall.

Windsor, C., and N. Ashkanasy. 1995. The effect of client management bargaining power, moral reasoning development and belief in a just world on auditor independence. *Accounting, Organizations and Society* 20(7/8): 701-720.

Witkin, H., and D. Goodenough. 1977. Field dependence and interpersonal behaviour. *Psychological Bulletin* 84: 661-689.